Takeo Oku

Solar Cells and Energy Materials

## Also of Interest

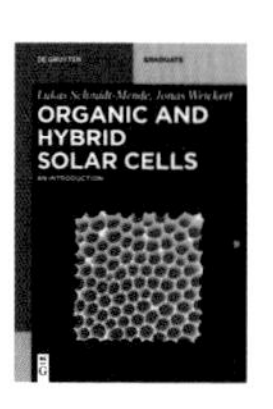


*Organic and Hybrid Solar Cells*
An Introduction
Schmidt-Mende, Weickert (Eds.), 2016
ISBN 978-3-11-028318-1, e-ISBN 978-3-11-028320-4

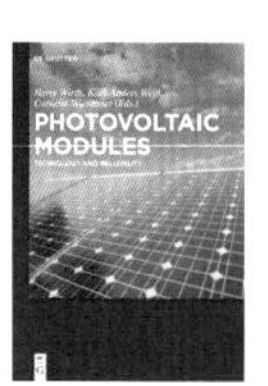


*Photovoltaic Modules*
Technology and Reliability
Wirth, Weiß, Wiesmeier, 2016
ISBN 978-3-11-034827-9, e-ISBN 978-3-11-034828-6

*Solar Energy and Technology*
Volume 1: English-German Dictionary / Deutsch-Englisch Wörterbuch
Mijic, 2016
ISBN 978-3-11-047575-3, e-ISBN 978-3-11-047717-7

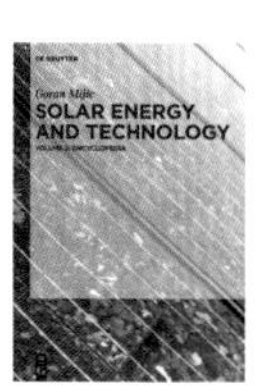


*Solar Energy and Technology*
Volume 2: Encyclopedia
Mijic, 2018
ISBN 978-3-11-047577-7, e-ISBN 978-3-11-047721-4

*Intelligent Materials and Structures*
Abramovich, 2016
ISBN 978-3-11-033801-0, e-ISBN 978-3-11-033802-7

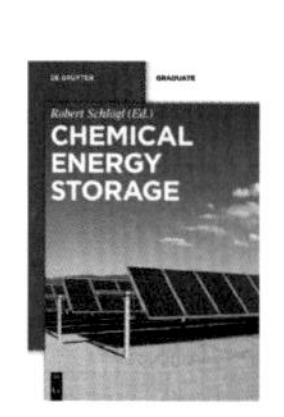


*Chemical Energy Storage*
Schlögl (Ed.), 2012
ISBN 978-3-11-026407-4, e-ISBN 978-3-11-026632-0

Takeo Oku

# Solar Cells and Energy Materials

DE GRUYTER

**Author**
Prof. Takeo Oku
The University of Shiga, Prefecture Japan
Department of Materials Science
Hassaka 2500, Hikone
Shiga 522-8533, Japan
oku@mat.usp.ac.jp

ISBN 978-3-11-029848-2
e-ISBN (PDF) 978-3-11-029850-5
e-ISBN (EPUB) 978-3-11-038106-1
Set-ISBN 978-3-11-041209-3

**Library of Congress Cataloging-in-Publication Data**
A CIP catalog record for this book has been applied for at the Library of Congress.

**Bibliographic information published by the Deutsche Nationalbibliothek**
The Deutsche Nationalbibliothek lists this publication in the Deutsche Nationalbibliografie; detailed bibliographic data are available on the Internet at http://dnb.dnb.de.


Typesetting: PTP-Berlin, Protago-TEX-Production GmbH, Berlin
Printing and binding: Hubert & Co. GmbH & Co. KG, Göttingen
♾ Printed on acid-free paper
Printed in Germany

www.degruyter.com

# Preface

"Let there be light". These words appear at the beginning of Genesis in the Old Testament. Advanced cosmological science has likewise shown that the universe started from a Big Bang 13.7 billion years ago. Light is an elemental root of our universe. The "flow of time" that we feel in what we perceive to be the present is an effect of the close relationship between light and matter, and time does not flow by itself. Quantum teleportation utilizing the connected spins of quantum particles at a distance results from principles that still remain unknown to contemporary scientists. Light has also not yet been entirely clarified by the advances of contemporary science, and several aspects of it remain unexplained.

On the other hand, light is essential to our everyday life in contemporary society. Light contains energy and information simultaneously, and most new electronic apparatuses, such as mobile phones, laptop computers, DVD players or satellite communications, function in ways closely connected to light.

After the big earthquake in Japan in 2011, all nuclear reactors in Japan were stopped, and some of them had already been closed previously. Saving electricity is promoted both in both industries and ordinary homes, and the world's energy problem has become a very familiar one in our everyday lives. Although Japan's energy self-sufficiency was 58 % in 1960, it had been significantly reduced to only 4 % by 2014. In other words, 96 % of the energy used in Japan depends on the import of materials such as oil, coal, liquefied natural gas and uranium, a fact that suggests improvements are needed to the current imbalance in supply and demand. Energy problems resulting from dependence on fossil fuels, which are limited energy resources, will soon be an inescapable issue all over the world. The author would be very pleased if the present book is helpful in solving future energy problems.

I would like to acknowledge many of my colleagues for their support, as follows: Atsushi Suzuki, Tsuyoshi Akiyama, Kenji Kikuchi, Taisuke Matsumoto, Akihiro Takeda, Akihiko Nagata, Kazuya Fujimoto, Junya Nakagawa, Hiroki Kidowaki, Nariaki Kakuta, Kazumi Yoshida, Masato Kanayama, Ryosuke Motoyoshi, Haruto Maruhashi, Masahito Zushi, Yuya Ohishi, Kohei Suzuki, Takuya Kitao, Taishi Iwata, Tatsuya Noma, Kazuma Kumada, Satoru Hori, Makoto Iwase, Yuma Imanishi, Yuji Ono, Kengo Kobayashi, Shiomi Kikuchi, Katsuhisa Tokumitsu, Mie Terada, Kouichi Hiramatsu, Masashi Yasuda, Yasuhiro Shirahata, Youichi Kanamori, Mikio Murozono, Yasuhiro Yamasaki, Mika Nakamura, Masahiro Yamada, Sakiko Fukunishi, Kazufumi Kohno, Misaki Fukaya, Yoshikazu Takeda, Akio Shimono, Eiji Ōsawa, Yasuhiko Hayashi, Tetsuo Soga, Tatsuo Oku, Ichihito Narita, Masaki Kuno, Naruhiro Koi, Atsushi Nishiwaki, Tadachika Nakayama, Katsuaki Suganuma and Brian D. Josephson. Thanks for excellent collaborative works, useful discussion, providing samples and experimental help. In addition, some figures were drawn after revision of original figures from some books and websites and I would like to offer my appreciation for

DOI 10.1515/9783110298505-001

this material. I would also like to give thanks for the financial support I received from The University of Shiga prefecture, Satellite Cluster Program of the Japan Science and Technology Agency, and a Grant-in-Aid for Scientific Research. Finally, I'd like to thank warmly Ria Fritz, Karin Sora, Anne Hirschelmann, Kelly Bescherer, Stefan Sossna, and Julia Lauterbach, who handled the production of this book.

November 2016, Takeo Oku

# Contents

# Table for physical constants

| Physical constants | Symbol | Values | SI units |
|---|---|---|---|
| Velocity of light | $c$ | 2.99792458 | $10^{8}$ m s$^{-1}$ |
| Planck constant | $h$ | 6.62607 | $10^{-34}$ J s |
| Dirac constant | $\hbar = h/2\pi$ | 1.05457 | $10^{-34}$ J s |
| Gravitational constant | $G$ | 6.67384 | $10^{-11}$ m$^{3}$ s$^{-2}$ kg$^{-1}$ |
| Electron charge | $e$ | 1.60218 | $10^{-19}$ A s (C) |
| Electron mass | $m_e, m_0$ | 9.10938 | $10^{-31}$ kg |
| e Proton mass | $m_p$ | 1.67262 | $10^{-27}$ kg |
| Neutron mass | $m_n$ | 1.67493 | $10^{-27}$ kg |
| Electron energy | $m_e\, c^2$ | 0.5110 | MeV |
| Compton wavelength | $\lambda_c$ | 2.4263 | $10^{-12}$ m |
| Boltzmann constant | $k, k_B$ | 1.38065 | $10^{-23}$ J K$^{-1}$ |
| Magnetic permeability | $\mu_0 = 4\pi \times 10^{-7}$ | 1.25664 | $10^{-6}$ H m$^{-1}$ (N A$^{-2}$) |
| Dielectric constant | $\varepsilon_0 = 1/\mu_0\, c^2$ | 8.85419 | $10^{-12}$ F m$^{-1}$ (N V$^{-2}$) |
| Avogadro constant | $N_A$ | 6.02214 | $10^{23}$ mol$^{-1}$ |
| Gas constant | $R = k\, N_{\mathrm{A}}$ | 8.31446 | J K$^{-1}$ mol$^{-1}$ |

| Physical constants | Symbol | Values and units |
|---|---|---|
| Ångström | Å | 0.1 nm = $10^{-10}$ m |
| Electron volt | eV | $1.60218 \times 10^{-19}$ J |
| Wavelength of 1 eV photon | $\lambda$ | 1239.84 nm |
| Standard atmosphere | atm | $1.01325 \times 10^{5}$ Pa |

DOI 10.1515/9783110298505-002

# Periodic table

| 1 | 2 | 3 | 4 | 5 | 6 | 7 | 8 | 9 | 10 | 11 | 12 | 13 | 14 | 15 | 16 | 17 | 18 |
|---|---|---|---|---|---|---|---|---|---|---|---|---|---|---|---|---|---|
| 1 H Hydrogen 1.008 | | | | | | | | | | | | | | | | | 2 He Helium 4.003 |
| 3 Li Lithium 6.941 | 4 Be Bery llium 9.012 | | | | | | | | | | | 5 B Boron 10.81 | 6 C Carbon 12.01 | 7 N Nitrogen 14.01 | 8 O Oxygen 16.00 | 9 F Fluorine 19.00 | 10 Ne Neon 20.18 |
| 11 Na Sodium 22.99 | 12 Mg Magnesium 24.31 | | | | | | | | | | | 13 Al Aluminum 26.98 | 14 Si Silicon 28.09 | 15 P Phosphorus 30.97 | 16 S Sulfu r 32.07 | 17 Cl Chlorine 35.45 | 18 Ar Argon 39.9 5 |
| 19 K Potassium 39.10 | 20 Ca Calcium 40.08 | 21 Sc Scandium 44.96 | 22 Ti Titanium 47.87 | 23 V Vanadium 50.94 | 24 Cr Chromium 52.00 | 25 Mn Manganese 54.94 | 26 Fe Iron 55.85 | 27 Co Cobalt 58.93 | 28 Ni Nickel 58.69 | 29 Cu Copper 63.55 | 30 Zn Zinc 65.41 | 31 Ga Gallium 69.72 | 32 Ge Germanium 72.64 | 33 As Arsenic 74.92 | 34 Se Selenium 78.96 | 35 Br Bromi ne 79.90 | 36 Kr Krypton 83.80 |
| 37 Rb Rubidium 85.47 | 38 Sr Strontium 87.62 | 39 Y Yttrium 88.91 | 40 Zr Zirconium 91.22 | 41 Nb Niobium 92.91 | 42 Mo Molybdenum 95.94 | 43 Tc Technetium (99) | 44 Ru Ruthenium 101.1 | 45 Rh Rhodium 102.9 | 46 Pd Palladium 106.4 | 47 Ag Silver 107.9 | 48 Cd Cadmium 112.4 | 49 In Indium 114.8 | 50 Sn Tin 118.7 | 51 Sb Antimony 121.8 | 52 Te Tellurium 127.6 | 53 I Iodine 126.9 | 54 Xe Xenon 131.3 |
| 55 Cs Caesium 132.9 | 56 Ba Barium 137.3 | 57-71 Lutetium ♦ | 72 Hf Hafnium 178 .5 | 73 Ta Tantalum 180.9 | 74 W Tungsten 183.8 | 75 Re Rhenium 186.2 | 76 Os Osmium 190.2 | 77 Ir Iridium 192.2 | 78 Pt Platinum 195.1 | 79 Au Gold 197.0 | 80 Hg Mercury 200.6 | 81 Tl Thallium 204.4 | 82 Pb Lead 207.2 | 83 Bi Bismuth 209.0 | 84 Po Polonium (210) | 85 At Astatine (210) | 86 Rn Radon (222) |
| 87 Fr Francium (223) | 88 Ra Radium (226) | 89-103 Lawrencium ♦♦ | 104 Rf Rutherfordium (267) | 105 Db Du bnium (268) | 106 Sg Seaborgium (271) | 107 Bh Bohrium (272) | 108 Hs Hassium (277) | 109 Mt Meitnerium (276) | 110 Ds Darmstadtium (281) | 111 Rg Roentgenium (280) | 112 Cn Copernicium (285) | 113 Nh Nihonium (284) | 114 Fl Flerovium (289) | 115 Mc Moscovium (288) | 116 Lv Livermorium (293) | 117 Ts Tennessine (294) | 118 Og Oganesson (294) |

Symbol for element →
Element →
Atomic mass (u) →

| | | | | | | | | | | | | | | | |
|---|---|---|---|---|---|---|---|---|---|---|---|---|---|---|---|
| ♦ | 57 La Lanthanum 138.9 | 58 Ce Cerium 140.1 | 59 Pr Praseodymium 140.9 | 60 Nd Neodymium 144.2 | 61 Pm Promethium (145) | 62 Sm Samarium 150.4 | 63 Eu Europium 152.0 | 64 Gd Gadolinium 157.3 | 65 Tb Terbium 158.9 | 66 Dy Dysprosium 162.5 | 67 Ho Holmium 164.9 | 68 Er Erbium 167.3 | 69 Tm Thulium 168.9 | 70 Yb Ytterbium 173.0 | 71 Lu Lutetium 175.0 |
| ♦♦ | 89 Ac Actinium (227) | 90 Th Thorium 232.0 | 91 Pa Protactinium 231.0 | 92 U Uranium 238.0 | 93 Np Neptunium (237) | 94 Pu Plutonium (239) | 95 Am Americium (243) | 96 Cm Curium (247) | 97 Bk Berkelium (247) | 98 Cf Californium (252) | 99 Es Einsteinium (252) | 100 Fm Fermium (257) | 101 Md Mendelevium (258) | 102 No Nobelium (259) | 103 Lr Lawrencium (262) |

| Metal | | | | | | Metalloid | Nonmetal | | | Unknown chemical properties |
|---|---|---|---|---|---|---|---|---|---|---|
| Alkali metal | Alkaline earth metal | Lanthanide | Actinide | Transition metal | Post -transition metal | | Polyatomic nonmetal | Diatomic nonmetal | Noble gas | |

DOI 10.1515/9783110298505-003

# 1 Energy

## 1.1 What is energy?

The word "energy" can have the following meanings:
1. The capacity for work that a certain system potentially has.
2. The ability to do physical work.
3. A useful resource for human society.
4. A resource required for physical or mental activity.

In the field of physics, energy generally refers to a quantity of work, as in definition (1). Heat, light, electromagnetic waves, and mass are also forms of energy. Within general usage, definitions (2) and (3) are more commonly used. There are many types of energy resources, and exhaustive energy and renewable energy have often been compared. Recently, a transition from exhaustive energy to renewable energy has begun taking place across the world.

The measurement used for energy in the International System of Units (SI unit) is the joule (J). The electron volt (eV) and kilowatt hour (kWh) are also used in the field of solar cells, as is shown in Table 1.1.

Table 1.1: Unit of energy.

| Item | Symbol of quantity |
|---|---|
| Energy | $E$ |
| Dimension | kg $m^2$ $s^{-2}$ |
| Kind | scalar |
| SI unit | J (Joule) |
| CGS unit | erg = $10^{-7}$ J |
| MKS system of units | kgf m |
| Planck unit | Planck energy $E_P = 1.956\times 10^9$ J |
| Atomic unit | Hartree $E_h = 4.360\times 10^{-18}$ J |
| Kilo watt hour (kWh) | 3.6 MJ |
| Electron volt (eV) | $1.602\times 10^{-19}$ J |

There are many types of energy, including: physical energy, kinetic energy, potential energy, elastic energy, chemical energy, ionization energy, heat energy, light energy, electric energy, acoustic energy, nuclear energy, mass energy and dark energy. Resources that are useful for industry, transportation and human life are generally referred to as "energy resources", which include oil, coal, natural gas, nuclear power energy, water power, solar heat and so on. Recently, a distinction has been made between energy resources that are exhaustive forms of energy and those that are re-

DOI 10.1515/9783110298505-004

newable. A development towards the increased use of renewable energy sources is currently in progress.

## 1.2 Fermions and bosons

An atom consists of a nucleus with positive charge and electrons with negative charge. The nucleus consists of protons with positive charge and electrically neutral neutrons. An electron is believed to be an elementary particle, and measures less than $10^{-18}$ m in diameter. Elementary particle is a general term for particles that cannot be further divided. Electrons do not orbit around the nucleus in the usual sense of the word, even though textbook figures often illustrate them as if they did. Electron clouds are stochastically distributed around the nucleus, which contributes to the size of the atom (diameter: ~ 0.2 nm). Electron clouds also exist like waves, which can be observed as a particle when measured. However, it is difficult to define the size of electron clouds. When atoms connect through chemical bonding to form molecules, or they are ionized, the size of atomic clouds change naturally and the size of atoms also becomes different.

The nucleus consists of protons and neutrons, and measures ~ $10^{-15}$ m (1 fm) in diameter. Mesons transmit the force of protons at a minute scale. According to the standard model, protons and neutrons consist of up and down quarks, and there are six types of quarks with three stages of generation in nature.

An electron is one of the six particles referred to as leptons. A proton consists of two up quarks and one down quark, and a neutron consists of one up quark and two down quarks, as shown in Fig. 1.1. These quarks are believed to be elementary particles at present, though superstring theory has also been proposed as a further

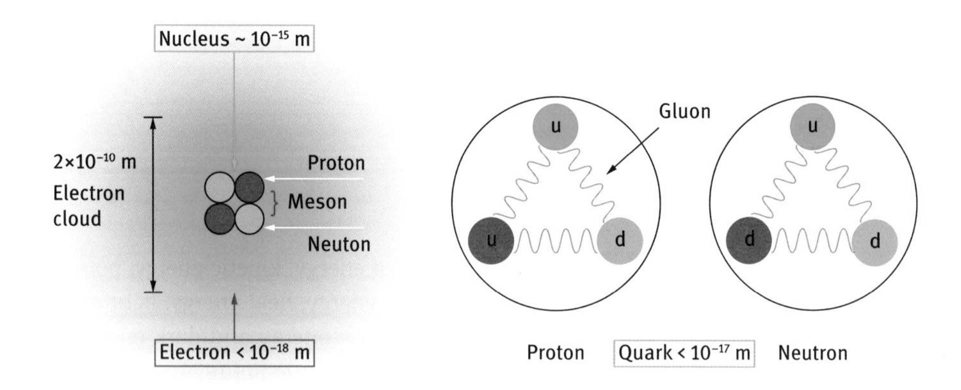


Fig. 1.1: Structure of atom, proton and neutron.

theory. Superstring theory indicates that elementary particles are a certain kind of string, and that quarks and leptons can be formed by the vibration of the strings. This theory is also called the quantum theory of gravity because of its inclusion of gravity.

Fermions are quantum particles with a spin angular momentum of half-integers such as 1/2, 3/2 and 5/2, as listed in Table 1.2. Fermions are guided by the Pauli Exclusion Principle, which indicates that two particles cannot occupy the same quantum state. Fermi-Dirac statistics apply to identical particles with half-integer spins in a system with thermodynamic equilibrium. The particles classified as fermions are quarks and leptons such as electrons, muons and neutrinos.

Table 1.2: Fermions and bosons.

| Fermions | First generation | | | Second generation | | | Third generation | | |
|---|---|---|---|---|---|---|---|---|---|
| | Charge / Spin / Mass | | | Charge / Spin / Mass | | | Charge / Spin / Mass | | |
| Quarks | Up u | | | Charm c | | | Top t | | |
| | +2/3 | +1/2 | ~ 2.3 MeV | +2/3 | +1/2 | ~ 1.3 GeV | +2/3 | +1/2 | ~ 173 GeV |
| | Down d | | | Strange s | | | Bottom b | | |
| | −1/3 | +1/2 | ~ 4.8 MeV | −1/3 | +1/2 | ~ 95 MeV | −1/3 | +1/2 | ~ 4.2 GeV |
| Leptons | Electron $e^-$ | | | Muon $\mu^-$ | | | Tau $\tau^-$ | | |
| | −1 | +1/2 | 0.511 MeV | −1 | +1/2 | 106 MeV | −1 | +1/2 | 1.78 GeV |
| | Electron neutrino $\nu_e$ | | | Muon neutrino $\nu_\mu$ | | | Tau neutrino $\nu_T$ | | |
| | 0 | +1/2 | < 2.2 eV | 0 | +1/2 | < 170 keV | 0 | +1/2 | < 16 MeV |

| Bosons | Force | Charge | Spin | Mass |
|---|---|---|---|---|
| Photon $\gamma$ | Electromagnetic | 0 | 1 | 0 |
| Z boson $Z^0$ | Weak | 0 | 1 | 91.2 GeV |
| W boson $W^\pm$ | Weak | ± 1 | 1 | 80.4 GeV |
| Gluon g | Strong | 0 | 1 | 0 |
| Gaviton G | Gravty | 0 | 2 | 0 |
| Higgs boson H | Mass | 0 | 0 | 126 GeV |

On the other hand, bosons are quantum particles with an integer spin angular momentum, as listed in Table 1.2. A photon is a particle with a spin of 1. Bosons can occupy the same quantum state even in the case of more than one particle in one system. Bose-Einstein statistics apply to identical particles with an integer spin in systems with thermodynamic equilibrium. Examples of bosons include gauge particles, which carry the forces of elementary particles, such as photons, weak bosons and gluons. A graviton is an undiscovered boson with a spin of 2. A Higgs boson, which causes mass in elementary particles is a boson with a spin of 0. Cooper pairs, which are related to the phenomenon of superconductivity, obey Bose-Einstein statistics.

Neutrino is a general name for electrically neutral leptons, and neutrinos come in three flavors: electron neutrinos, muon neutrinos and tau neutrinos, associated with the electron, muon and tau, respectively. Although several quadrillion neutrinos pass through the human body each second, nobody feels them as they pass. Neutrinos almost never interact with matter, and it is quite difficult to observe them.

## 1.3 Important physical constants in the universe

The most important physical constants in our universe are the following:
- Velocity of light $c$ ($3.00 \times 10^8$ m s$^{-1}$)
- Planck constant $h$ ($6.63 \times 10^{-34}$ J s)
- Gravitational constant $G$ ($6.67 \times 10^{-11}$ m$^3$ S$^{-2}$ kg$^{-1}$)

The Planck constant is a universal constant at the quantum scale. The energy of light ($E$) is proportional to the frequency ($\nu$) of light, and the proportionality constant is a Planck constant.

$$E = h\nu \tag{1.1}$$

The velocity of light and the gravitational constant are large-scale constants valid across the universe, while the Planck constant is a constant at an extremely small scale.

## 1.4 Four fundamental forces of nature

- Gravity: The universal gravitation ($F$) between $m_1$ and $m_2$ at a distance of $r$ is expressed as follows:

$$F = G\frac{m_1 m_2}{r^2} \tag{1.2}$$

Although gravitation interacts at a distance in a similar way to the electromagnetic force, gravitational force is very weak. Stars with high mass density attract and confine light, and can potentially form black holes.
- Electromagnetic force: The electrostatic force ($F$) between $q_1$ and $q_2$ with a distance of $r$ is expressed by Coulomb's law as follows:

$$F = \frac{1}{4\pi\varepsilon_0}\frac{q_1 q_2}{r^2} \tag{1.3}$$

$\varepsilon_0$ is a dielectric constant of a vacuum. Magnetic force functions similarly, and gravitational and electromagnetic forces depend on $r^2$. Various forms of energies central to life depend on the electromagnetic force, such as chemical reactions or bioenergy.

- Weak force: The weak force found by Fermi works at the elementary particle scale ($10^{-18}$ m) and causes radioactive decay such as beta decay, in which a beta-ray (an electron) and an associated neutrino are emitted from an atomic nucleus.
- Strong force: The strong force is about 100 times stronger than electromagnetic force according to the theory of nucleus force and mesons. The interaction range of the strong force extends to about the size of nucleus ($10^{-15}$ m) and its potential is expressed as follows:

$$U(r) \sim -\frac{g^2}{4\pi} \frac{e^{-r/\lambda}}{r} \tag{1.4}$$

Where $m$ is the mass of a meson, a particle has a Compton wavelength ($\lambda = h/mc$), and $g^2/4\pi$ is a bonding constant. As expressed by the exponential $e^{-r/\lambda}$, the force only acts at a close distance and other repulsive forces also act around the center of the nucleus.

These are the four forces that exist in the universe and their interaction ranges are different, as listed in Table 1.3. The forces interact through the distortion of fields and the exchange of particles. Gravitational and electromagnetic forces act over an infinite range. These four forces can be used in various ways as energy resources.

Table 1.3: The four fundamental forces as gauge bosons.

| **Particles mediating** | **Photon** $\gamma$ | **Weak boson** W, Z | **Gluon** g | **Graviton** G |
|---|---|---|---|---|
| Transmission force | Electromagnetic force | Weak force | Strong force | Gravitation |
| Strength scale ratio | $10^{-2}$ | $10^{-5}$ | 1 | $10^{-40}$ |
| Spin | 1 | 1 | 1 | 2 |
| Mass | 0 | 80, 91GeV | 0 | 0 |
| Acts on | Electric charge | Flavor | Color charge | Mass, energy |
| Functional ranges | Infinite | $10^{-18}$ m | $10^{-15}$ m | Infinite |
| Functional places | Atoms and molecules | Inside of nucleus | Inside of nucleus | Universal space |
| Energy source | Fossil fuels | Geothermal energy | Sun, nuclear energy | Hydropower, tide power |

## 1.5 The mass of light

Particles with an electric charge, such as electrons and protons, absorb or discharge photons and the kinetic energy of the photons is exchanged for the transmission of electromagnetic force. Light is a quantum particle in an electromagnetic field. Its static mass $m_0$ is 0 and its spin is 1. Energy ($E$) is expressed by $h\nu$ ($\nu$: frequency), and the direction of the electric field vector is vertical to the direction of the movement of light as a result of $m_0$.

However, light is not actually static, but, rather, moves at the velocity of light. A static mass of zero is an expedient value without any physical meaning, and light has energy as expressed in Eq. (1.1):

$$E^2 = c^2p^2 + m_0^2c^4 \tag{1.5}$$

where c is the velocity of light and p is momentum. E indicates static energy for $p = 0$, and Eq. (1.1) can be applied to photons and phonons for $m_0 = 0$. This equation has two solutions as expressed in Eq. (1.2).

$$E = \pm(c^2p^2 + m_0^2c^4)^{1/2} \tag{1.6}$$

Here, the minus sign corresponds to an antiparticle or antimatter such as a positron for an electron or an antihydrogen atom for a hydrogen atom.

The kinetic energy of ordinary light is $h\nu$, and the mass of moving light, which interacts with gravity, can be expressed as $h\nu/c^2$. However, this mass is also an expedient value without a physical meaning, and the mass is generally expressed as $E/c^2$ by using energy.

Light waves detected as particles are photons, and photons transmit electromagnetic force. In the interaction between two objects at a distance, charged particles such as protons and electrons always repeated emit and absorb photons. When the charged particles are close to each other, the force acts through the frequent emission and absorption of photons. Since the photons have no mass, the electromagnetic force acts over an infinite range.

According to the theory of relativity, four-dimensional spacetime is considered as a combination of three-dimensional space and one-dimensional time. What connects spatial coordinates and the axis of time is "information transmitted by photons". Since the velocity of light is not infinite, it takes some time for photons to reach a certain distance. Therefore, when the photon arrives, the information carried by the photon has already become old.

The energy of photons for a number of photons ($N$) is expressed by the following equation. Here, $\lambda$ is the wavelength of light, and $P$(W) expresses the power of light.

$$E = h\nu = \frac{hc}{\lambda} \tag{1.7}$$

$$N = \frac{E}{h\nu} = \frac{Pt\lambda}{hc} \tag{1.8}$$

## 1.6 The materialization of light and antimatter

If the density of the atomic nucleus (~ $10^{17}$ kg m$^{-3}$) is transformed into energy, the energy is calculated to be ~ $10^{34}$ J m$^{-3}$ ($10^{25}$ J mm$^{-3}$). When such a very high energy density is obtained, energy can materialize. There are few forms of matter with such huge energy density that exist in our universe. Matter's density is generally low because the atomic nucleus is surrounded by an electron cloud. However, black holes and neutron stars have densities close to that of an atomic nucleus.

It has been shown that a quantum particle of radiation generates an electron-positron pair from the occurrence of positive and negative electron pairs when high energy cosmic-rays collide with the nucleus inside a cloud chamber [1]. An energy of $2mc^2$, equivalent to twice the mass of an electron (or positron), is necessary for the pair generation according to Einstein's equation

$$E = mc^2 \tag{1.9}$$

In this way, it was first demonstrated that light can be transformed into matter.

Electrons and positrons are generated together from high energy photons such as gamma rays, and they revert to photons through pair annihilation when they collide, as shown in Fig. 1.2(a). This indicates the occurrence of probable modulation according to quantum mechanics and energy equivalence with mass for the theory of relativity. In relativistic quantum theory, a vacuum is not an empty space, but rather a space filled with particle pairs in a process of generation and annihilation. Sakharov indicated that a smaller amount of matter in comparison to antimatter could be produced through a small violation of charge parity (CP) symmetry.

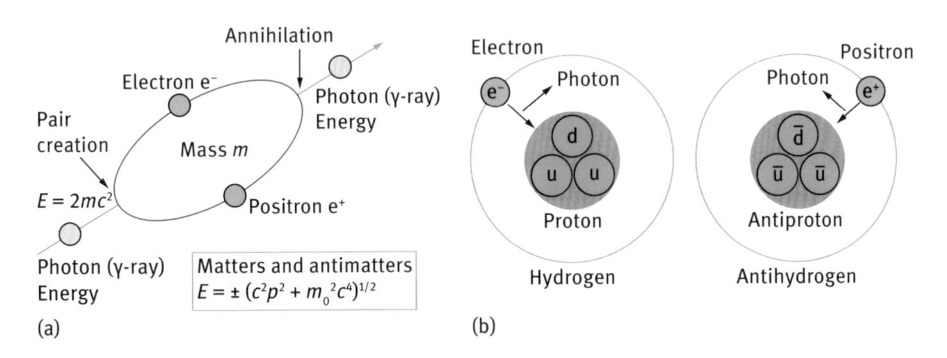


Fig. 1.2: The materialization of light and the generation of hydrogen.

Dirac predicted the existence of antimatter theoretically, and Anderson found positron antimatter in 1932 [2]. Positrons have the same weight as electrons, a positive charge, and combine with electrons. An electron and positron pair is formed from high energy light such as gamma rays, and, after pair annihilation, becomes a photon again.

In 1996, researchers at the European Organization for Nuclear Research (CERN) succeeded in synthesizing antihydrogen atoms, as shown in Fig. 1.2(b). An antihydrogen atom consists of antiparticles and is the form of antimatter with the simplest structure. Recently, the generation of a large amount of antihydrogen atoms has been reported [3].

On the other hand, Einstein's finding that matter changes into light is utilized in nuclear reactors. Matter also changes into light in the sun. The nuclear fission of uranium occurs in nuclear reactors, the nuclear fusion of protons occurs in the sun, and in both cases we utilize the huge energies generated.

Theoretically, antimatter is a completely clean energy source [4]. When matter and antimatter annihilate each other, all the mass is transformed into energy (photons). The pair annihilation is 100 % efficient, and comparatively more energy can be produced than in nuclear fission and nuclear fusion reactions. Although antimatter could be used as fuel without leaving behind any residue because no reaction products are produced, this is extremely difficult to do in reality. However, it's possible to consider a method in which a layer resulting from the first reaction blocks the remaining reaction and antimatter is stopped from continuing the reaction, through a phenomenon comparable to the the Leiden frost effect.

## 1.7 Bose-Einstein condensation and freezing light

Usually, at the high temperature that is room temperature, both bosons and fermions behave as classical particles and can be distinguished from each other. Fermions behave claustrophobically, and two fermions cannot occupy identical quantum states in the same location [5]. Electrons, protons and neutrons are fermions. On the other hand, bosons behave gregariously, and bosons of a particular species tend to gather together in identical quantum states if given the opportunity. Photons are bosons. Composite particles such as atoms are also either bosons or fermions. An atom made of an even number of protons, neutrons and electrons is a boson.

When wavelengths of thermal de Broglie waves increase to the distances between atomic particles at low temperatures, these particles cannot be distinguished, and bosons condense [6]. Bose-Einstein condensation (BEC) is a phenomenon in which the wave functions of Bose particles (bosons) are extended and overlap at very low temperatures, and all bosons have the same quantum state, as shown in Fig. 1.3. Bose-Einstein condensates (BECs) have a huge size of 10 μm, which is almost the same as that of human cells. The probability of occupying the same state for a number of N bosons increases according to the equation (2N/(N+1)).

To freeze light, BEC as a macroscopic quantum phenomenon is necessary. Similar phenomena such as superconductivity and superfluidity have been found, and BEC with ideal Bose gas was first achieved in 1995 [5, 7], Atoms in a gaseous condensate experience a small amount of mutual repulsion or attraction, depending on the species.

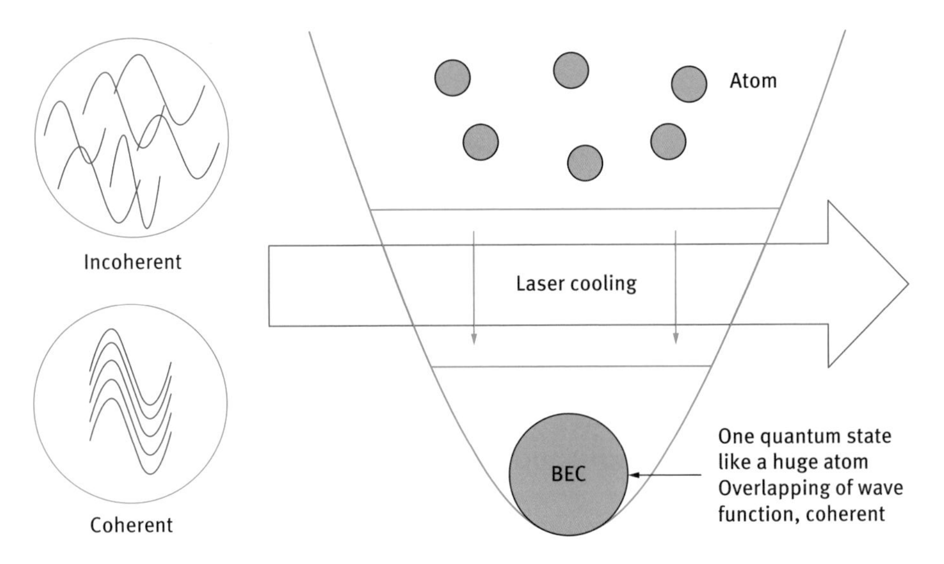


Fig. 1.3: Schematic illustration of BEC.

Importantly, repulsion stabilizes condensates, whereas attraction destabilizes them. Consequently, repulsive atoms form BECs consisting of millions of atoms.

Other BEC-related systems include lasers, superconductivity, superfluidity and excitons. Since photons (bosons) have characteristics that place them in the same quantum state, laser light is produced by the coherent phase of the photon wave. Superconductivity and superfluidity are phenomena in which electron pairs and helium atoms are Bose-condensed. Excitons are pairs of a hole and an electron which are Bose-condensed. Such phenomena related to quantum coherence have been observed for $C_{60}$ fullerene [8]. In addition to the BEC, macroscopic quantum coherence has also been achieved by using a superconducting quantum interference device (SQUID) [9, 10].

Although producing fermions in the laboratory was difficult originally, most matter around us consists of fermions. To make an equivalent Fermi condensate requires pairing off reluctant fermions so that their combined spin is an integer. A method that makes pairs of them like bosons was discovered in 2003 [11]. Fermi condensation could be related to phenomena that play a role in life [12].

Recently, a method for freezing light was developed [5, 13]. The method is described below. First, Bose-Einstein Condensate (BEC) is prepared as the light freezing medium. Before the light pulse reaches the cloud of BEC atoms that will freeze it, all the atoms' spins are aligned, and a coupling laser beam renders the BEC transparent to the pulse [14]. The BEC atoms greatly slow and compress the pulse, and the atoms' states change in a wave that accompanies the slow light. When the pulse is fully inside the cloud, the coupling beam is turned off, halting the wave and the light. Then, the light vanishes at zero velocity. Later, the coupling beam is turned on again, regenerating the light pulse and setting the wave and the light back in motion.

The light pulse comes to a grinding halt and turns off. However, the information that was in the light is not lost. That information has already been imprinted on the atoms' states. When the pulse halts, the imprint is simply frozen in place. The frozen pattern imprinted on the atoms contains all information about the original light pulse. This phenomenon is equivalent to a hologram of the pulse written on the atoms of gas. The hologram is read out by turning the coupling laser on again. The light pulse reappears and sets off in slow motion again, along with the wave of the atoms' state. It was reported that light could be stored for 1 ms by using the above method [15, 16].

## 1.8 Quantum brain theory and light

Quantum brain theory clarifies the elemental process that brings forth consciousness even at the level of a quantum effect on the basis of the minute structure of brain cells. Theoretically describing the physical elemental processes that take place in the organization of the brain as a macroscopic condensate is unreasonable in quantum mechanics; quantum field theory can handle a specific physics phenomenon in a system with infinite degrees of freedom.

Quantum brain theory was developed to propose a relationship between brain areas in the cranium and their functions by using first principle calculations from quantum field theory [17]. It has now developed into the field of quantum brain dynamics (QBD) [18]. It was theoretically argued that water inside of cells should behave like macroscopic condensates because the ground state of an electric dipole field degenerates infinitely and electric dipoles have a macroscopic quantum state with a size of several tens of microns [17]. It was also theoretically reported that microtubles should be produced from protein molecules, and coherent light (super-radiance) should be radiated [19, 20]. This light could be related to polaritons, which have the characteristics of both light and matter. Based on the Higgs mechanism for a gauge field in the cranium, critical temperature $T_C$ is expressed as follows:

$$T_C = \frac{2\pi\hbar^2}{mk_B}\left[\frac{n}{\zeta(3/2)}\right]^{2/3} \tag{1.10}$$

where quantum fluctuation energy is equal to thermal energy, mass is 13.6 eV, $k$ is the Boltzmann constant, $m_p$ is the weight of a photon, $n$ is the particle density, and $\zeta$ is the Riemann zeta function. Since the $T_C$ increases to room temperature, tunneling (evanescent) photons which are stable at around 300 K (close to human body temperature) appear as BECs [21]. These photons are a macroscopic quantum condensate and quantum particles of an unprogressive wave mode by the tunneling effect, and the momentum becomes an imaginary number. These tunneling photons generated in the brain could be closely related to consciousness. The coherent length $\xi = \hbar/m_p c$

is calculated to be several tens of microns. According to this theory, these evanescent photons are strongly related to the mind, and the vibration mode measured at the surface of cranium corresponds to brain waves (observed as an electroencephalogram). It has also been reported that the most important preservation site of light is deoxyribonucleic acid (DNA), and that DNA is a source of bio-photon emission [22]. It is believed that the phenomena of light could also be related to both consciousness and life-related phenomena, as described previously [23–25].

The holographic principle is a concept proposed by Gerard't Hooft, a 1999 Nobel laureate in physics [26, 27]. According to the holographic principle, the universe can be explained as a gigantic hologram. The universe where we live is a 4-dimensional system with time, and all information on time and space in the universe is recorded on a 3-dimensional boundary [28]. relation between information and energy was also indicated from the holographic principle, as the universal entropy boundary ($I_{UEB}$), expressed as the next equation [29].

$$I_{UEB} \leq \frac{2\pi ER}{\hbar c \ln 2} \tag{1.11}$$

This is the upper information limit when all energy $E$ is included within the radius $R$. From eqs. (1.9) and (1.11), mutual transformations betweeninformation, energy and mass are possible. The information projected from a 3-dimensional boundary would be transformed to energy, and a part of it would be materialized as atoms. Presently, the consciousness of a human being can be regarded as a form of information in science, and consciousness aould have energy. The holographic information limit that a certain space can hold might be related with the consciousness [23–25].

## 1.9 The materialization of vacuum

Our universe consists of positive energy. Negative energy is exceedingly unstable, and there are few cases of few negative energy in nature. Negative energy is also called antimatter or antiparticles. These are generated in particle accelerators such as the Large Hadron Collider (LHC), and become light again by combining with matter after an extremely short time. However, negative energy appears throughout the whole universe at a quantum level, and the vacuum is an inexhaustible treasure house for finding negative energy particles.

Heisenberg's uncertainty principle is expressed as follows.

$$\Delta t \Delta E \geq \frac{\hbar}{2}, \quad \Delta x \Delta p \geq \frac{\hbar}{2} \tag{1.12}$$

$X$, $p$, $E$ and $t$ are position, momentum, energy and time, respectively. Wave functions have two components of phase and amplitude, and two conjugate physical quantities cannot be determined simultaneously from the equation. The uncertainty principle indicates that if the time is confirmed, energy information becomes vague. When the time $\Delta t$ is exceedingly short, the energy $\Delta E$ become extremely huge, which results in the materialization of the energy even in a vacuum, as shown in Fig. 1.4 [30].

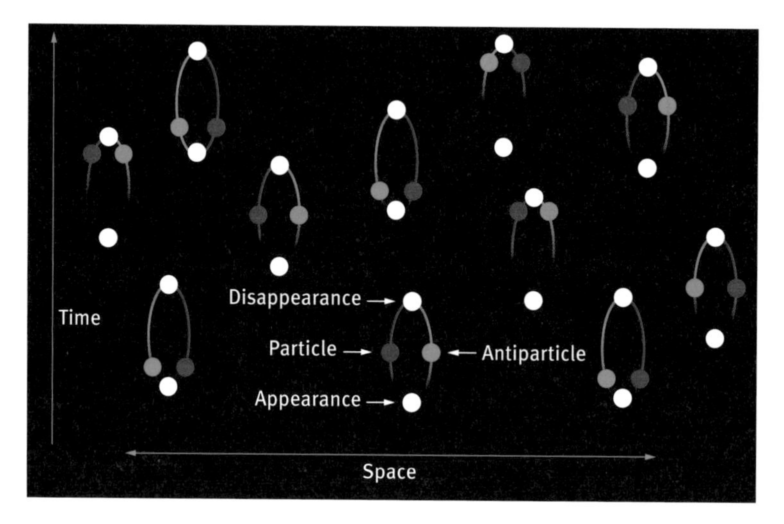


Fig. 1.4: The materialization of vacuum.

The vacuum experiences fluctuation and vibrations, and has an extremely small energy called zero-point energy in a certain region size $L$.

$$E = \frac{h^2}{8mL^2} \tag{1.13}$$

A pair of particle and antiparticle (positive and negative energies) appears at all times from the space of zero-point energy. After a very short time, the particles and antiparticles combine again, and return to the zero-point energy. The time and size are too small to detect negative energy or antiparticles. However, negative energy has been found indirectly by the Casmir effect, where an attractive force from the negative energy acts between two metal plates in a vacuum. The formation of the pair of particle and antiparticle indicates statistical fluctuation for quantum dynamics and the equivalence of energy and mass for theory of relativity. According to relativistic quantum theory, the vacuum is not a space that contains nothing, but rather a space filled with appearing and disappearing pairs of particles and antiparticles.

## 1.10 The energy constitution of the entire universe

In February 2003, NASA announced the observation results of the universe's temperature and light obtained by the investigation satellite Wilkinson Microwave Anisotropy Probe (WMAP). NASA determined basic data about the universe's constitution, expansion speed and geometry with a very high accuracy. The energy composition of the entire universe is shown in Fig. 1.5, which was determined with the high accuracy of 5 % or less error. The detailed constitution of the universe as far as it is understood is summarized in Table 1.4 [31, 32]. In addition, the geometry of the universe was determined to be flat with a space curvature of 0 [33], and the age of the universe was determined to be 13.7 billion years.

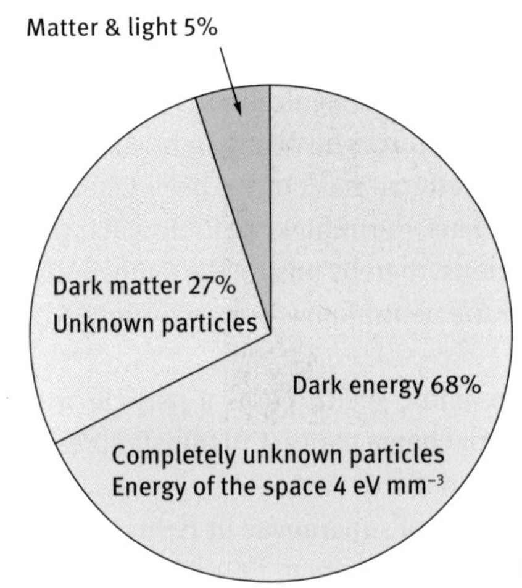


Fig. 1.5: The energy composition of the entire universe.

Table 1.4: The constitution of the universe.

| Material | Representative particles | Particle energy (eV) | Number of particles | Contribution to mass of universe |
|---|---|---|---|---|
| Baryonic matter | Protons, Electrons | $10^6$ to $10^9$ | $10^{78}$ | 4.9 % |
| Radiation | Photons | $10^{-4}$ | $10^{87}$ | 0.005 % |
| Dark matter | Supersymmetric particles? | $10^{11}$ | $10^{77}$ | 26.8 % |
| Dark energy | "Scalar" particles? | $10^{-33}$ | $10^{118}$ | 68.3 % |

Although the forms of energy in the universe which have been clarified are matter and radiation light, as shown in Table 2.1, these make up only 5 % of all the universe's energy. These correspond to quarks and leptons, and gauge particles transferring the forces are also included. However, dark matter makes up 27 % of the universe's energy and the particles which compose it are unknown. A further 68 % is dark energy that has not been clarified at all. That is, about 95 % of the entire universe is composed of unclarified forms of energy. This is the biggest mystery faced by cosmological and particle physics in the 21st century.

Although a leading candidate for hot dark matter in dark matter is neutrinos, the proportion of neutrinos in the universe is very small (0.3 %) [32]. It is important to note that cold dark matter occupies about 1/4 of the universe. Although cold dark matter has not yet been identified, several candidates have been considered [32].

The first candidate for cold dark matter is the still unknown elementary particle that supersymmetry theory predicts, the neutralino [34] These are an amalgam of superpartners, photons (which transmit electromagnetic force), Z bosons (which transmit weak nuclear force) and other particle types. The neutralino is very heavy, has zero charge and is unaffected by electromagnetic forces involving light. Since the neutralino has almost no interaction with other matter, it has not yet been detected. When an occasional neutralino hits an atomic nucleus, the unlucky particle will transfer a small amount of its kinetic energy to the nucleus, thereby raising the temperature of the material slightly. At present the search for the neutralino is being continued by researchers worldwide.

The second candidate for dark matter is an axion, predicted as a particle with strong force, which was theoretically introduced for charge parity (CP) transfer preservation [35]. However, the axion has also not yet been discovered.

Dark energy was discovered from the observation of supernovae in 1998 [36], and *Science* reported on it as the most important scientific discovery in 1998 [37]. As part of this discovery, the fact that the expansion of the universe is accelerating was clarified [38] and the energy responsible for this was named dark energy. This dark energy is completely burying all cosmic space with an energy of 4 eV $mm^{-3}$ [39].

The cosmological constant Einstein predicted is believed to be one of the candidates for dark energy [40, 41]. Quintessence, a cosmological constant with time dependence, has also been proposed [39]. This energy is based on the concept that energy exists in space which had been thought to be a vacuum. This concept is introduced from a basic principle of quantum mechanics combined with special relativity. Empty space without matter and light is actually filled with elementary particles that pop in and out of existence too quickly to be detected directly, which are called 'virtual particles'. According to the uncertainty principle of quantum theory, all things, even emptiness, are wavering, and the product of time and energy is larger than Planck's constant. In other words, the generation and disappearance of pair particles with high energy for a short period can occur without contradicting the energy preservation rule. An observed effect that the virtual particles cause on real space is the Casimir effect [39]. If particles with positive or negative energy are controlled, positive or negative energy can be observed. If the dark energy is due to a quantum gravity effect, the energy level would be expected to have the energy density of Planck's energy. However, the problem posed by the fact that the energy in fact has the smaller value of 123 orders remains unsolved. Although some possible candidates to explain dark energy have been presented in recent years, its essence remains entirely unclarified.

## 1.11 Cosmological constant

As described in the previous section, Einstein's cosmological term is proposed as a candidate for the dark energy that occupies 73 % of the universe. The Einstein equation, which connects the structure of time and space and the gravity of matter is as follows:

$$R_{\mu\nu} - \frac{1}{2}g_{\mu\nu}R + \Lambda g_{\mu\nu} = \frac{8\pi G}{c^4}T_{\mu\nu} \tag{1.14}$$

$R_{\mu\nu}$: Ricci tensor showing the distortion of time and space by the metric tensor, $g_{\mu\nu}$: metric tensor prescribing distance of time and space, $R$: $R = g_{\mu\nu}R_{\mu\nu}$, $\Lambda$: cosmological constant, $G$: Newton's gravity constant, $c$: velocity of light, $T_{\mu\nu}$: energy-momentum tensor showing the energy and momentum of matter. This equation prescribes mass, energy, time and space. Based on the general coordinate transformation covariance, scalar and vector (electromagnetic force etc.) quantities are tensors of the 0th and 1st rank, respectively, and the time and space geometry of this equation are expressed by a metric tensor of the second rank.

In this equation, $\Lambda g_{\mu\nu}$ is called the cosmological term, referring to negative pressure and antigravity. Matter particles such as gases have positive pressure; the kinetic energy of atoms and radiation pushes outward on the container. However, negative pressure behaves in the opposite way and interaction between atoms overcomes the kinetic energy, which causes the gas to implode. Implosive gas has a negative pressure. Note that the direct effect of negative pressure – implosion – is the opposite of its gravitational effect – repulsion.

A cosmological term in Einstein equation is the cosmological constant $\Lambda$, and it has a regular value. Since the cosmological term is not sufficient for a complete description of the accelerating expansion of the universe, the energy of a scalar field (invariable field to space rotation) dependent on time is introduced as a cosmological term dependent on time, which is called quintessence [42]. Quantum field theory predicts quintessence is conceivable as one candidate for dark energy. Although there have been many inquiries into dark energy itself, more detailed research will be necessary in the future.

## 1.12 Bibliography


[1] Blackett PMS. Positive electron. *Nature*. 1933; 132: 917–919.

[2] Anderson CD. The positive electron. *Phys Rev*. 1933; 43: 491–494.

[3] Amoretti M, Amsler C, Bonomi G, Bouchta A, Bowek P, Carraro C, Cesar CL, Charlton M, Collier MJT, Doser M, Filippiniq V, Fine KS, Fontanaq A, Fujiwara MC, Funakoshi R, Genovaq P, Hangstk JS, Hayano RS, Holzscheiter MH, Jørgensen LV, Lagomarsino V, Landua R, Lindelöf D, Lodi Rizzini E, Macri M, Madsen N, Manuzio G, Marchesottiq M, Montagnaq P, Pruys H, Regenfus C, Riedler P, Rochet J, Rotondiq A, Rouleau G, Testera G, Variola A, Watson TL, van der Werf DP. Production and detection of cold antihydrogen atoms. *Nature*. 2002; 419: 456–459.

[4] Frazer G. Antimatter: The ultimate mirror. Cambridge University Press, Cambridge, UK. 2002.
[5] Hau LV. Frozen Light. *Sci Amer*. 2001; 285(1): 52–59.
[6] Collins GP. The coolest gas in the universe. *Sci Amer*. 2000; 283(6): 68–75.
[7] Anderson MH, Ensher JR, Matthews MR, Wieman CE, Cornell EA. Observation of Bose-Einstein condensation in a dilute atomic vapor. *Science*. 1995; 269: 198–201.
[8] Arndt M, Nairz O, Vos-Andreae J, Keller C, van der Zouw G, Zeilinger A. Wave-particle duality of $C_{60}$ molecules. *Nature*. 1999; 401: 680–682.
[9] van der Wal CH, ter Haar ACJ, Wilhelm FK, Schouten RN, Harmans CJPM, Orlando TP, Lloyd S, Mooij JE. Quantum superposition of macroscopic persistent-current states. *Science*. 2000; 290: 773–777.
[10] Friedman JR, Patel V, Chen W, Tolpygo SK, Lukens JE. Quantum superposition of distinct macroscopic states. *Nature*. 2000; 406: 43–46.
[11] Greiner M, Regal CA, Jin DS. Emergence of a molecular Bose–Einstein condensate from a Fermi gas. *Nature*. 2003; 426: 537–540.
[12] Regal CA, Greiner M, Jin DS. Observation of resonance condensation of Fermionic atom pairs. *Phys Rev Lett*. 2004; 92: 040403-1-4.
[13] Fleischhauer M, Yelin SF, Lukin MD. How to trap photons? Storing single-photon quantum states in collective atomic excitations. Optics Commun. 2000; 179: 395–410.
[14] Hau LV, Harris SE, Dutton Z, Behroozi CH. Light speed reduction to 17 metres per second in an ultracold atomic gas. *Nature*. 1999; 397: 594–598.
[15] Liu C, Dutton Z, Behroozi CH, Hau LV. Observation of coherent optical information storage in an atomic medium using halted light pulses. *Nature*. 2001; 419: 490–493.
[16] Lukin MD, Imamoğlu A. Controlling photons using electromagnetically induced transparency. *Nature*. 2001; 413: 273–276.
[17] Stuart CI, Takahashi Y, Umezawa H. On the stability and non-local properties of memory. *J Theoretical Biol*. 1978; 71: 605–618.
[18] Jibu M, Hagan S, Hameroff SR, Pribram KH, Yasue K. Quantum Optical Coherence in Cytoskeletal Microtubules: Implications for Brain Function. *BioSystems*. 1994; 32: 195–209.
[19] del Giudice E, Doglia S, Vitiello MMG. Electromagnetic field and spontaneous symmetry breaking in biological matter. *Nucl Phys B*. 1986; 275: 185–199.
[20] Jibu M, Hagan S, Hameroff SR, Pribram KH, Yasue K. Quantum optical coherence in cytoskeletal microtubules: Implications for brain function. *BioSystems*. 1994; 32: 195–209.
[21] Jibu M, Pribram KH, Yasue K. From conscious experience to memory storage and retrieval: The role of quantum brain dynamics and Boson condensation of evanescent photons. *Int J Modern Phys B*. 1996; 10: 1735–1754.
[22] Rattemeyer M, Popp FA, Nagl W. Evidence of photon emission from DNA in living systems. *Naturwissenschaften*. 1981; 68: 572–573.
[23] Oku T. A study on consciousness and life energy based on quantum holographic cosmology. *J Intl Soc Life Info Sci*. 2005; 23(1): 133–154.
[24] Oku T. Consciousness-information-energy medicine: Health science based on quantum holographic cosmology. *J Intl Soc Life Info Sci*. 2007; 25(1): 140–163.
[25] Oku T. Science towards reality and meaning. *J Intl Soc Life Info Sci*. 2008; 26(1): 65–70.
[26] 't Hooft G. Dimensional reduction in quantum gravity. in Salam-festschrifft. Ed. Aly A, Ellis J, Randjbar-Daemi S. World Scientific, Singapore. 1993.
[27] 't Hooft G. Nobel Lecture: A confrontation with infinity. *Rev Modern Phys*. 2000; 72: 333–339.
[28] Susskind L. The world as a hologram. *J Math Phys*. 1995; 36: 6377–6396.
[29] Bekenstein J.D. Holographic bound from second law of thermodynamics. *Phys Lett B*. 2000; 481: 339–345.
[30] Krauss LM, Turner MS. A cosmic conundrum. *Sci Amer*. 2004; 291(3): 70–77.

[31] NASA/WMAP. http://map.gsfc.nasa.gov/
[32] Cline DB. The search for dark matter. *Sci Amer*. 2003; 288(3): 28–35.
[33] De Bernardis P, Ade PAR, Bock JJ, Bond JR, Borrill J, Boscaleri A, Coble K, Crill BP, De Gasperis G, Farese PC, Ferreira PG, Ganga K, Giacometti M, Hivon E, Hristov VV, Iacoangeli A, Jaffe AH, Lange AE, Martinis L, Masi S, Mason PV, Mauskopf PD, Melchiorri A, Miglio L, Montroy T, Netterfield CB, Pascale E, Piacentini F, Pogosyan D, Prunet S, Rao S, Romeo G, Ruhl JE, Scaramuzzi F, Sforna D, Vittorio N. A flat universe from high-resolution maps of the cosmic microwave background radiation. *Nature*. 2000; 404: 955–959.
[34] Jungman G, Kamionkowski M, Griest K. Supersymmetric dark matter. *Phys Rep*. 267; 1996: 195–373.
[35] Ogawa I, Matsuki S, Yamamoto K. Interactions of cosmic axions with Rydberg atoms in resonant cavities via the Primakoff process. *Phys Rev D*. 1996; 53: R1740–R1744.
[36] Perlmutter S, Aldering G, Della Valle M, Deustua S, Ellis RS, Fabbro S, Fruchter A, Goldhaber G, Goobar A, Groom DE, Hook IM, Kim AG, Kim MY, Knop RA, Lidman C, McMahon RG, Nugent P, Pain R, Panagia N, Pennypacker CR, Ruiz-Lapuente P, Schaefer B, Walton N. Discovery of a supernova explosion at half the age of the Universe. *Nature*. 1998; 391: 51–54.
[37] Glanz J. Astronomers see a cosmic antigravity force at work. Science 1998; 279: 1298–1299.
[38] Hogan CJ, Kirshner RP, Suntzeff NB. Surveying space-time with supernovae. *Sci Amer*. 1999; 280: 28–33.
[39] Ostriker JP, Steinhardt PJ. The quintessence Universe. *Sci Amer*. 2001; 284: 36–43.
[40] Krauss LM. The end of the age problem, and the case for a cosmological constant revisited. *Astrophys J*. 1998; 501: 461–466.
[41] Peebls PJE. Evolution of the cosmological constant. *Nature*. 1999; 398: 25–26.
[42] Ostriker JP, Steinhardt PJ. The observational case for a low-density Universe with a non-zero cosmological constant. *Nature*. 1995; 377: 600–602.

# 2 Solar energy

## 2.1 Energy problems and entropy on Earth

The society and technology that human beings have built for themselves are strongly dependent on the consumption of energy. However, earth's energy resources have been consumed for a long time, and we will have to face energy resource problems and find solutions now if we wish to develop a sustainable human society. If we continue to use fossil fuels at the current rate, they will be completely exhausted within the 21$^{st}$ century. Therefore, there is an urgent need to shift towards the use of renewable energies such as solar energy, nuclear fusion, wind power, biomass, and geothermal energy instead of continuing to use fossil fuels. Renewable energy resources are abundant and have a low impact on the natural environment [1]. On the other hand, they provide low energy density and strongly fluctuate depending on the place and time. Therefore, creating a stable supply of energy is difficult given the current state of technology and the social system we live in, and there are many difficulties in practically using these energy resources.

The first law of thermodynamics states that there is conservation of energy, which is indicated as follows:

$$\Delta U = w + q \tag{2.1}$$

An increase of internal energy ($\Delta U$) in a certain system is equal to the sum of work ($W$) and heat ($q$). In other words, the first law of thermodynamics implies that no perpetual motion machine can exist, as shown in Fig. 2.1(a). If fossil fuels and nuclear energy are mainly used to maintain a civilization that consumes a gigantic amount of energy in a closed system, these limited energy resources will be exhausted, as shown in Fig. 2.2(a).

Furthermore, the second law of thermodynamics indicates that entropy ($S$: a thermodynamic index representing a degree of randomness) increases unidirectionally for natural processes in a closed system, as shown in Fig. 2.1(b).

$$\Delta S > 0 \tag{2.2}$$

Various compounds are discharged by the gigantic amount of fuels consumed within the closed system on the earth. If the compounds were broken down naturally, this would present no problem. However, if these compounds are harmful to nature and exceed the quantities that can be dealt with naturally, they produce environmental pollution and health damage to human beings. One example of this can be seen in global warming, which is caused by the increase of carbon dioxide, as shown in Fig. 2.2(b).

Even if various efforts are made to solve the problem from within the closed system on the earth, other problems will appear because through such efforts the earth's entropy will inadvertently be increased. If a civilization functions by consuming inner

DOI 10.1515/9783110298505-005

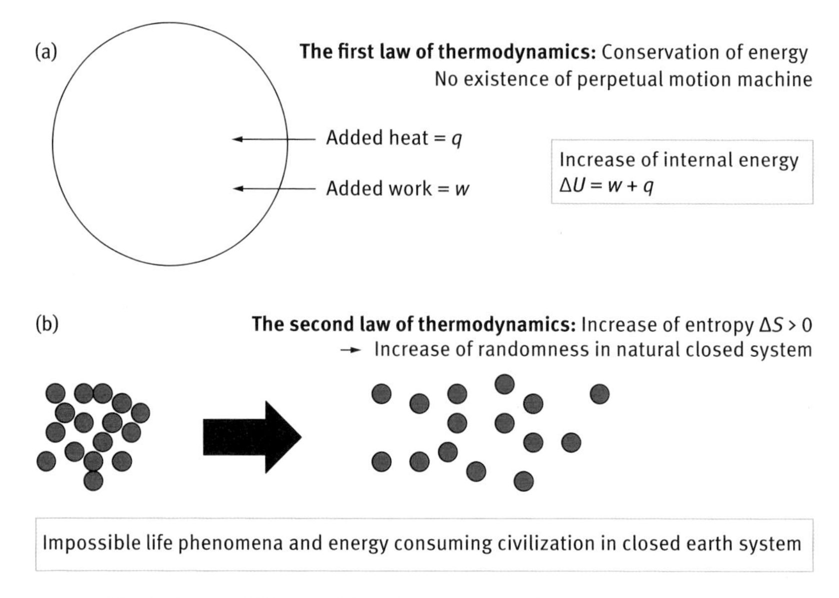


Fig. 2.1: The (a) first and (b) second law of thermodynamics.

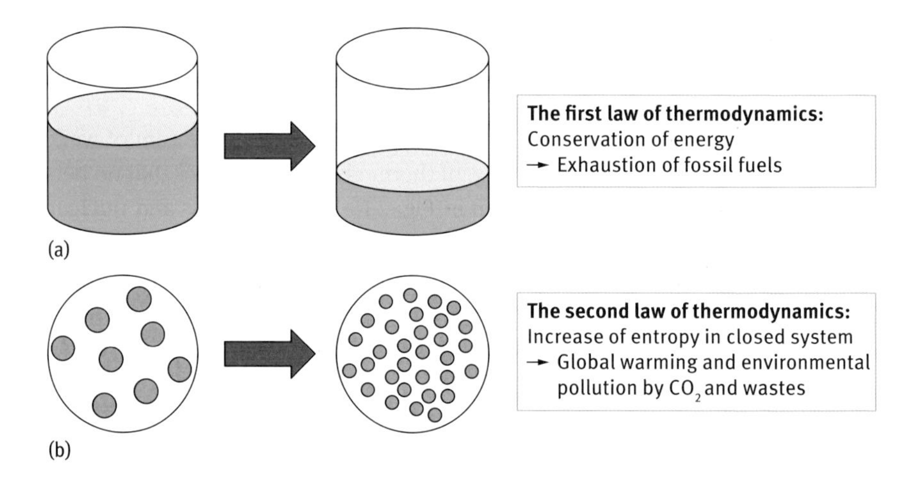


Fig. 2.2: The (a) first and (b) second law of thermodynamics and their predictions on Earth.

energy in the closed system on the earth, the civilization must break down according to the first and second law of thermodynamics.

## 2.2 The energy circulation of photons and electrons

Primitive life emerged around 3.5 billion years ago, and it continues to thrive in spite of the first and second laws of thermodynamics. Why has the phenomenon of life managed to continue for 3.5 billion years? Because most of the energy that maintains life relies on photosynthesis, which uses sunlight as an energy resource. In other words, the phenomenon of life is organized not only within the closed system on earth but also within a system opened to the sun. If the earth were a closed system, the phenomenon of life would not be able to continue.

Life-supporting photosynthesis is a chemical reaction that synthesizes carbohydrate ($C_6H_{12}O_6$) from water ($H_2O$), carbon dioxide ($CO_2$) and visible light containing 45 % of the sunlight. Then, oxygen molecules are also generated through the following reaction:

$$CO_2 + H_2O + 8 \text{ photons} \rightarrow (C_6H_{12}O_6)_{1/6} + O_2 \tag{2.3}$$

The Gibbs standard free energy ($\Delta G°$) and entropy ($\Delta S°$) obtained by the photochemical reaction are as follows:

$$\Delta G^\circ = 114\,\text{kcal}\,CO_2\,\text{mol}^{-1} = 480\,\text{kJ}\,CO_2\,\text{mol}^{-1} \tag{2.4}$$

$$\Delta S^\circ = -43.6\,\text{J}\,\text{K}^{-1}\,CO_2\,\text{mol}^{-1} \tag{2.5}$$

The reaction corresponds to the process of 4 electrons from 1 $CO_2$ molecule, and the free energy is 120 kJ per 1 mol electrons, which corresponds to 1.24 eV. The theoretical decomposition voltage of $H_2O$ is 1.23 V and therefore photosynthesis provides energy which corresponds to the decomposition voltage of $H_2O$ by visible sunlight. Furthermore, photosynthesis provides negative entropy, as shown in eq. (2.5).

In this way, animals use positive free energy and negative entropy from photosynthesis as food. They also take the oxygen abandoned by plants to breathe, and life is preserved through the act of breathing. Excretions are decomposed into photosynthesis ingredients ($CO_2$ and $H_2O$) by bacteria and in other ways, and the material cycle comes full circle.

Animals use the high energy electrons preserved in photosynthesis productions as food, and they also use the oxygen released through photosynthesis to breathe. Then, with the high energy of electrons returned to oxygen, the obtained Gibbs free energy is used for life activity. The process of fossil fuel combustion is similar to the above process. Both life phenomena and the combustion of fossil fuels are represented as electron flows. Energy circulation on the earth is expressed as an electron flow, and the driving force of the electron flow is photons from the sunlight. Therefore, energy circulation on the earth can be understood through just two kinds of fundamental particles *i.e.* photons and electrons. Carbon is a necessary element for life to form various compounds, and carbon dioxide plays a role in carbon circulation.

The environmental problems of our present civilization result from exceeding the permissible limits of nature for the cultural cycle by burning the fossil fuels (photosynthesis products) preserved across the earth. To maintain a civilization with gigantic energy consumption in spite of the first and second laws of thermodynamics, opening our cultural cycle to the sunlight is important, as shown in Fig. 2.3. If sunlight is used as an energy resource, future civilization could be maintained.

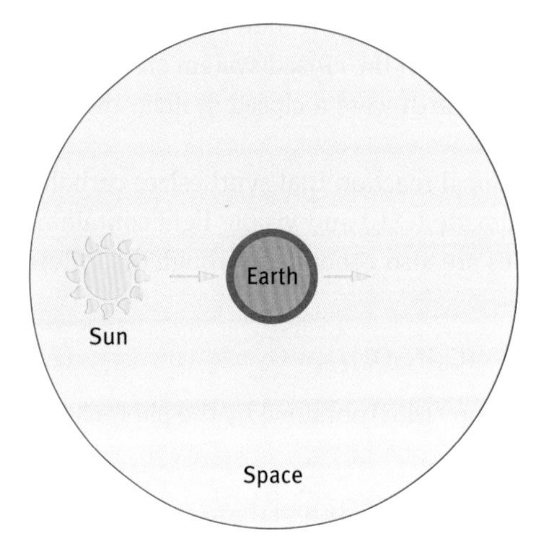


Fig. 2.3: The earth is opened system.

## 2.3 Homeostasis of life and civilization

Due to the first law (energy conservation) and second law (entropy increase) of thermodynamics, our civilization's activity will break down as a result of gigantic energy consumption in a closed system. Why have similar activities of life continued constantly for several billions? It is because, as already described, the most important energy resource used in sustaining the phenomenon of life is sunlight energy in an opened system.

The exchange of energy and matter is observed in the phenomenon of life , and stable equilibrium states are maintained even under non-equilibrium conditions. This is called homeostasis, which is also a system in dynamical equilibrium. An equilibrium system is expressed as $A \leftrightarrow B$, which indicates an equilibrium state between A state and B, and this can be realized even in a closed system. Although life and civilization on earth are dynamical equilibrium systems, equilibrium states are still maintained. Such a non-equilibrium steady state cannot be achieved in a closed system, but can be achieved in an opened system.

The present energy and environmental problems have resulted from the disintegration of a stable state of civilization. This is because social activity is performed only within a closed system. When energy consumption is small, no problem results from this. However, when the scale of society's energy consumption is as huge as it is today, the problem becomes obvious. Our civilization should focus on opening the system of energy production to the universe in order to preserve homeostasis and life, as shown in Fig. 2.4.

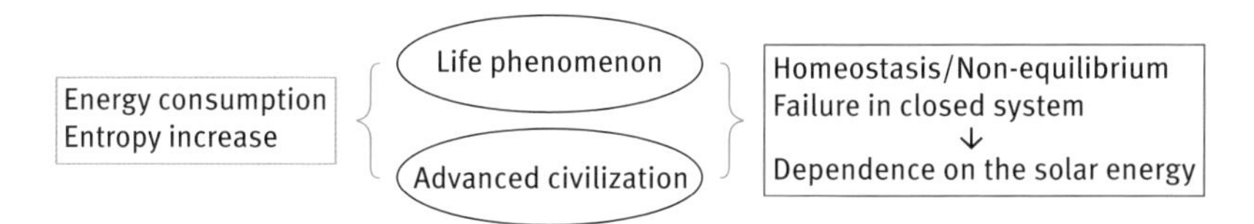


Fig. 2.4: The homeostasis (dynamical equilibrium) of life and civilization.

The use of solar energy is not only an alternative energy resource instead of fossil fuels but also provides an answer to one of the essential problems posed to the existence of civilization and human beings. In fact, nature can provide for a civilization with such a huge energy consumption, as listed in Table 2.1. However, if civilization is to continue to function in the same way, we should construct a social system opened to sunlight like photosynthesis. If human beings humbly learn the meaning of our existence from nature, it will be possible to preserve and develop a higher civilization by obtaining the requisite minimum energy from sunlight.

Table 2.1: The supply and demand of energy in the world (2010).

| Energy | Quantity of energy ($10^{18}$ J $y^{-1}$) | Power (TW) |
|---|---|---|
| Sunlight in outer space | 5,450,000 | 173,000 |
| Sunlight on the earth | 3,820,000 | 121,000 |
| Photosynthesis production | 4,000 | 127 |
| Food | 16 | 0.5 |
| Energy demand | 480 | 15.2 |
| Biomass waste | 128 | 4.1 |

## 2.4 Global warming

Throughout the history of earth, it is thought that the global warming and cooling of the climate have been repeated many times. However, the phenomenon that all of the earth's climate has become warm is remarkable from the second half of $20^{th}$ century, and this phenomenon is the global warming of the earth.

The "global warming of the earth" includes the increase of average temperatures of air and sea and various secondary problems caused by the temperature increase such as changes in ecological systems and the erosion of the coastline by rising seas.

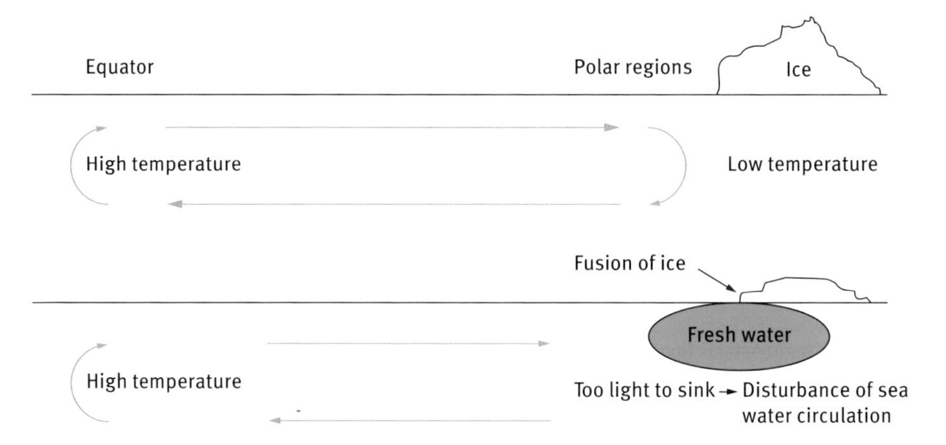


Fig. 2.5: Balanced air temperature by sea water circulation and fusion of ice in polar regions.

The worst effects of global warming should be considered, and one of these is shown in Fig. 2.5.
From the results of the direct observation of air, estimates about the past several tens of thousands of years and long-term simulations, it is believed that the cause of the global warming is gases such as carbon dioxide ($CO_2$) and methane ($CH_4$). Combustion is a reaction in which carbon and hydrogen, consisting of fossil fuels and fire wood, react with oxygen in air, and the carbon and hydrogen form oxides in that process. The reaction of methane, which is the main ingredient in natural gas, is expressed as follows:

$$CH_4 + 2O_2 \rightarrow CO_2 + 2H_2O \tag{2.6}$$

The reactions of carbohydrates $C_nH_{2(n+1)}$, such as kerosene and gasoline, are expressed as follows:

$$C_nH_{2(n+1)} + \{(3n+1)/2\}O_2 \rightarrow nCO_2 + (n+1)H_2O \tag{2.7}$$

small amount of nitrogen in the carbohydrate forms nitride compounds such as $N_2O$, $NO_3^-$ and $NO_2^-$, which provide environmental pollution such as global warming and eutrophication.

The concentration of carbon dioxide in the air during earlier periods can be estimated from the $CO_2$ concentration enclosed in the ice of the South Pole. Although the $CO_2$ concentration has remained almost constantly around 280 ppm, it rapidly increased after the mid-18$^{th}$ century as a result of burning coal in the Industrial Revolution. When oil appeared in the first 20$^{th}$ century, the quantity of fossil fuels increased. Through rapid progress in scientific technology such as automobiles and airplanes, the quantity of fossil fuels used accelerated more and more. The current $CO_2$ concentration is ~ 400 ppm, and by the end of the 21$^{st}$ century it will be over 560 ppm, which is twice the levels that existed before the Industrial Revolution.

Atmospheric temperature increases in response to the artificial discharge of global warming gasses. From 1990 to 2100, the average increase in temperature is predicted to be in the range of 1.1 ~ 6.4 °C. These values are extraordinarily abnormal in comparison to atmospheric temperatures during past ten thousand years. The average rate of temperature increase in the North Pole regions is twice that of 100 years ago, and the average area of sea ice reduced ~ 2 % per year. $CO_2$ and $CH_4$ emissions from the earth's environment are promoted by global warming, and will accelerate. We should think about these problems seriously.

## 2.5 Solar light and Earth

Solar energy from the sun and the spectrum of solar light are shown in Fig. 2.6 and 2.7(a), respectively. The spectrum outside of the atmosphere of the earth is called Air Mass 0 (AM-0) [2]. The spectrum on the ground perpendicular to the surface after absorption and scattering by $H_2O$ and $O_2$ in the atmosphere is called AM-1. The sunlight incidence is actually below 90 °, and AM-1.5 (incidence angle of 41.8 °) is generally used for evaluation of solar cells.

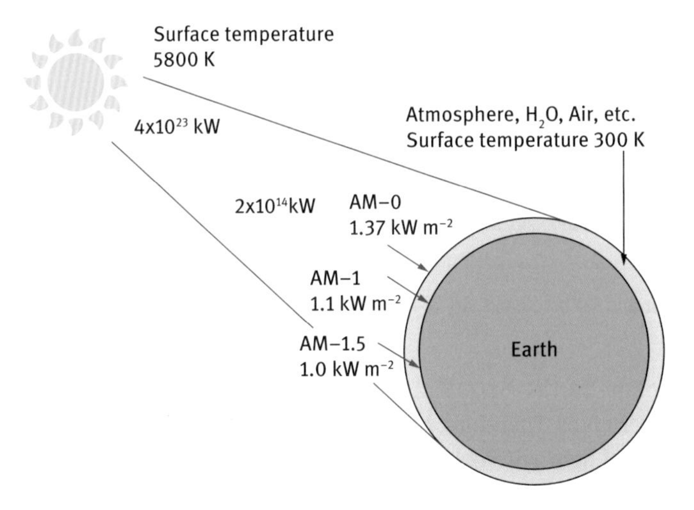


Fig. 2.6: Energy from the Sun to the Earth.

If the sensitivity of human eyes evolved by responding to the highest radiation energy of sunlight as shown in Fig. 2.7(b), the wavelength of light is calculated by Wien's displacement law as follows.

$$\lambda_{\max} T = \frac{hc}{5k} \tag{2.8}$$

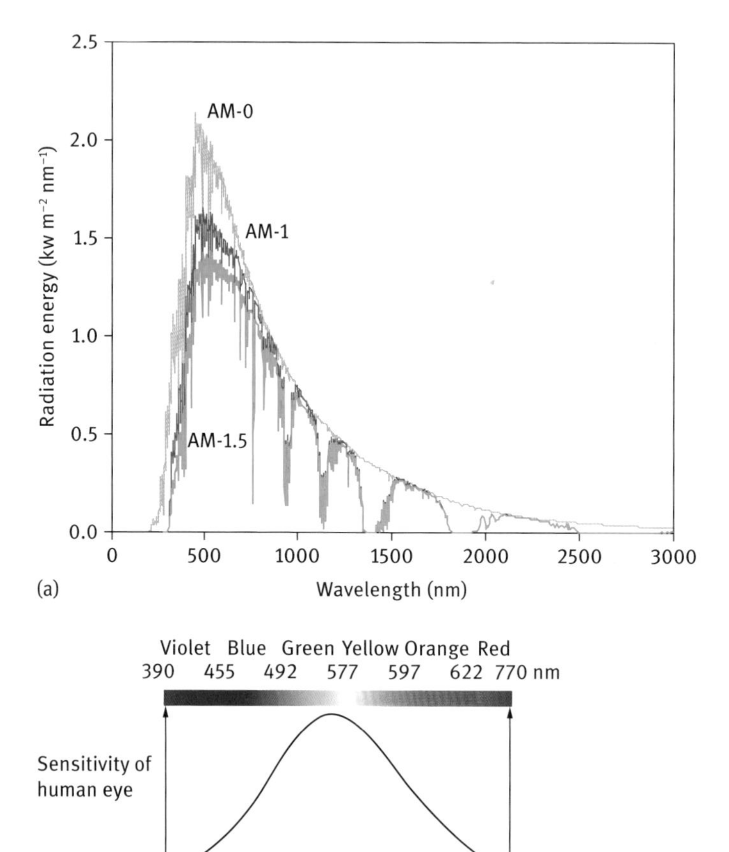


Fig. 2.7: (a) Spectra of sunlight and (b) the sensitivity of human eyes.

The surface temperature of the Sun is ~ 5800 K, and the wavelength of the light it emits is ~ 500 nm, which is a green color. Therefore, human's eyes are sensitive to green. Incidentally, the energy gaps of light with colors of red, blue and violet correspond to 1.8, 2.6 and 2.9 eV, respectively. Since the light visible to humans ranges from red to violet, semiconductors with energy gaps in the range of 1.8 ~ 2.9 eV should be selected to obtain visible light. Si and Ge are infrared, and diamond and GaN are ultraviolet. Therefore, if photons are emitted from these semiconductors, they are normally invisible to human eyes.

## 2.6 Renewable energy

Renewable energy refers to various energies that are supplied from nature at a rate which is faster than social and industrial consumption, which are energy resources supplied regularly from natural power such as sunlight, wind power, wave activated power, hydropower, tide power, ocean thermal energy conversion, geothermal energy and biomass, as listed in Table 2.1 and 2.2. These energies are utilized for electric power generation, hot-water supply systems, air conditioning and fuel. Because of a sudden rise in prices, the exhaustion of underground resources and a goal of preventing global warming, the utilization of renewable energies has recently increased, and one third of newly built power plants depend on renewable energies. Wind power generation in particular has increased rapidly. The wind power generated met 2.3 % of world's electric power demands in 2010, and will reach ~ 10 % by 2020.

Table 2.2: Various renewable energies and problems.

| Energy | Scale | Problems and others |
|---|---|---|
| Hydropower | Small ~ Large | Dependence on weather, geographical location and influence on natural environment |
| Wind power | Medium | Supporting power, effects on birds and noise |
| Solar cells | Small ~ Large | Strong dependence on time, weather and season. Development of storage battery |
| Artificial photosynthesis | Small | Under development |
| Biomass | Small ~ Medium | Bio-waste and cost |

In contrast with renewable energies, exhaustible energies are mainly fossil fuels: oil, coal, natural gas, oil sand, shale gas and methane hydrate. In addition, uranium is also a limited resource, and thermal power generation and nuclear power generation are also included in the category of exhaustible energies.

Fossil fuels such as petroleum, coal and natural gas have been formed from the anaerobic decomposition of buried dead organisms such as fauna plankton and dead plants originating in ancient photosynthesis by exposure to heat and pressure in the Earth's crust over millions of years. They are organic materials in which harmful compositions, such as carbonic acid gas dispersed in atmosphere, had been fixed in fossils deep underground for many years by sunlight energy and organisms. Mankind is now exhausting these fossil fuels after only several hundred years of use.

Nuclear power plants and resources utilize the huge energy emitted from nuclear fusion, nuclear fission and atomic decay, which harness energy from the nuclear fusion of deuterium and tritium, the nuclear fission of uranium and plutonium, and radioactive decay of $^{60}Co$, respectively. Although uranium is a limited natural resource, deuterium and tritium are almost unlimited. In addition, although fuel cells are not an energy resource, they are included in the new energies.

## 2.7 Solar energy plan

Solar energy is almost infinite compared with our life span. All of the energy consumed in the world over one year is 14 TW, which is equal to the sunlight energy irradiated on the Earth during 40 min. If commercial solar cells are set in the Gobi Desert, the electricity needed for all of human society can be obtained. If all of the supply of energy power is achieved through solar power generation, various problems, such as the exhaustion of fossil fuels, global warming by carbon dioxide and radioactive waste, can be avoided.

A grand plan for a large-scale switchover from fossil fuels to solar energy power generation has been proposed in the US [3]. There is 650 thousand $km^2$ of land suitable for solar power plants in the southwest of the US. 2.5 % of the solar energy available in that area converted to electricity could provide a supply for all of the energy consumed in the US. If the extensive lands are covered with solar cell modules and solar heating troughs, electricity can be transferred to all over the country by direct current power transmission. By 2050, a substantial switch from fossil fuels and nuclear power plants to solar power plants could supply 69 % of the US's electricity and 35 % of the total energy. By 2100, renewable energy such as solar energy, wind power, biomass and geothermal energy could generate 100 % of the US's electricity and 90 % of total energy. A world energy vision in Germany also shows that 90 % and 70 % of energy in 2100 will be renewable energy and solar energy, respectively. A roadmap towards solar power generation by 2050 has been created by the New Energy and Industrial Technology Development Organization in Japan, and various subjects and problems have been indicated.

The merits of solar cells are as follows: 1. Use a resource that is almost infinite and free, 2. Environmentally clean, 3. No noise without moving parts, 4. Unattended operation, 5. Easy maintenance, 6. Long lifetime (~ 30 years), 7. Multiple use of lands. However, there are several demerits such as high cost, power dependence on sunlight irradiation and small energy density.

The basic technologies necessary to achieve these plans for solar energy plans already exist to a certain extent. The largest barrier for these plans is the price. The price for solar power generation is ~ 4 times more expensive compared with thermal and nuclear power generation. Therefore, reducing the cost of the present silicon solar cells is very important. To reduce the cost of single-crystal Si solar cells, various-type solar cells such as poly-crystalline Si, thin film Si, $CuInSe_2$, dye-sensitized $TiO_2$ and organic thin films have been developed.

Desertec Foundation promoted a Desertec project that generated electric power in a desert using solar energy and wind power and transmitted if to consumer countries (Desertec, http://www.desertec.org/). The core of the Desertec network is an international network consisting of specialists on renewable energy and politicians. A concentrated solar thermal power system, a solar cell power generation system and a wind power system are arranged around the Sahara, and the electricity is transmitted to Europe and Africa by a super-grid high voltage direct current cable.

## 2.8 Bibliography

[1] Kaneko M, Nemoto J. Bio-photochemical batteries (in Japanese). Kogyo Chosakai Pub. 2008.
[2] National Renewable Energy Laboratory. Reference Solar Spectral Irradiance. http://rredc.nrel.gov/solar/spectra/am1.5/
[3] Zweibel K, Mason J, Fthenakis V. A solar grand plan. *Sci Amer*. 2008; 298: 64–73.

# 3 Basics of solar cells

## 3.1 Properties of semiconductors

Solar cells consist of semiconductors. Materials that transmit electricity, such as metals, are conductors, and materials that intercept electricity are insulators. Semiconductors have intermediate properties between metals and insulators. By utilizing these properties, semiconductors generate electricity from light, generate light from electricity, amplify electrical signal and promote switch action.

A main semiconductor that drives computers and generates electricity from light is silicon (Si), and the structure of Si is shown in Fig. 3.1(a) [1]. The atomic arrangement of Si is the same as that of diamond, and the interatomic distances are longer compared with those of carbon in diamond. Germanium (Ge) had been also used for first stage transistors, and the structure is the same as that of Si except for the lattice constant. The transistor was developed by Brattain, Shockley and Bardeen in AT&T Bell Laboratories, and the Nobel award in physics was given to them in 1956. Because of its semiconductor properties and technological merits, Si became a mainstream semiconductor material. SiGe semiconductors are also being studied recently, as they provide good electrical conductivity, small electricity consumption and low noise. GaAs has been also used for various optical device applications, and a model of its structure is shown in Fig. 3.1(b) [2].

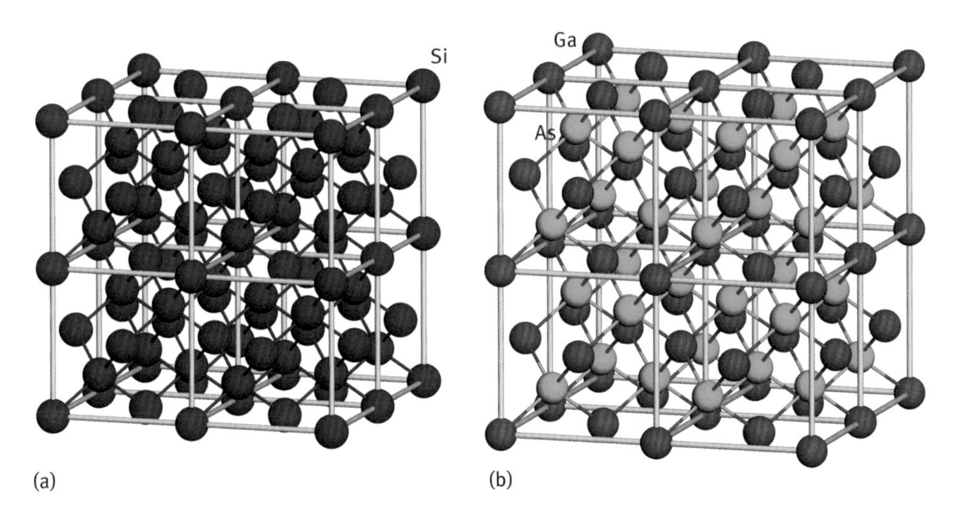


Fig. 3.1: Structure model of (a) Si and (b) GaAs.

Energy gaps (bandgaps, $E_g$) are important in the use of semiconductors. As shown in Fig. 3.2 and 3.3, electrons exist at a conduction band and a valence band. On the

DOI 10.1515/9783110298505-006

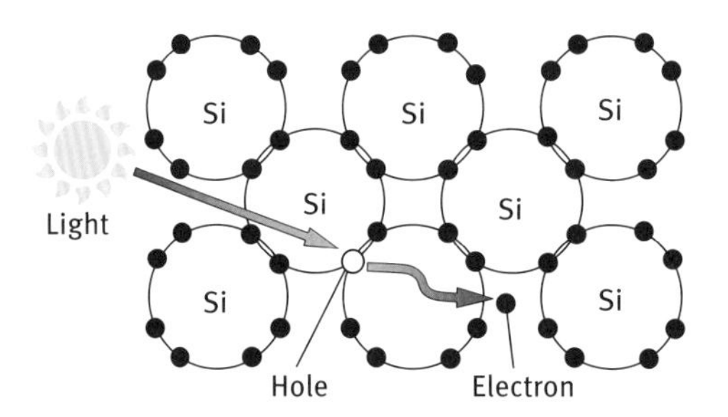


Fig. 3.2: Generation of electron and hole by the light excitation of a Si.

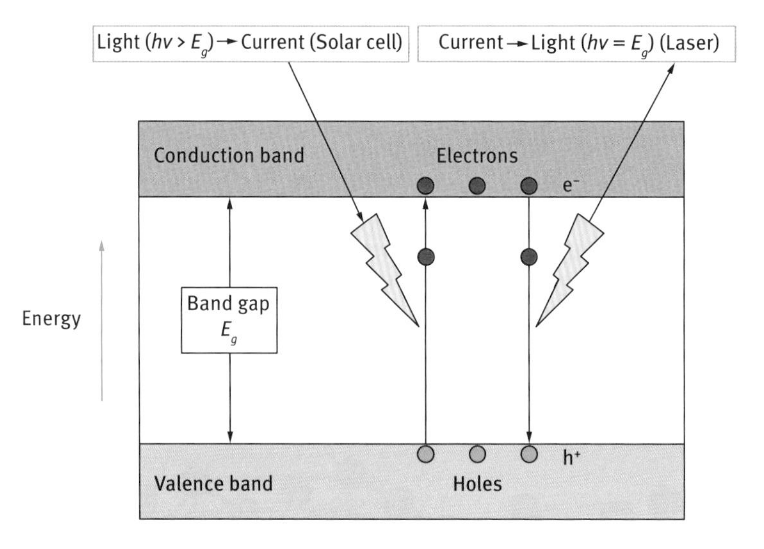


Fig. 3.3: The emission and absorption of light.

other hand, the region where electrons have no energy levels is called the energy gap (bandgap), which is a very important value that greatly affects the semiconductor's properties

When light above the bandgap energy ($E_g$) is irradiated onto semiconductors, electrons on the valence band receive the light energy and are excited to the conduction band. Holes are generated by the excitation of electrons, and current occurs by these flows of electrons and holes. The light energy of a photon is related with its wave length as follows:

$$E(\mathrm{J}) = h\nu = \frac{hc}{\lambda}, \quad E_g\ (\mathrm{eV}) = \frac{1240}{\lambda\ (\mathrm{nm})} \tag{3.1}$$

## 3.2 *pn* junction

When electrons on the conduction band fall down to the valence band to combine with holes, light with the energy $E_g$ is emitted. Interfacial structures called *pn* junctions are fabricated by doped semiconductors, and emission or excitation is carried out effectively. Various applications, such as laser diodes and solar cells, can be achieved by using semiconductors.

When light is irradiated onto solar cells, electrons and holes are generated, as shown in Fig. 3.4. These electrons and holes are moved to *n*-type and *p*-type Si, respectively, and then the current occurs.

Coincidence of the Fermi level and band slope by internal electric field are important for solar cells. Although the Fermi levels of *n*-type and *p*-type Si are different, the Fermi levels of both coincide at the interface of the *pn*-junction, and the energy band

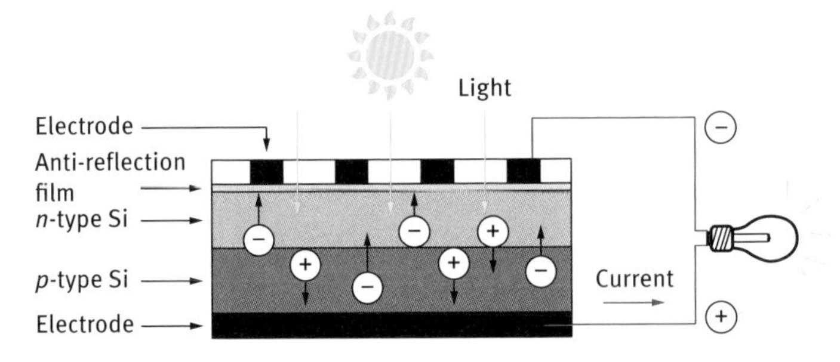


Fig. 3.4: The basic structure of a solar cell.

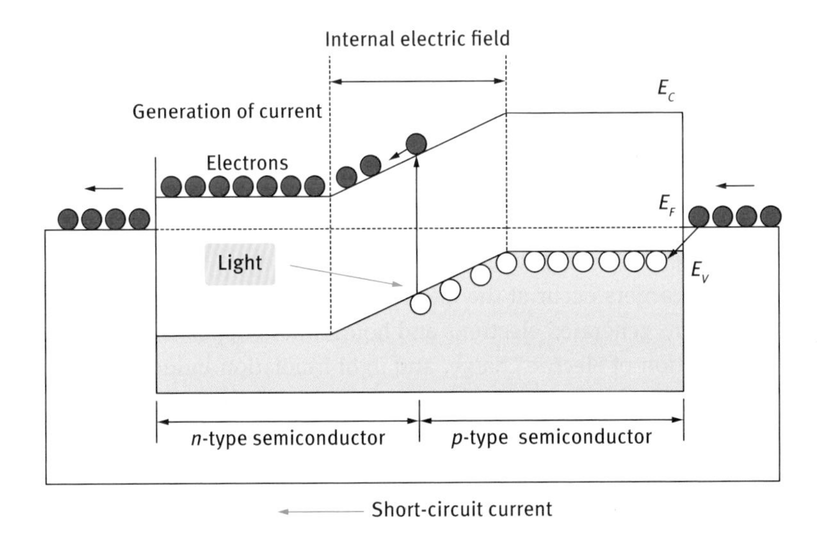


Fig. 3.5: Light irradiation at the *pn*-junction and the band structure of a solar cell.

is sloped by the internal electric field, as shown in Fig. 3.5. Then, electrons and holes move in the directions with lower energies, and current flows by the voltage.

Figure 3.6 shows the structure of a Si solar cell and its cross-section [3]. A thin n-type layer is formed around the p-type Si by impurity diffusion, and metal electrodes are attached. Photo-carriers are generated around the *pn* junction with as internal electric field from light. An electron and hole pair is separated by the internal electric field around the *pn* junction, and voltage occurs at the electrodes. When the terminals of solar cells are in condition of opened circuit under light irradiation, no current flows. Then electromotive force occurs, and its voltage is called open-circuit voltage ($V_{OC}$). When no voltage is applied, its current per area is called short-circuit current density ($J_{SC}$).

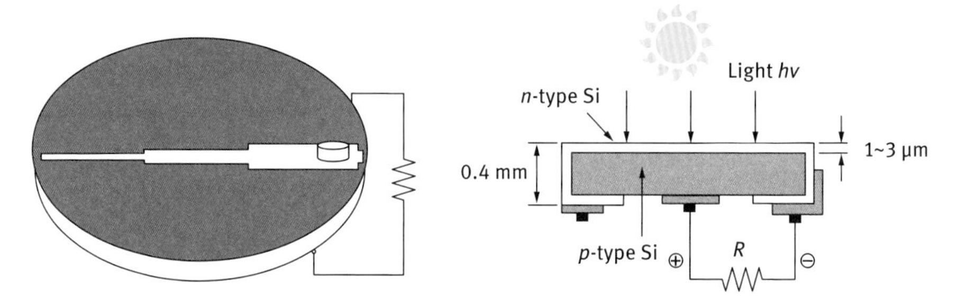


Fig. 3.6: Si solar cell wafer and its cross-section.

## 3.3 Photovoltaic effect

When light is irradiated on the semiconductor, photoconductive phenomena occur. If photo-generated carriers are inhomogeneous by an internal electric field at the *pn* junction, the density-distribution equilibrium of electrons and holes generated by diffusion or drift effect are violated, and the electromotive force occurs. This phenomenon is called the photovoltaic effect.

Strong internal electric fields exist at interfaces and surfaces, such as the *pn* junction and grain boundaries, as a result of electron affinity and Fermi-level difference. When photo-generated carriers occur at the interface or surface of the semiconductor under light irradiation, generated electrons and holes drift to opposite directions, which causes a polarization of electric charge, and light irradiation induced voltage occurs. Energy band diagrams of various semiconductor interfaces and the photo-generated voltage effect are shown in Fig. 3.7 [3]. Such interfacial potentials of *pn* junction, hetero-junction and Schottky barrier are utilized in current solar cell devices.

Interfacial electric fields are also generated by a difference in chemical potential at interfaces between semiconductors (or metals) and conducting liquid. The generated voltage is applied to an electrochemical reaction such as electrolysis. When photoactive semiconductors or dyes are used for electrodes or electrolytes, respectively, photoelectric or photochemical effects occur. An energy band diagram of the interface of *n*-type semiconductors and electrolytes is shown in Fig. 3.7. The electrolyte closed to the *n*-type semiconductor is negatively charged by electrons from the semiconductor surface, and a *p*-type reversed layer forms at the surface of the *n*-type semiconductor. When photons with higher energy compared with the bandgap are irradiated at the interface, the electromotive force is generated at the liquid and semiconductor electrode by the polarization of generated carriers. This phenomenon is applied to photocells and photosensitized electrolysis reactions.

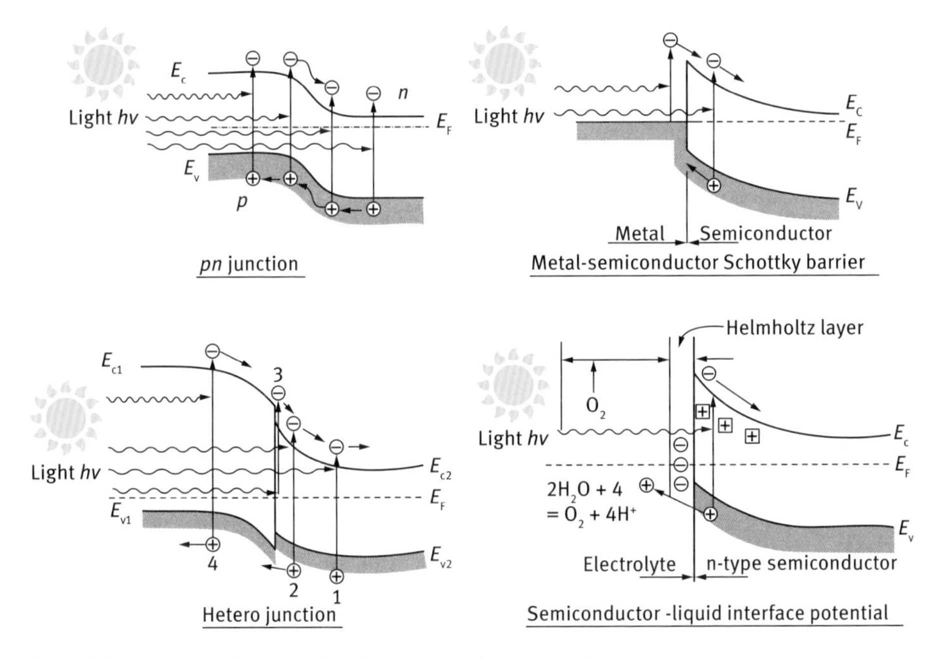


Fig. 3.7: Energy band diagrams of various semiconductor interfaces.

The bandgap energies of semiconductors are closely related with open-circuit voltage and short-circuit current, as shown in Fig. 3.8 [4]. If the bandgap energy increases, the open-circuit voltage increases, but the short-circuit current decreases. On the other hand, if the bandgap energy decreases, the short-circuit current increases because of photovoltaic behavior resulting from a long wavelength. Since, theoretically, the limit efficiency is proportional to the product of the open-circuit voltage and the short-circuit current, efficiency reaches a maximum at the bandgap of ~ 1.5 eV. Actually, for a single *pn* junction, GaAs with a bandgap of ~ 1.4 eV provides high conversion efficiency, as shown in Fig. 3.9.

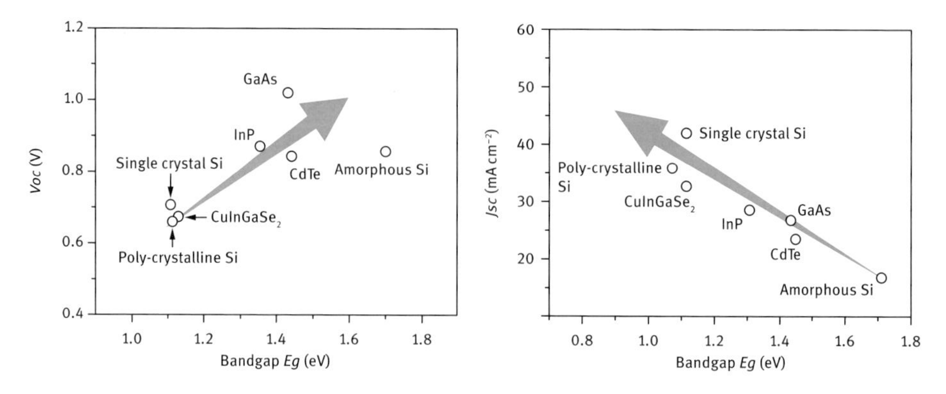


Fig. 3.8: Open-circuit voltages and short-circuit current depending on bandgap.

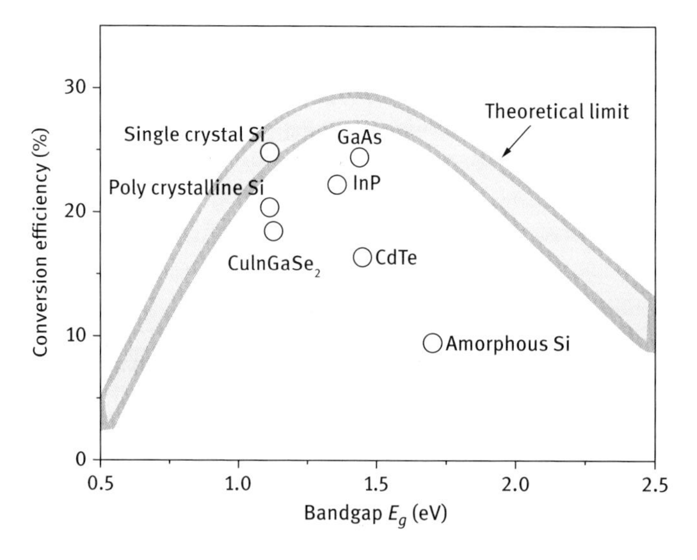


Fig. 3.9: Bandgap and conversion efficiencies.

## 3.4 Energy loss and the requirements for high efficiency

The theoretical limits of conversion efficiencies are determined by the materials of semiconductors and strongly depend on the bandgap energy, which is related to the absorbable wavelength of light. The theoretical limit of Si crystal solar cells is ~ 27 %, as shown in Fig. 3.10, and several main sources of loss are indicated.

The first is transmission loss of light with a longer wavelength compared with that of Si. Light with lower energy than the bandgap energy of Si cannot be absorbed. The loss corresponding to spectrum mismatch is 44 %.

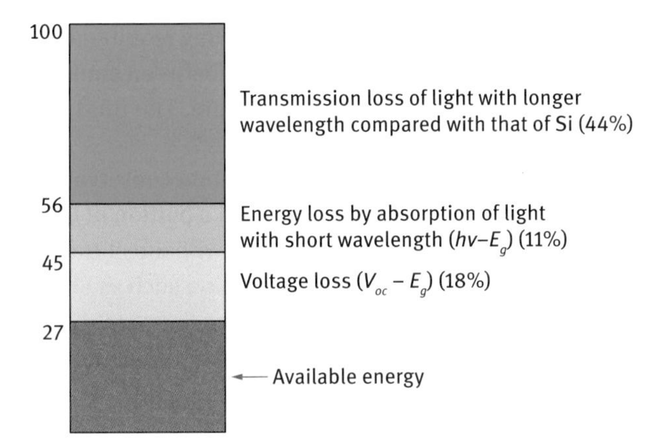


Fig. 3.10: Factors that theoretically limit the conversion efficiency of Si solar cells.

The second is energy loss during the absorption of light. A portion of the light absorbed with a shorter wavelength transforms into heat energy, and the loss can be estimated by $hv - E_g$ (11 %).

The third is voltage loss, which can be estimated by $V_{OC} - E_g$. The open-circuit voltage becomes the maximum by light absorption corresponding to the bandgap energy. The difference between the bandgap energy and the open-circuit voltage is the loss of 18 %.

Other factors that lead to decreased efficiency are reflection loss, recombination loss at defects and surfaces, and the details are indicated in Fig. 3.11.

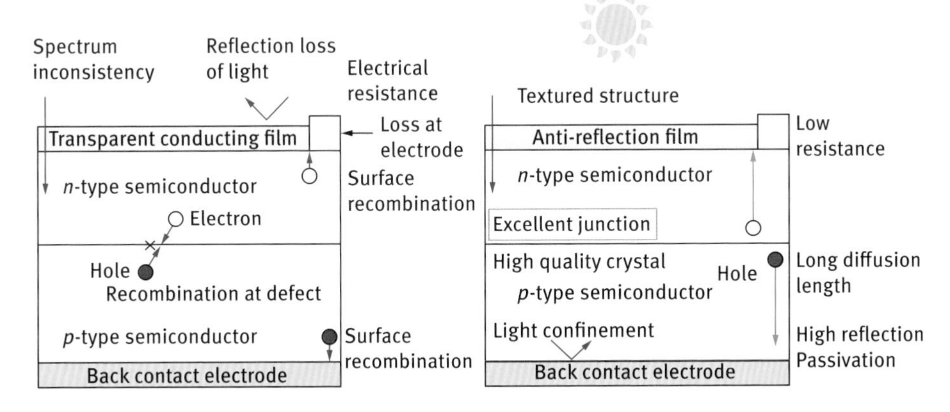


Fig. 3.11: Factors of efficiency loss and improvement for high efficiency.

To improve the conversion efficiency of solar cells, all of the following requirements for the semiconductors should be fulfilled. The first is consistency between sunlight and bandgap energies. The second is direct-transition band structure. The third is a large optical absorption coefficient.

In order for the optical transition to satisfy energy and momentum conservation laws, a phonon is sometimes created during indirect transition, and a portion of light energy becomes heat energy. Semiconductors with band structures with direct transition such as GaAs are advantageous over those with indirect transition such as Si.

A large optical absorption coefficient is also desirable. Since the Si crystal has a band structure of indirect transition and a small optical absorption coefficient, the substrate thickness should be increased. On the other hand, materials with large optical absorption coefficients such as $CuInSe_2$,which has long carrier diffusion lengths, absorb light efficiently even for thin films.

Necessary factors for efficiency improvement are as follows: high crystal quality, excellent *pn* junction, high carrier mobility, the confinement of light in a semiconductor by reducing surface reflection, anti-reflection thin films, surface textured structure, light confinement by forming high reflection films such as Al at the back contact and the formation of passivation films to reduce the surface recombination.

## 3.5 Characterization of solar cells

The energy conversion efficiency ($\eta$) of solar cells is expressed by an energy ratio of the electric output power from the solar cell and the light energy from the sun, as follows:

$$\eta = \frac{[\text{Electric output power from the solar cell}]}{[\text{Light energy introduced in the solar cell}]} \times 100\ (\%) \tag{3.2}$$

For the International Electrotechnical Commission (IEC), a nominal efficiency ($\eta$) of solar cells on the ground is defined as a ratio of maximum electric output power to incident light power (100 mW $cm^{-2}$) under AM1.5 condition. The relation between the nominal efficiency, the maximum output voltage ($V_{max}$), the maximum output current density ($J_{max}$), open-circuit voltage ($V_{OC}$) and short-circuit current density ($J_{SC}$) is described below.

The maximum output power of a solar cell under the connection of the optimum load resistance ($R_L$) is indicated by an intersection point of $V_{max}$ and $J_{max}$ in Fig. 3.12, and the area indicated by a rectangle is the output power. For actual measurements of the official efficiency of solar cells, a solar simulator that simulates a natural solar spectrum is used. Irradiated light is normalized as AM-1.5 and AM-0 at 100 mW $cm^{-2}$ for solar cells on the ground and in space, respectively. Based on the irradiated condition, the maximum power $P\ (V_{max}, I_{max})$ can be measured, and the nominal efficiency

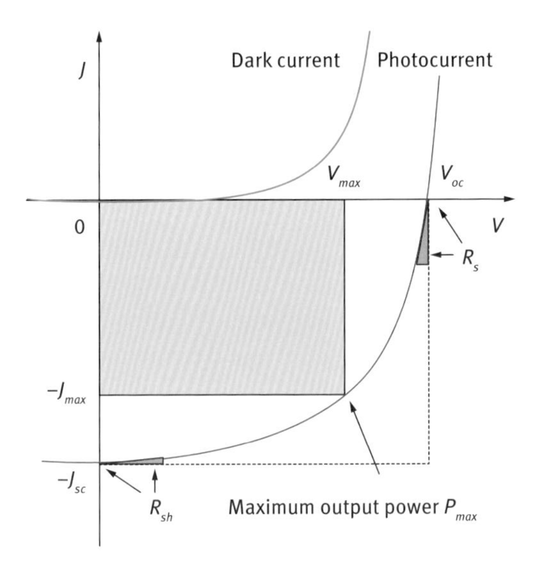


Fig. 3.12: Current density-voltage characteristics of solar cell.

can be calculated as follows:

$$\eta = \frac{V_{OC} \cdot J_{SC} \cdot FF}{100\ (\mathrm{mW\ cm^{-2}})} \times 100\ (\%) = V_{OC}\ (\mathrm{V}) \cdot J_{SC}\ (\mathrm{mA\ cm^{-2}}) \cdot FF\ (\%) \tag{3.3}$$

$$FF = \frac{V_{\mathrm{max}} \cdot J_{\mathrm{max}}}{V_{OC} \cdot J_{SC}} = \frac{P_{\mathrm{max}}}{V_{OC} \cdot J_{SC}} \tag{3.4}$$

A fill factor (*FF*) is the rectangle's area divided by the area of $V_{OC} \times J_{SC}$, which is an important index indicating the properties of solar cells. The product of $V_{OC}$, $J_{SC}$, and *FF* values measured during experiments under the normalized input power of 100 mW cm$^{-2}$ is the nominal efficiency. A solar simulator that has a coincidence with sunlight (AM-1.5) is used for the current-voltage measurements, and an example of the measurement is shown in Fig. 3.13. A comparison of solar cell properties is listed in Table 3.1. $V_{OC}$ and $J_{SC}$ values depend on the bandgap energies.

Table 3.1: Comparison of solar cell properties.

| Materials | $E_g$ (eV) | $J_{SC}$ (mA cm$^{-2}$) | $V_{OC}$ (V) | *FF* | $\eta$ (%) |
|---|---|---|---|---|---|
| Single crystal Si | 1.12 | 42.2 | 0.706 | 0.828 | 24.7 |
| GaAs | 1.42 | 28.2 | 1.022 | 0.871 | 25.1 |
| CdTe | 1.56 | 26.1 | 0.840 | 0.731 | 16.0 |

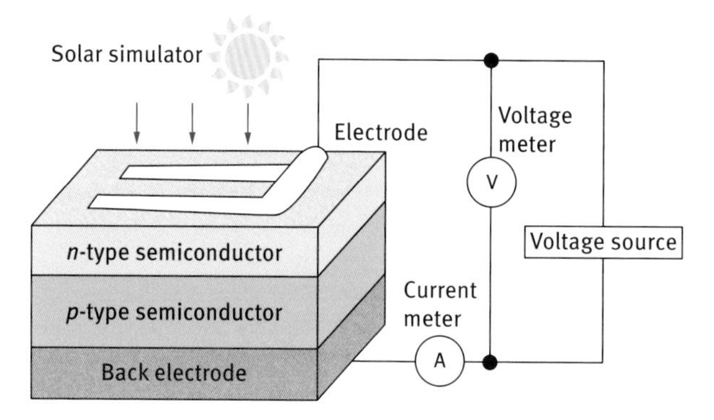


Fig. 3.13: Measurement of a solar cell.

## 3.6 Internal and external quantum efficiency

Internal quantum efficiency (IQE) is expressed as *the number of generated carriers/the number of incident photons*. When 80 carriers are generated from 100 incident photons, the internal efficiency is 80 %. External quantum efficiency (EQE) is expressed as *internal efficiency × actual obtained number of electrons for the cell*, and various loss factors, such as surface reflection, electric resistance and recombination in the actual solar cell, are considered. The external quantum efficiency is also expressed as *the number of current electrons on the external circuit / number of photons absorbed in the semiconductor*. Therefore, the EQE value is generally lower than the IQE value. The EQE is also expressed as incident photon-to-current conversion efficiency (IPCE). The IQE and EQE are strongly dependent on the wavelength of incident light, and these values are often used in the evaluation of solar cells. The efficiency ($\eta$) of the EQE is expressed as follows:

$$\eta = \frac{\frac{hc}{e\lambda}}{\Phi_0} \times 100 \tag{3.5}$$

where the intensity of incident light is $\Phi_0$ [W m$^{-2}$] and the wavelength of light is $\lambda$.

## 3.7 Series and shunt resistances

Series resistance ($R_s$) and shunt resistance ($R_{sh}$) should be considered in current photovoltaic devices [5, 6]. Series resistance is an electrical resistance that occurs when the current flows through each circuit, and low series resistance is a desirable quality for well-functioning devices. The series resistances in the circuit result from the electrical resistances of metal electrodes, ohmic contact at the metal/semiconductor interface, and bulk semiconductors.

The shunt resistance is an electrical resistance at the *pn* junction, and a high shunt resistance is desirable for reducing the leak current at the junction. An equivalent circuit is shown in Fig. 3.14.

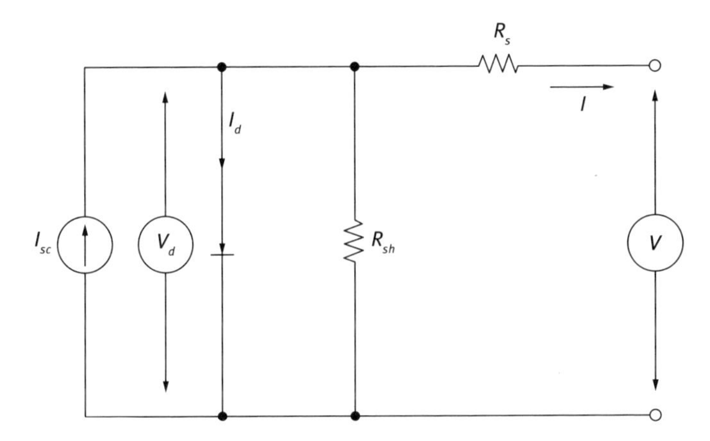


Fig. 3.14: Equivalent circuit of a photovoltaic device.

These series and shunt resistances can be estimated by the following simple calculations:

$$R_S = -\frac{\Delta V_{OC}}{\Delta J_{OC}} \tag{3.6}$$

$$R_{sh} = -\frac{\Delta V_{SC}}{\Delta J_{SC}} \tag{3.7}$$

Although $R_S$ values do not significantly affect the open-circuit voltage ($V_{OC}$), they reduce the short-circuit current density ($J_{SC}$). On the other hand, while $R_{sh}$ values do not significantly affect the short-circuit current density, they reduce the open-circuit voltage.

In contemporary solar cells, $R_S$ values are below ~ 0.5 Ω, and their influence is small. However, when light irradiation intensity is large, for example in large scale cells or light-concentrating solar cells, it should be noted that the $R_S$ values affect the efficiencies.

When there are many defects at the *pn* junction with a large leakage current, the $R_{sh}$ values should be noted. In contemporary solar cells, $R_{sh}$ values are above ~ 1 kΩ, and their influence is small. However, when light irradiation intensity is small, it should be noted that the $R_{sh}$ values affect the efficiencies.

The influence of the $R_S$ value on the power output can be estimated as follows: for a Si *pn* junction solar cell, current-voltage characteristics are calculated from an assumption of $J_{SC}$ = 30 mA cm$^{-2}$, $V_{OC}$ = 0.51 V, $R_{sh} = \infty$ and $\Phi_0$ = 100 mW cm$^{-2}$. When $R_S$= 0 Ω, the efficiency and fill factors are $\eta$ = 12.2% and $FF$ = 0.8, respectively. When

$R_S = 1, 5$ and $20\,\Omega$, the efficiencies are $\eta = 11.5, 8.5$ and $3.0\,\%$, respectively. The $R_S$ values in recent Si solar cells are below $0.5\,\Omega$.

The influence of the $R_{sh}$ value on the power output can also be estimated. When $R_S = 0\,\Omega$ and $R_{sh} = \infty$, the efficiency and fill factors are $\eta = 12.2\,\%$ and $FF = 0.8$, respectively. When $R_{sh} = 500, 100$ and $50\,\Omega$, $\eta = 11.7\,\%$, $11.1\,\%$ and $8.6\,\%$, respectively. The $R_{sh}$ value affects $V_{OC}$ values. $R_s$ and $R_{sh}$ are important values for understanding and designing solar cells.

## 3.8 Bibliography


[1] Kitano A, Sakata M, Moriguchi K, Yonemura M, Munetoh S, Shintani A, Fukuoka H, Yamanaka S, Nishibori E, Takata M. Structural properties and thermodynamic stability of Ba-doped silicon type-I clathrates synthesized under high pressure. *Phys Rev B*. 2001; 64: 045206-1-9.

[2] Laaksonen K, Komsa HP, Arola E, Rantala TT, Nieminen RM. Computational study of $GaAs_{1-x}N_x$ and $GaN_{1-y}As_y$ alloys and arsenic impurities in GaN. *J Phys Cond Matt*. 2006; 18: 10097–10114.

[3] Hamakawa Y. Kuwano Y. Solar energy engineering. Baifukan Co. Ltd. (In Japanese). 1994.

[4] The National Institute of Advanced Industrial Science and Technology. Book of solar cells (In Japanese). Nikkan Kogyo Shimbun Ltd. 2007.

[5] Handy RJ. Theoretical analysis of the series resistance of a solar cell. *Solid-State Electron*. 1967; 10: 765–775.

[6] Sahai R, Milnes AG. Heterojunction solar cell calculations. *Solid-State Electron*. 1970; 13: 1289–1299.

# 4 Inorganic solar cells

## 4.1 Comparison of solar cells

The characteristics and application fields of various solar cells are shown in Table 4.1 [1]. Perovskite solar cells have recently begun undergoing rapid development as one form of next-generation solar cells.

Table 4.1: The characteristics and application fields of various solar cells.

| Cell type | Materials | Efficiency (%) | Radiation resistance | Reliability | Cost | Application |
|---|---|---|---|---|---|---|
| Bulk-type | Single crystal Si | 25.0 | △ | ◎ | ○ | Ground power, space |
| | Polycrystal Si | 20.8 | △ | ◎ | ○ | Ground electric power |
| Thin film | Amorphous Si | 13.4 | △ | △ | ◎ | Consumer |
| | Thin film poly-Si | 12.3 | △ | ○ | ◎ | Ground electric power |
| | HIT | 25.6 | △ | ○ | ○ | Ground electric power |
| | $CuInGaSe_2$ | 21.0 | ◎ | ○ | ○ | Ground electric power |
| | CdTe | 21.5 | ○ | ○ | ◎ | Ground electric power |
| | $CuZnSnS_{4-y}Se_y$ | 12.6 | △ | △ | ◎ | Consumer |
| High-efficiency | Light-concentrating multi-junction | 46.0 | ◎ | ○ | ○ | Ground electric power |
| Space | GaAs | 29.1 | ○ | ◎ | △ | Space |
| | InP | 22.1 | ◎ | ◎ | △ | Space |
| | Multi-junction | 38.9 | ○ | ◎ | △ | Space |
| New materials | Dye-sensitized | 11.9 | – | △ | ◎ | Consumer |
| | Organic (Carbon) | 11.7 | – | △ | ◎ | Ground electric power |
| | Perovskite | 22.1 | – | △ | ◎ | Consumer |

In developing next-generation solar cells, the goal is to realize high efficiencies and low cost simultaneously. Therefore, thin films, no environmental pollution, non-toxic materials, rich resources, stability and durability are desirable. The potential conversion efficiencies of solar cells are shown in Fig. 4.1. To reach the thermodynamic limit of 74 %, a circulation machine against the asymmetry of time, which is a basic physical law, is needed [2]. A maximum efficiency of 68 % would be obtainable through a method with time symmetry. Such methods are used in tandem and hot-carrier cells.

DOI 10.1515/9783110298505-007

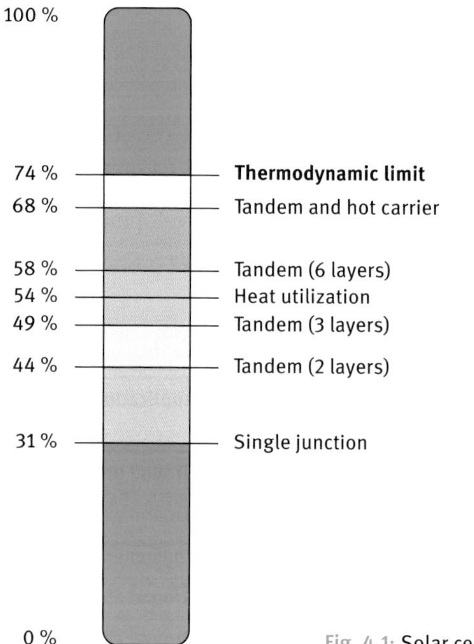


Fig. 4.1: Solar cells and the efficiency limit.

## 4.2 Amorphous Si

The atomic arrangement of amorphous silicon (a-Si) is disordered, unlike crystalline silicon with ordered atomic arrangement. Since the interaction between light and the Si network is large, the a-Si is able to absorb more light in comparison to crystalline Si. Therefore, thin films can be used for a-Si solar cells, and a thickness below 1 µm can generate electricity. In addition, transparent solar cells can be produced by using glass and plastic substrates.

The basic structure of a-Si thin film solar cell is a *pin*-type structure, as shown in Fig. 4.2. *pn*-junctions are formed by continuously depositing *p*-type, *i*-type and *n*-type a-Si layers on the substrate, by plasma and chemical vapor deposition methods using silane gas ($SiH_4$). During the deposition process, multiple cells can be connected in the series, and the desired voltage can be obtained. The *i*-layer is an intrinsic, pure semiconductor, which has no impurities, such as those of *n*- and *p*-type, and its Fermi level exists on the center of the bandgap of Si. Since the structural quality of a-Si are inferior to crystalline Si, a non-doped *i*-type semiconductor is inserted between the doped *pn*-junction layers. A high electric field is applied to the depletion region of *i*-Si, and photo-generated carriers are transported by the high electric field, which is a carrier-drift type device. The doped layers form the internal voltage in the *i*-layer. Since the photo-generated carriers in the doped layers recombine, the photo-current cannot be obtained in the dead layer. Then, wide-bandgap materials with small absorption coefficients are desirable for creating a light-incident window for the p-type layer. The

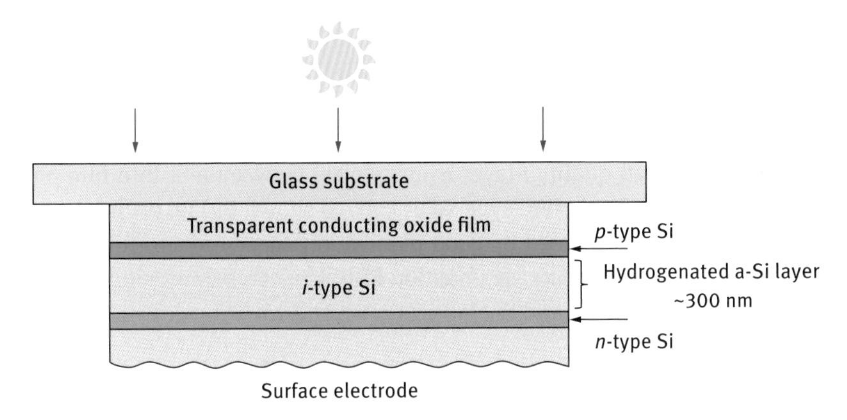


Fig. 4.2: The structure of a-Si solar cells.

energy band diagram of a-Si is shown in Fig. 4.3. a-Si has a wider bandgap of ~ 1.8 eV in comparison to that of crystalline Si (1.1 eV), which results in a high $V_{OC}$ and low $J_{SC}$, as listed in Table 4.2. A high efficiency was obtained for multi-junction solar cells with microcrystalline Si layers [3].

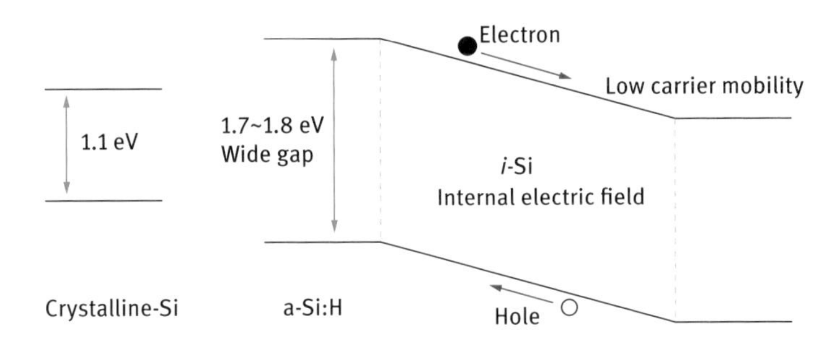


Fig. 4.3: Energy diagram of amorphous Si.

Table 4.2: Photovoltaic properties of Si solar cells.

| **Materials** | $J_{SC}$ **(mA cm$^{-2}$)** | $V_{OC}$ **(V)** | ***FF*** | $\eta$ **(%)** |
|---|---|---|---|---|
| Single crystal Si | 42.7 | 0.706 | 0.828 | 25.0 |
| Poly crystal Si | 38.1 | 0.654 | 0.795 | 19.8 |
| Amorphous Si | 19.4 | 0.887 | 0.741 | 12.7 |

## 4.3 HIT

Solar cells with heterojunction with intrinsic thin-layer (HIT) achieved fairly high conversion efficiencies for mass production, and an efficiency of 25.6 % was achieved, as listed in Table 4.1. A high quality *i*-layer is introduced between a-Si thin film on the single crystal Si, which is a hybrid-type solar cell, as shown in Fig. 4.4 [4]. High open-circuit voltages can be obtained by good passivation effect and control of surface recombination. The characteristic degradation following temperature increase is smaller than that of crystalline Si, and the fabrication cost is low.

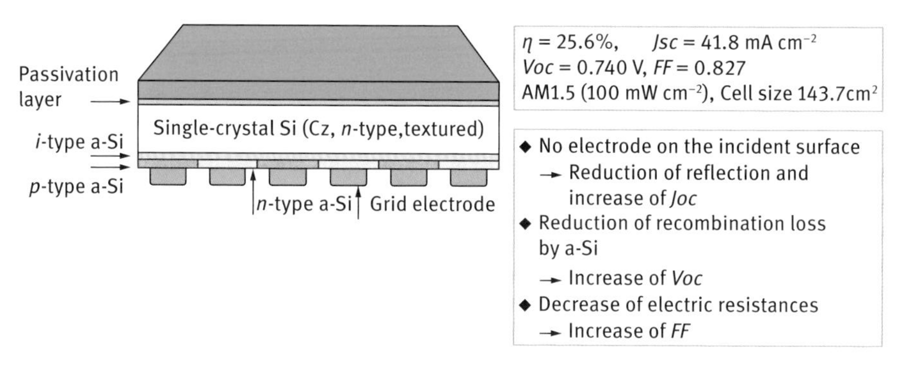


Fig. 4.4: Structure of HIT solar cell.

## 4.4 CdTe

Cadmium telluride (CdTe) is expected to be a solar cell material with high conversion efficiency and low production cost. Crystal [5] and the device structures of CdTe solar cell are shown in Fig. 4.5. The bandgap of the CdTe semiconductor is 1.44 eV, which is the most suitable value for single-junction solar cells, as shown in Fig. 4.5. The optical absorption coefficient is so high that photo-conversion is possible with only a

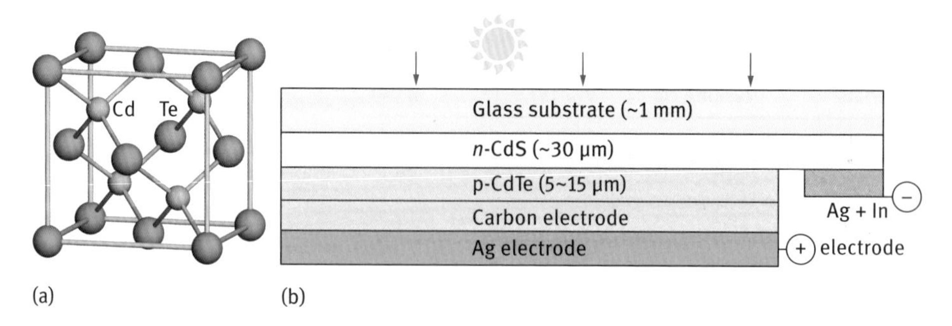


Fig. 4.5: (a) Crystal structure and (b) device structure of CdTe solar cell.

thickness of a few μm. In addition, CdTe solar cells are fabricated by a screen printing method which is a low cost method. Although Cd is poisonous material, it becomes non-poisonous by reacting with Te. Since the dangerousness of Cd-defluvium became low by the guaranteed recycling of the productions, the market for power plants is expanding. Flexible, high-efficiency, low-cost CdTe solar cells are also being fabricated [6].

## 4.5 CIGS

I-III-VI group compounds called the chalcopyrite-system are used as solar cell materials and consist of copper (Cu), indium (In), gallium (Ga), selenium (Se) and sulfur (S), as shown in Fig. 4.6(a). Typical compounds are $Cu(In,Ga)Se_2$ (CIGS) [7], $Cu(In,Ga)(Se,S)_2$ (CIGSS), $CuInS_2$ (CIS) and $Cu_2ZnSnS_{4-x}Se_x$ (CZTSS) [8]. There are various types of materials and fabrication methods, and they can be applied to create both high-quality and low-cost solar cells. They have polycrystalline structures, and a flexible solar cell with a large area can be fabricated, as shown in Fig. 4.6(b). Since their bandgap energies can be controlled by controlling the compositions and elements, they are also expected to be able to be used in multi-junction solar cells for mass production.

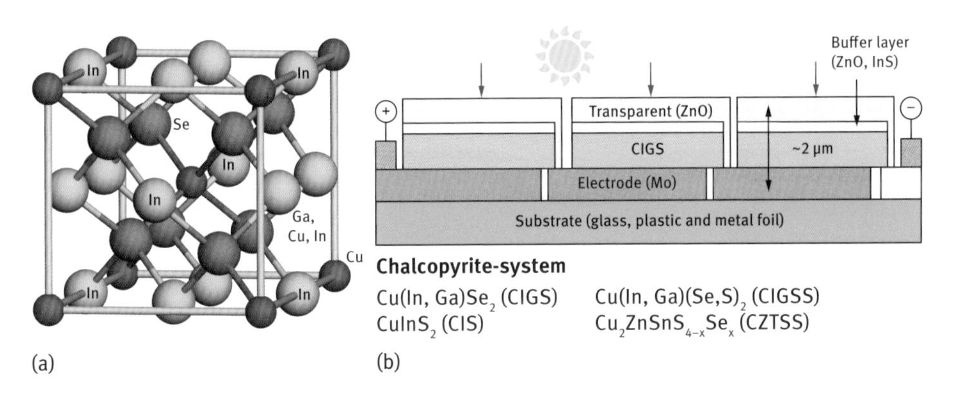


Fig. 4.6: (a) Crystal structure and (b) device structure of a CIGS solar cell.

CIGS solar cells consists of a $Cu(In,Ga)Se_2$ compound, and have attracted a great deal of attention because of their following characteristics: 1. High efficiency (21 %), 2. High optical absorbance and a thickness of several μm is enough for the cells, 3. Stable for a long time and 4. Dark black color. Various flexible substrates such as ceramics, metal foils and polymers are also used for high efficiency solar cells.

## 4.6 Spherical Si

Solar cells are expected to be used as clean energy devices instead of fossil fuels. Although they are clean energy devices that produce no greenhouse gases, the cost reduction of solar cells is an important issue. Spherical silicon (Si) solar cells [9–11] can reduce the consumption of Si in comparison to conventional crystal Si solar cells [12–14]. The basic structure of spherical Si solar cells is shown in Fig. 4.7. The Si spheres are *p*-type semiconductors, and an n-type element such as phosphorus is doped at the surface of Si spheres. Various directions of light incidence on the Si spheres can also be utilized for power generation. Flexible solar cells have also been manufactured using silicon spheres with a diameter of ~ 1 mm with a *pn* junction on aluminum substrates, as shown in Fig. 4.8(a) and (b) [15]. To improve the conversion efficiencies of the spherical Si solar cells [16], high quality silicon balls, optical confinement structures, inactivation of defects by passivation, formation of a texture structure on Si surface [17–19], optimization of structures of reflectors [20, 21], reduction in resistance, and high transmission of anti-reflection films are mandatory [22–25].

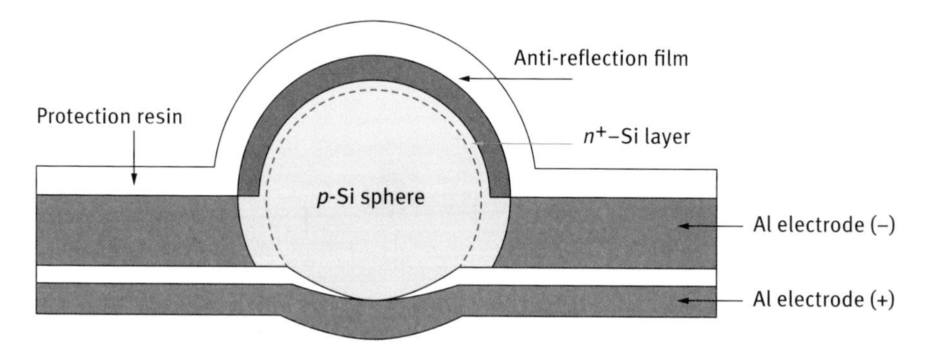


Fig. 4.7: Structure of basic spherical Si solar cells.

The *pn* junctions are formed at a depth of ~ 500 nm from the Si surface [26], and the Si spheres are arranged on hexagonal reflectors. Surrounding light around the Si spheres can be received from various directions by the reflectors. Since the reflection coefficient of Si is large, 30–50 % of light is reflected, and the refractive index and reflectivity increase in the short wavelength band. Therefore, anti-reflection (AR) thin films are added as coating on the Si spheres to suppress the reflection of incident sunlight. The properties of anti-reflection films depend on the refractive index and film thickness. For the present spherical Si solar cells, fluorine-doped tin oxides ($SnO_2$:F, FTO) are used, which are conductive and can easily be fabricated on the sphere. Two-types of fabrication methods for the Si spheres have been proposed, as shown in Fig. 4.9. Methods for the free dropping of melted Si are shown in Fig. 4.9(a) and powder melting on

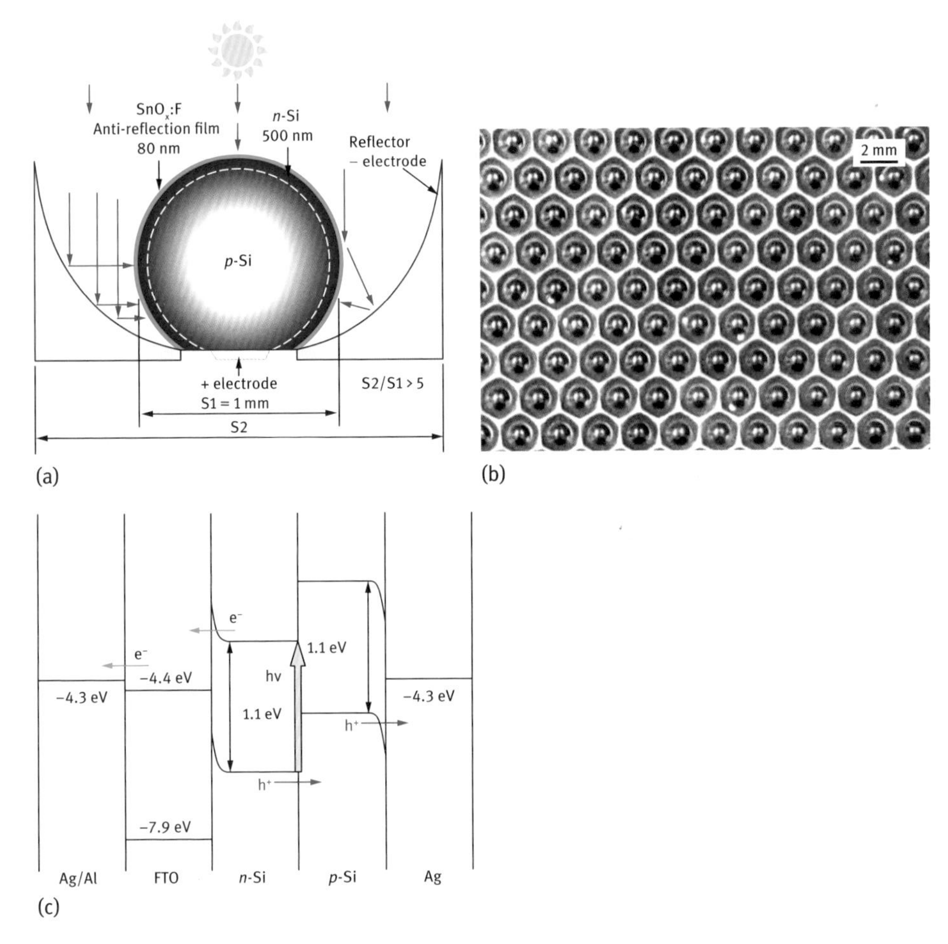


Fig. 4.8: (a) Schematic illustration of spherical Si solar cell. (b) Photograph of spherical Si solar cell. (c) Energy level diagram of present spherical Si solar cells.

the heated substrate, as shown in Fig. 4.9(b,c), is used for fabrication by utilizing the surface tension of the melted Si [27–32].

Here, the results on the microstructures and photoelectric conversion properties of spherical Si solar cells with anti-reflection (AR) F-doped tin oxide ($SnO_x$:F, FTO) thin films are shown. The annealing effects on conversion efficiencies were investigated, and the optical properties were investigated by optical absorption and fluorescence spectroscopy. Microstructure analysis of Si and AR thin films was carried out by X-ray diffraction (XRD) and transmission electron microscopy (TEM). Thermodyanamic calculations at the metal semiconductor interface were also performed and discussed. The combination of microstructure analysis and the photoelectric properties of the spherical Si will give us a guideline for the improvement of the efficiency of spherical Si solar cells.

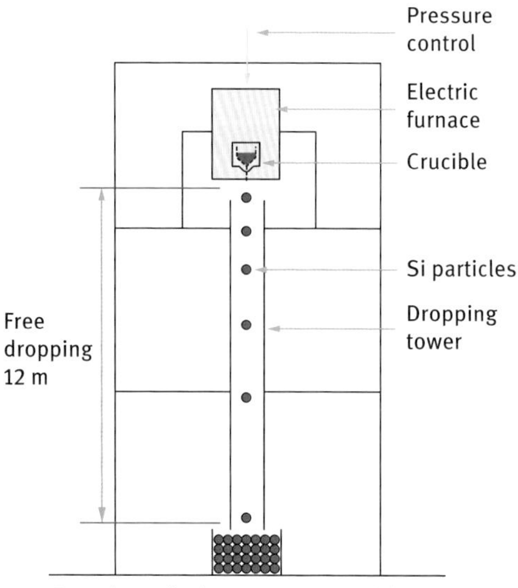


Nozzle for silicon powder supply
Spray gun
Silicon
powder
Liquid binder
with dopant
Outer frame
of the
cylindrical
Silicon
powder
Air slit
Bottom
plate
A rotatable
support rod
(b)

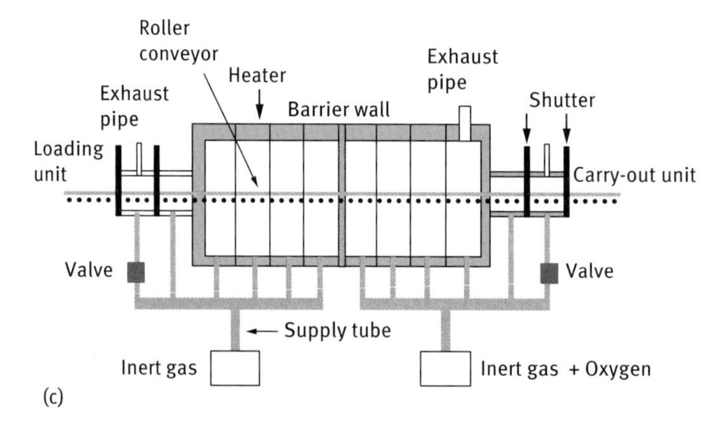


Fig. 4.9: (a) Free dropping and (b,c) powder melting methods for the production of Si spheres.

TEM images and the electron diffraction patterns of spherical Si taken along the [111] and [011] directions are shown in Figs. 4.10(a), (b), (c) and (d), respectively [33]. The TEM image in Fig. 4.10(a) indicates a clear Si crystal, and clear Kikuchi lines are observed in Fig. 4.10(b), which indicates that the spherical silicon has a structure with high crystallinity. The electron diffraction pattern taken along [011] in Fig. 4.10(d) also shows a high-crystallinity structure. Structural models of Si are shown in Fig. 4.11(a), observed along the [001], [011], [111], and [233] directions. The corresponding calculated electron diffraction patterns based on the structural models in Fig. 4.11(a) are shown in Fig. 4.11(b). The calculated [111] diffraction pattern agrees with the center of the observed diffraction pattern in Fig. 4.10(b). A reflection of 200 was not observed in the calculated [011] pattern of Fig. 4.11(b), which is due to the extinction rule of the Si structure (space group *Fd*3*m*). On the other hand, the 200 reflection is observed in Fig. 4.10(d), which is due to the dynamic diffraction effect in the Si crystal.

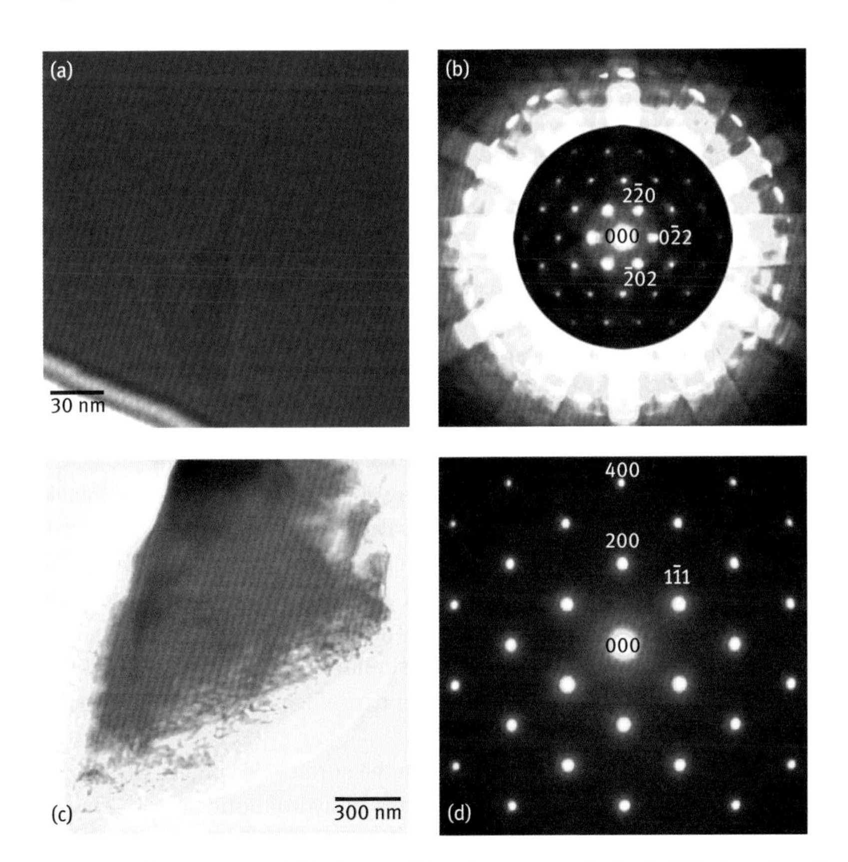


Fig. 4.10: (a) TEM image and (b) electron diffraction pattern of spherical Si taken along the direction of [111]. (c) TEM image and (d) electron diffraction pattern of spherical Si taken along the direction of [011].

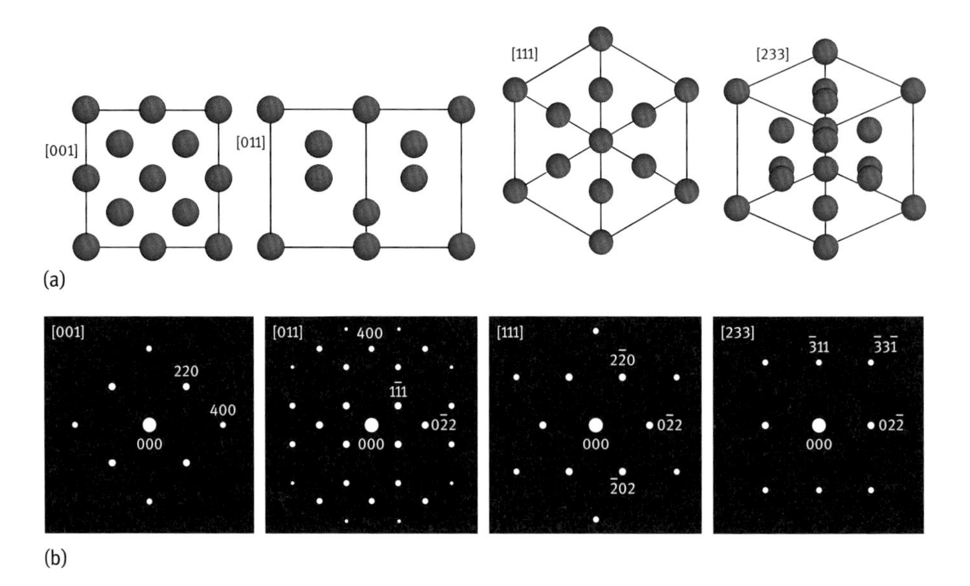


Fig. 4.11: (a) Structural models of Si observed along the directions of [001], [011], [111], and [233]. (b) Calculated electron diffraction patterns based on the structure models of (a).

An HREM image taken along the [011] direction Si crystal is shown in Fig. 4.12(a). In the present work, all HREM images were taken close to the Scherzer defocus, which is an optimum defocus value for the electron microscope to investigate the atomic structures in detail [34]. To observe atomic arrangements more clearly, image processing was carried out by using Fourier filtering. Si atoms are observed as clear dark dots in the image, and a unit cell of Si is indicated in the HREM image. To investigate the observed HREM image in detail, HREM images were calculated based on the projected structural model of Si along the [011] direction, as shown in Fig. 4.12(b). The image calculations were carried out for various defocus values (under defocus) and crystal thicknesses to determine the imaging conditions of the observed image. The calculated image has a crystal thickness of 3.0 nm and a defocus value of 50 nm, which agrees with the observed HREM image in Fig. 4.12(a). Although HREM images calculated at 10 and 90 nm defocus values also show similar contrast, the actual HREM images were nearly taken under Scherzer defocus, and a defocus value of 50 nm would have been more appropriate.

To compare the center of spherical silicon with the surface, Si spheres of 0.5 mm in diameter were prepared by dissolving the spheres in hydrofluoric acid to remove the surface regions of Si spheres. A TEM image and an electron diffraction pattern of spherical Si taken along the [233] direction are shown in Figs. 4.13(a) and (b), respectively. Image contrast such as network structures is observed in the TEM image in Fig. 4.13(a), which could be due to lattice defects such as dislocation loops in the Si crystal. Fig. 4.13(c) and (d) are another type of image contrast and electron diffraction

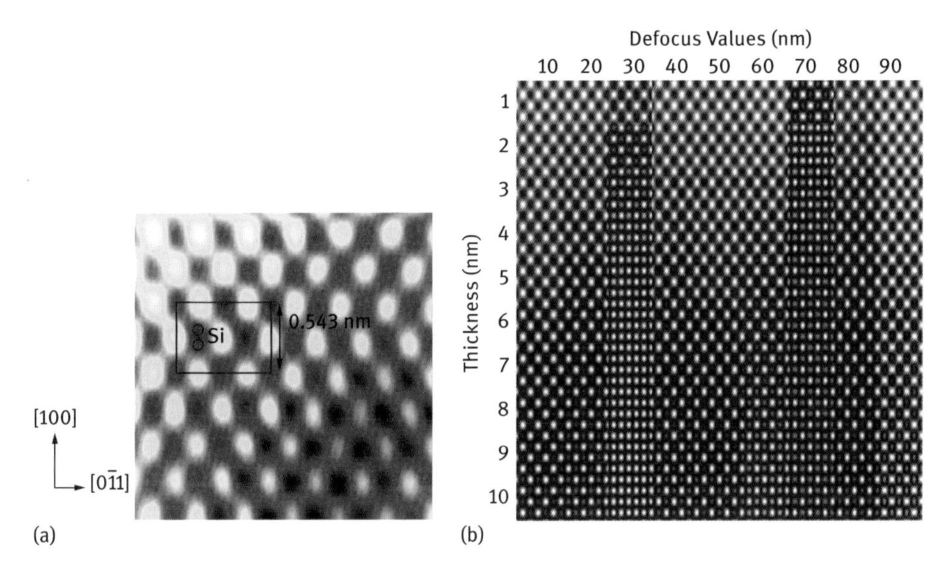


Fig. 4.12: (a) HREM image of spherical Si taken along [011]. (b) Calculated HREM images of Si along [011] as functions of crystal thickness and defocus values of objective lens.

pattern, respectively, which were taken along the [001] direction of Si. In Fig. 4.13(d), split reflections corresponding to 400 and 440 are observed in the electron diffraction pattern as indicated by arrows, which could also be due to defects in the crystal. An HREM image of the center of the Si sphere is shown in Fig. 4.14(a), which was taken along the [011] direction of the Si structure. Two-directional lattice fringes corresponding to (111) planes are imaged, and a dark contrast is observed around the center of the image. The Fourier transform of Fig. 4.14(a) is shown in Fig. 4.14(b), and separation of 111 reflections is observed as indicated by two arrows.

A filtered inverse Fourier transform of Fig. 4.14(b) is shown in Fig. 4.14(c), and lattice fringes are clearly observed [33]. Fig. 4.14(d) is an inverse Fourier transform of Fig. 4.14(b) using 000, Si 111, and Si 111 reflections. Enlarged HREM images of parts of Figs. 4.14(c) and (d) are shown in Figs. 4.14(e) and (f), respectively. Si atoms are observed as dark dots, as indicated by circles in Fig. 4.14(e), and an edge-on dislocation is observed, as indicated by arrows in Figs. 4.14(e) and (f), which could be due to the lattice distortion generated during Si growth. The lattice distortion of Si is also observed as diffused streaks of Si 111 and Si 111 reflections in the Fourier transform, as indicated by arrows in Fig. 4.14(b). These types of defects can form during the crystal growth of the Si spheres.

XRD patterns of spherical Si with anti-reflection $SnO_x$:F films and center parts of Si spheres of 0.5 mm in diameter are shown in Figs. 4.15(a) and (b), respectively. The lattice constants at the surface and center of the Si spheres were calculated to be 0.5421 and 0.5440 nm, respectively. The surface and center of the Si spheres here are doped with phosphorous and boron for *n*- and *p*-type semiconductors, respectively.

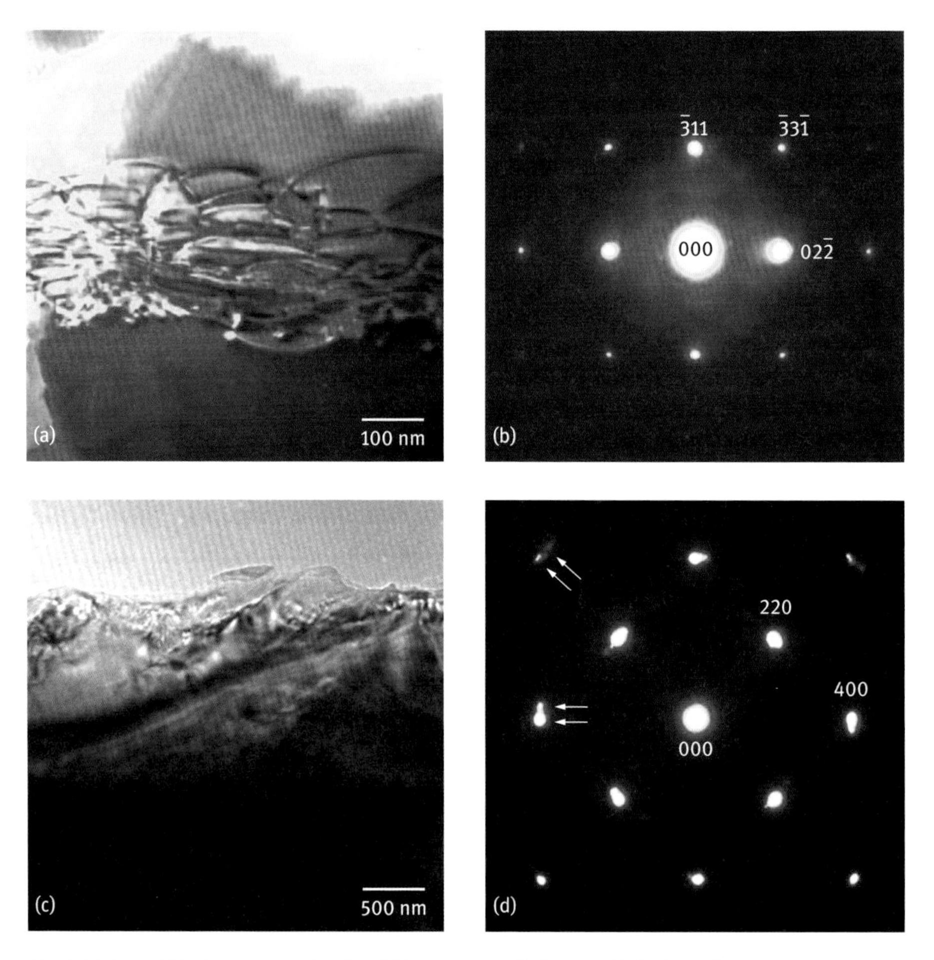


Fig. 4.13: (a, c) TEM images of spherical Si taken along [233] and [001]. (b, d) Corresponding electron diffraction patterns of (a) and (c), respectively.

Since the sizes of atomic radii of tetrahedral covalent bonds for Si, boron, and phosphorous are 0.117, 0.088, and 0.110 nm, respectively, the lattice constant of the surface is predicted to be larger than that of the center of Si spheres. However, the lattice constants measured by XRD indicate the opposite results. The results of TEM observation in Figs. 4.10, 4.13, and 4.14 indicate that the surface and center of the Si spheres have highly crystalline structures and various defects, respectively. The XRD results also indicate that the surface had a smaller lattice constant and higher crystallinity compared to the Si spheres' centers' larger lattice constant due to crystal defects, which agrees with the TEM results.

XRD patterns of spherical silicon with AR $SnO_x$:F thin films before and after annealing at 650 °C for 4 h are shown in Fig. 4.15(a). The Miller indices of Si and $SnO_2$ (space group $P4_2/mnm$) [35, 36] are indicated in the figure, and the measured grain

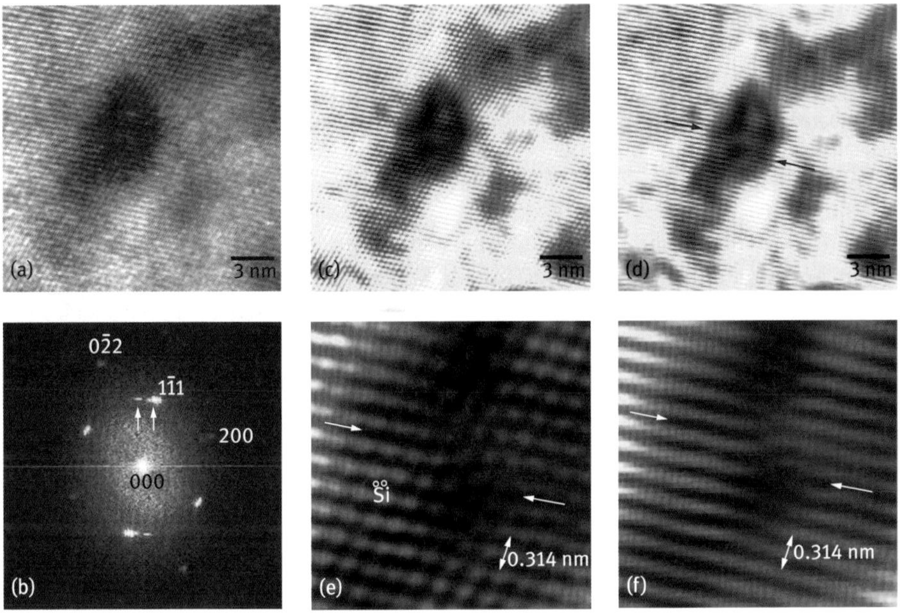

Fig. 4.14: (a) HREM image of center of spherical silicon taken along [011]. (b) Fourier transform of (a). (c) Fourier filtered HREM image of (a). (d) Inverse Fourier transform of (c) using 111 reflection. (e, f) Enlarged HREM images of a part of (c) and (d), respectively.

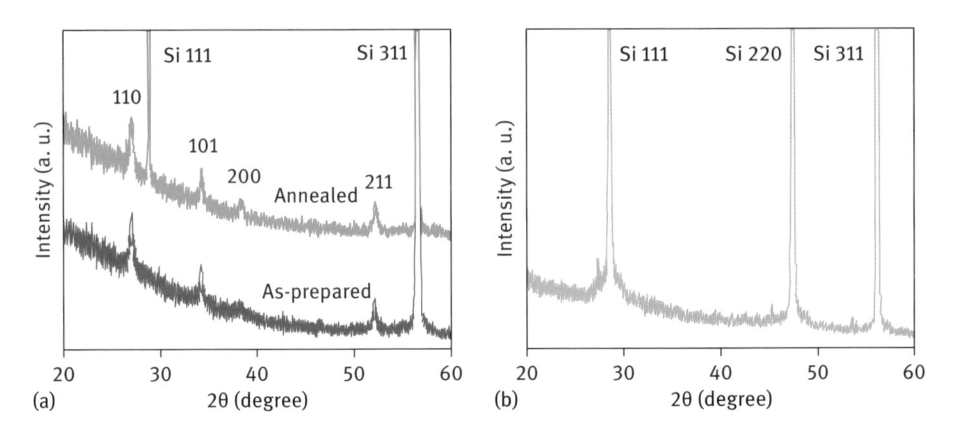


Fig. 4.15: XRD patterns of (a) spherical Si with $SnO_x$:F films and (b) center of spherical Si.

sizes and lattice constants of the *a*-axis and *c*-axis for $SnO_2$ before and after annealing are summarized in Table 4.4. The grain sizes were calculated using Scherrer's formula. After annealing, the grain size of $SnO_2$ increased, which indicates that the crystallinity of the $SnO_x$:F thin films was improved by the annealing.

Table 4.3: Measured parameters of spherical silicon solar cells.

| Samples | $J_{SC}$ (mA cm$^{-2}$) | $V_{OC}$ (V) | $FF$ | $\eta$ (%) | $R_s$ ($\Omega$) |
|---|---|---|---|---|---|
| As-prepared | 24.9 | 0.511 | 0.542 | 6.9 | 89.0 |
| Annealed | 26.3 | 0.593 | 0.717 | 11.2 | 59.3 |

Figure 4.16(a) shows the measured optical absorption of spherical Si with AR thin films before and after annealing at 650 °C for 4 h. Spherical Si absorbs light in the range of 300 to 1150 nm. The optical absorption of the spherical Si increased after annealing at 650 °C. Since the absorption range was shifted to a lower energy, a structural change could have occured after annealing. The refractive indices of the AR $SnO_x$:F films were measured to change from 1.8 to 1.9 after annealing, and the refractive indices of the substrate and antireflection films would influence reflectance and optical absorption. Since the refractive indices are different for different wavelengths of light, the effect of the reflectance reduction depends on the wavelength.

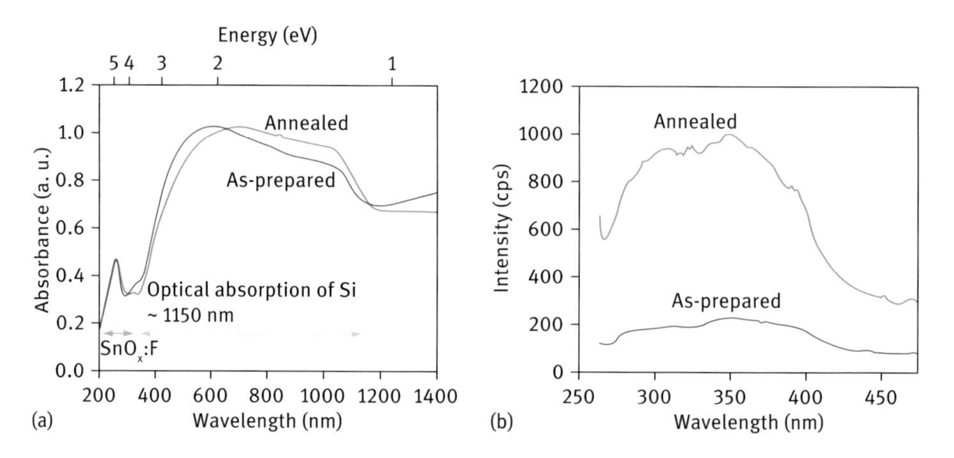


Fig. 4.16: (a) Absorption spectra and (b) fluorescence emission spectra of spherical Si with AR films.

Figure 4.16(b) shows the fluorescence emission spectra of spherical Si before and after annealing. The energy gap of $SnO_2$ is known to be 3.6 eV, which corresponds to ~ 340 nm. Fluorescence peaks of $SnO_x$:F are observed at ~ 350 nm in the present work, which could be due to F-doping and oxygen content in the $SnO_2$ structure. Fluorescence emission spectra due to impurities in Si might also be contained in the spectra [37]. The intensity of the fluorescence emission of $SnO_x$:F increased after annealing, which indicates the crystallinity of the $SnO_2$ structure was improved by annealing. This result agrees with the XRD results in Table 4.4.

A TEM image and an electron diffraction pattern of spherical Si at the Si/$SnO_x$:F interface are shown in Figs. 4.17(a) and (b), respectively. The thickness of the $SnO_x$:F

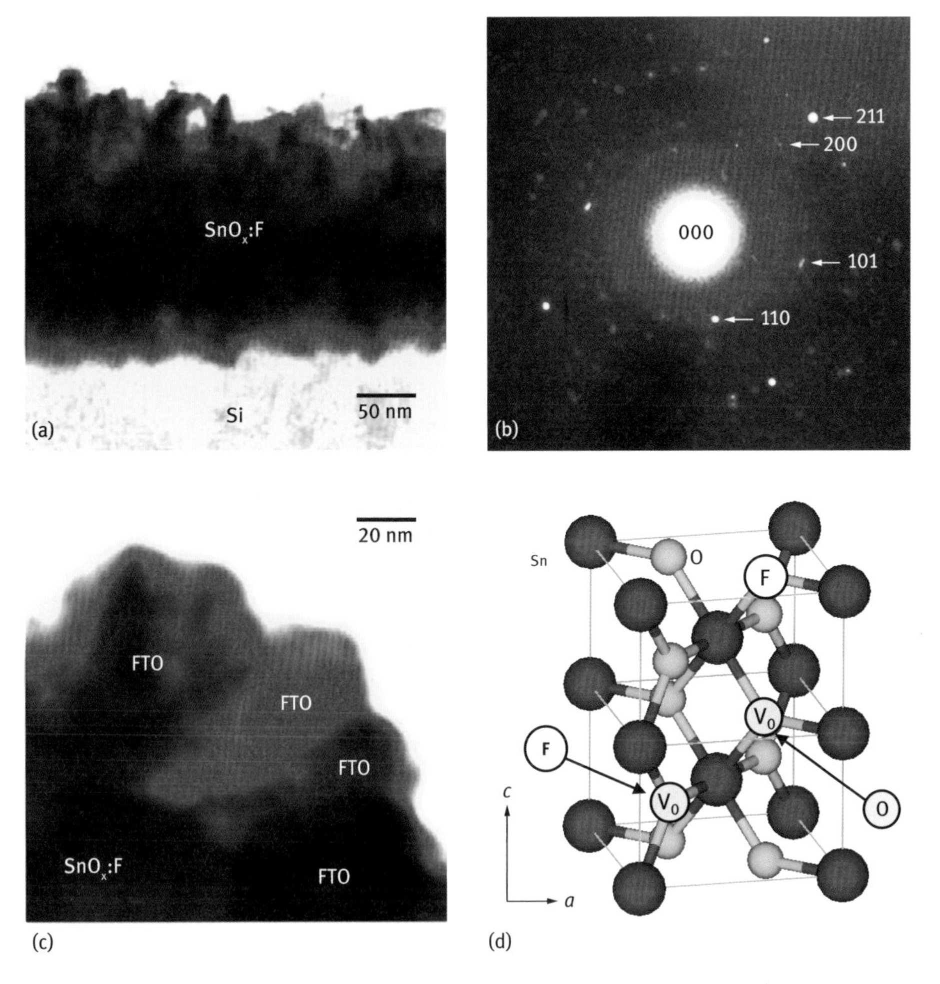


Fig. 4.17: (a) TEM image and (b) electron diffraction pattern of spherical Si at the Si/FTO interface. (c) Enlarged TEM image of the surface of FTO. (d) Schematic illustration of atomic diffusion in $SnO_2$ structure during annealing.

is ~ 200 nm, and the roughness at the Si/$SnO_x$:F interface is ~ 20 nm. The surface and interface roughness could be effective in forming a textured structure to promote the optical confinement of sunlight. Debye–Scherrer rings indexed with $SnO_2$ structure are observed in Fig. 4.17(b), which indicates a nanocrystalline structure without preferred crystal orientation.

An enlarged TEM image of the surface of the $SnO_x$:F is shown in Fig. 4.17(c). The grain sizes of the $SnO_x$:F are ~ 50 nm, which agrees with the XRD results in Table 5.3. $SnO_2$ crystals have high electrical resistance, and doped F and oxygen vacancies are electrical carriers for $SnO_x$:F [38–41]. To consider the change of the oxygen vacancy during annealing, a thermodynamic calculation was carried out, as shown in Fig. 4.18(a). The Gibbs free energy change ($\Delta G$) for $SnO_2$ oxidation is negative at 650 °C in atmosphere, which suggests that the number of oxygen vacancies in $SnO_2$ decreases after annealing, and F atoms could occupy oxygen sites. If the number of oxygen vacancies decreases, an increase in lattice constants may be expected. However, the lattice constants of $SnO_x$:F decreased after annealing, as indicated in Table 4.4, which could be due to the interstitial F atoms being removed or substituted for the oxygen sites. A schematic illustration of the structural change of $SnO_x$:F by annealing is shown in Fig. 4.17(d).

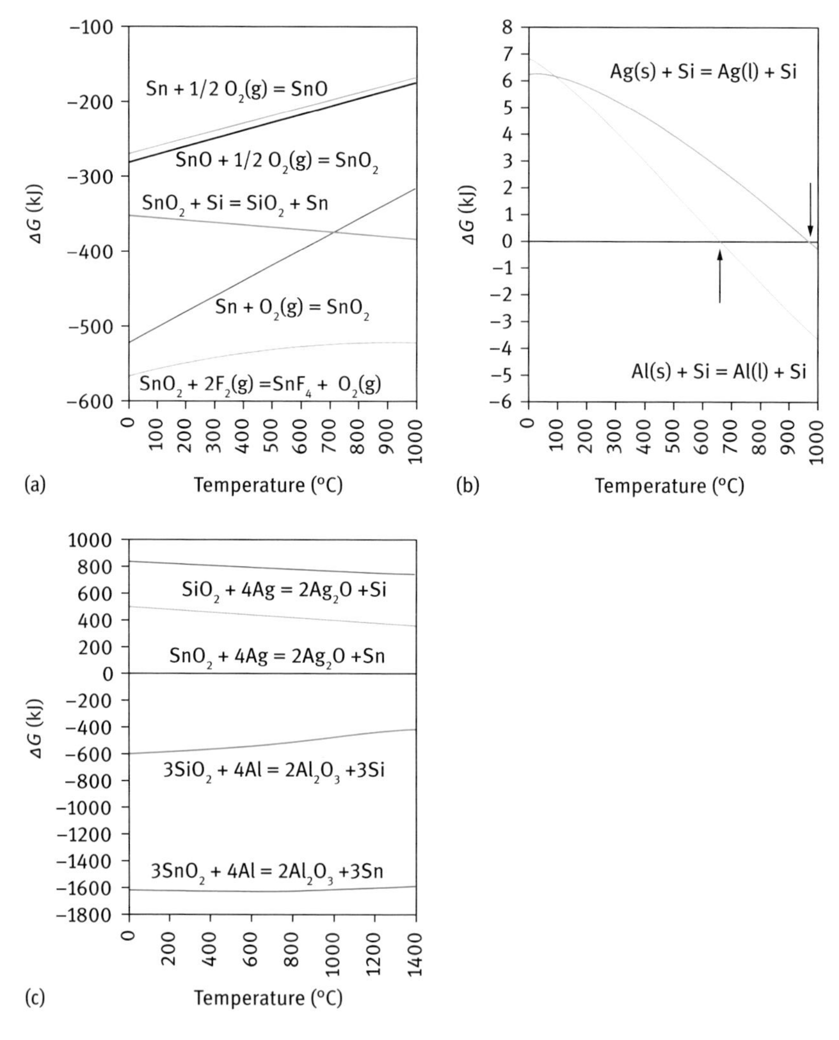


Fig. 4.18: Thermodynamical calculations of (a) $SnO_2$ with F, (b) electrode metals with Si, and (c) $SnO_2$ and $SiO_2$ with electrode metals.

The performance of the spherical Si solar cells was improved after annealing, as summarized in Table 4.3. If some atoms exist at interstitial sites, the electronic resistance increases. Although $SnO_2$ has no optical absorption for the absorption wavelength of Si, the impurities contained in $SnO_2$ could cause light dispersion and absorption [42]. Reduction of the interstitial atoms would decrease the electronic resistance and increase the optical absorption of Si, which would result in improvement of the conversion efficiency of the solar cells. Reduction of the lattice constants could also be due to the decrease in F, which exists at interstitial sites in $SnO_2$. The results indicate that the microstructures and properties of AR films depended on the annealing process, and that optimization of the formation process of AR films is mandatory for further performance improvement.

Table 4.4: Grain sizes and lattice constants of $SnO_x$:F.

| **Samples** | **Grain size (nm)** | **Lattice constant (nm)** | |
|---|---|---|---|
| | | ***a*-axis** | ***c*-axis** |
| As-prepared | 41.0 ± 14 | 0.4713 | 0.3182 |
| Annealed | 56.4 ± 32 | 0.4695 | 0.3170 |
| $SnO_2$ | – | 0.4737 | 0.3185 |

An energy level diagram of a spherical silicon solar cell is shown in Fig. 4.8(c). The spherical Si solar cell with a reflector cup is completed by mounting the spherical Si on a reflector cup consisting of a Ag-electroplated Al substrate, as illustrated in Fig. 4.8(a). The reduction of contact resistances at the metal electrode/semiconductor interface is important to improve the efficiency. As calculated in Figs. 4.18(b) and (c), Al is more reactive with Si, $SiO_2$, and $SnO_2$ than Ag. This indicates that Ag would be more suitable for forming low contact resistances with $SnO_2$, and Al would be suitable for contact with Si in order to remove the native silicon oxide layer. The eutectic temperatures of Ag–Si (Ag ~ 90 at.%), Al–Si (Al ~ 88 at.%), and Al–Ag (Ag ~ 42 at.%) are 835, 577, and 567 °C, respectively, which suggests that the Al–Ag alloy would be more reactive with Si and suitable for the formation of the ohmic contact materials.

Electronic carriers of electrons and holes are separated at the *pn* junction by light irradiation. Separated holes could be transferred from the p-Si to the Ag electrode, and separated electrons could be transferred from the n-Si to the Ag/Al electrode. As shown in Fig. 4.8(c), the charge separation is strongly dependent on the incident angle of light. To improve the efficiency, microstructure control of the anti-reflective coating by annealing and deposition conditions is important. In addition, there could be an energy barrier at the semiconductor/metal interfaces, as indicated by band bending in Fig. 4.8(c) [43, 44], which results in contact resistance. Heavy doping or control of the interfacial microstructure by annealing would reduce the contact resistance.

The microstructures and optical and photoelectric conversion properties of spherical Si solar cells with AR thin films were investigated and discussed. TEM observation and XRD results showed that the surface of the spherical Si had high crystallinity. The lattice constant of the center of the spherical Si was larger than that of the surface, which could be due to lattice distortion from defects such as dislocations at the center of Si spheres. The conversion efficiencies of the spherical Si solar cells coated with the $SnO_2$:F films were improved from 6.9 to 11.2 % by annealing. The optical absorption and fluorescence of the solar cells increased after annealing, and the lattice constants of $SnO_2$:F anti-reflection layers decreased, which could be due to the introduction of fluorine and the crystallization of the $SnO_x$:F structure. Thermodynamic calculations also supported fluorine doping in $SnO_2$, and suggested that Al-Ag alloys are suitable contact materials for Si and $SnO_2$.

## 4.7 $ZnO/Cu_2O$

Copper oxides ($Cu_2O$ and CuO) semiconductors are a promising alternative to silicon-based solar cells because they possess high optical absorption, non-toxicity, and have low production costs. The crystal structures of $Cu_2O$ and CuO are shown in Fig. 4.19(a) and (b), respectively [45, 46]. $Cu_2O$ is known to be a p-type semiconductor oxide and it is a suitable material for high efficiency solar cells because of its direct bandgap of 2.1 eV. A high efficiency of 5.38 % has been obtained for $Cu_2O$ solar cells fabricated by high-temperature annealing and pulsed laser deposition [47]. $Cu_2O$-based solar cells fabricated by electrodeposition and photochemical deposition have been reported, and zinc oxide (ZnO)/$Cu_2O$ thin-film solar cells prepared by electrodeposition have also been reported [48–53]. Electrodeposition is a low-temperature method for solar cell fabrication with a low process cost. A high conversion efficiency of 1.28 % has been reported for electrodeposited $ZnO/Cu_2O$ solar cells using an electrolyte containing KOH for $Cu_2O$ deposition [48]. Conversion efficiencies of 1.06 and 0.88 % have also been reported for p-$Cu_2O$/n-$Cu_2O$ [53] and $ZnO/Cu_2O$ [54] solar cells prepared by electrodeposition.

The fabrication of $ZnO/Cu_2O$ thin-film solar cells by electrodeposition and investigation of the effect of LiOH, KOH, and NaOH in the electrolytes of the $Cu_2O$ thin films are described here. ZnO is a good electron acceptor and has been used as an n-type semiconductor active layer for inorganic thin-film solar cells [55–57], as shown in Fig. 4.19(c) [58]. The $ZnO/Cu_2O$ solar cells prepared in this study were investigated using structural analysis, optical absorption, and photovoltaic measurements.

$Cu_2O$ and ZnO layers were prepared on a pre-cleaned FTO glass plate by electrodeposition using a platinum counterelectrode. Copper(II) sulfate ($CuSO_4$) and L-lactic acid were dissolved in distilled water. The pH of the electrolyte was adjusted to 12.5 by the addition of NaOH, KOH, or LiOH. The electrolyte temperature was kept at 65 °C during electrodeposition. The preparation of $Cu_2O$ layers was carried out at a current

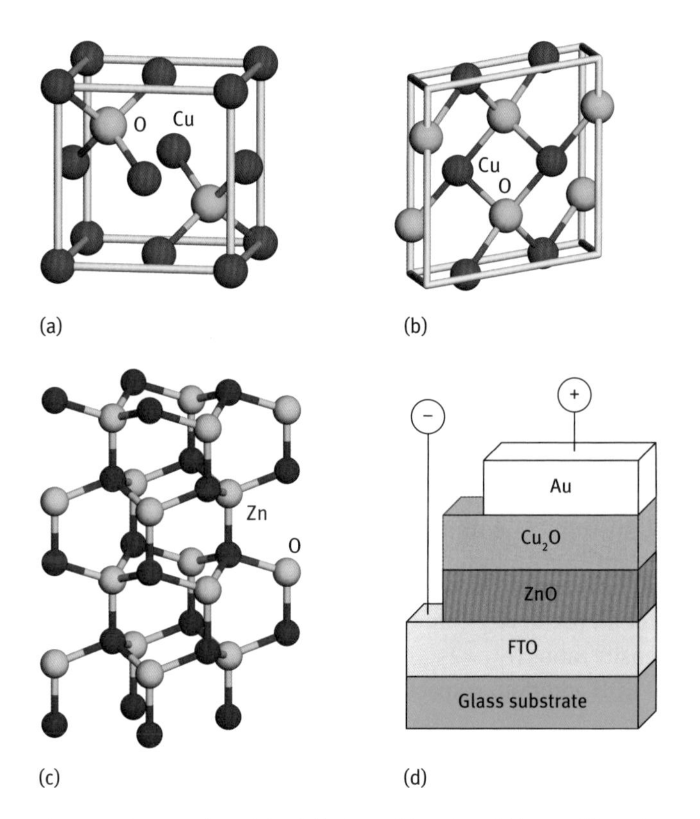


Fig. 4.19: Structural models of (a) $Cu_2O$, (b) CuO and (c) ZnO. (d) Device structure of a $FTO/ZnO/Cu_2O/Au$ heterojunction solar cell.

density of −3.0 mA $cm^{-2}$ and an electric charge of 3.0 C $cm^{-2}$ using a potentio/galvanostat. ZnO layers were also prepared at a current density of 0.5 mA $cm^{-2}$ and an electric charge of 0.12 C $cm^{-2}$ using an aqueous solution of zinc nitrate hexahydrate [$Zn(NO_3)_2 \cdot 6H_2O$]. Au metal contacts were deposited as top electrodes. The structure of the heterojunction solar cells is thus $FTO/ZnO/Cu_2O/Au$, which is shown in Fig. 4.19(d) as a schematic illustration [59].

The current density–voltage ($J$–$V$) characteristics of the $ZnO/Cu_2O$ structure under illumination at 100 mW $cm^{-2}$ obtained using an AM 1.5 solar simulator are shown in Fig. 4.2(a). The photocurrent was observed under illumination and the $ZnO/Cu_2O$ structure showed characteristic curves with regard to the short-circuit current and open-circuit voltage. The highest efficiency was obtained for the $ZnO/Cu_2O$ solar cell prepared using an electrolyte containing LiOH. A solar cell with a $FTO/ZnO/Cu_2O/Au$ structure provided a power conversion efficiency ($\eta$) of 1.43 %, a fill factor ($FF$) of 0.596, a short-circuit current density ($J_{SC}$) of 4.47 mA $cm^{-2}$, and an open-circuit voltage ($V_{OC}$) of 0.535 V. The measured parameters of these $Cu_2O$-based solar cells are summarized in Table 4.5.

Table 4.5: Measured parameters of the solar cells.

| PH reagent | $\eta$ (%) | *FF* | $V_{oc}$ (V) | $J_{sc}$ (mA cm$^{-2}$) | $R_s$ (Ω cm$^2$) | $R_{sh}$ (Ω cm$^2$) |
|---|---|---|---|---|---|---|
| LiOH | 1.43 | 0.596 | 0.535 | 4.47 | 26.0 | 1250 |
| NaOH | 0.698 | 0.425 | 0.445 | 3.69 | 51.3 | 476 |
| KOH | 0.591 | 0.413 | 0.415 | 3.45 | 51.3 | 400 |

Figure 4.20(b) shows the optical absorption of the solar cells. The $ZnO/Cu_2O$ structure shows a high absorption between 300 and 650 nm. In Fig. 5.23, an absorption peak at 540 nm for the $ZnO/Cu_2O$ structure was due to $Cu_2O$. The microstructures of the $FTO/ZnO/Cu_2O$ thin films were investigated by XRD, as shown in Fig. 4.21(a). Diffraction peaks corresponding to $Cu_2O$ were observed for the $Cu_2O$ thin films and they consisted of a cuprite phase with a cubic system (space group *Pn*3*m* and lattice parameter $a = 0.4250$ nm). The particle size was estimated using Scherrer's equation: $D = 0.9\,\lambda/B\cos\theta$, where $\lambda$, $B$ and $\theta$ represent the wavelength of the X-ray source, the full width at half maximum and the Bragg angle, respectively. The crystallite sizes of $Cu_2O$ were found to be 100 nm for all the devices, and $Cu_2O$ showed highly {111}-oriented crystallites. Diffraction intensity ratios ($I_{111} = I_{200}$) of $Cu_2O$ on the ZnO were 41, 35, and 12 for the LiOH, NaOH, and KOH addition, respectively.

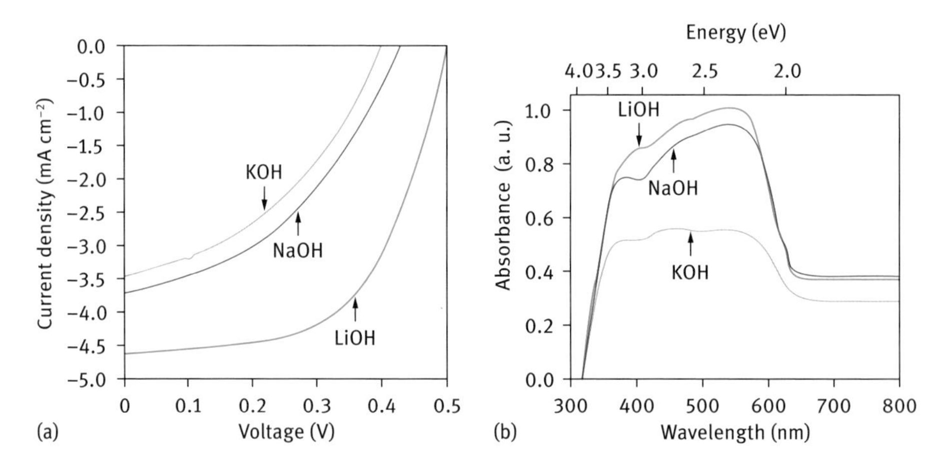


Fig. 4.20: (a) Measured *J*–*V* characteristic of $ZnO/Cu_2O$ solar cells under illumination. (b) Absorption spectra of $ZnO/Cu_2O$ thin films containing LiOH, NaOH, and KOH.

The IPCE spectra of the solar cells with a $ZnO/Cu_2O$ structure are shown in Fig. 4.21(b). The IPCE profiles agree well with the absorption spectra of the ZnO and $Cu_2O$ thin films and this indicates that excitons were effectively generated in the active layers

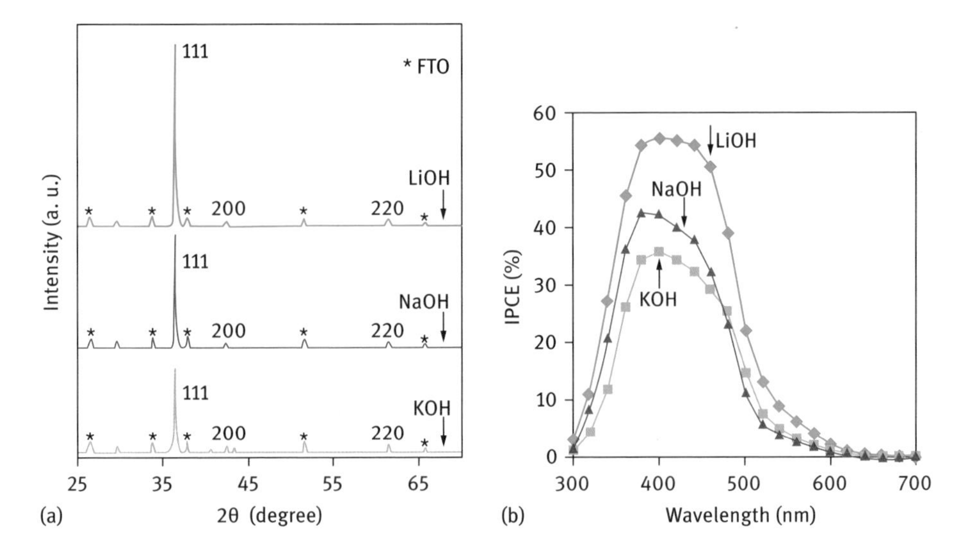


Fig. 4.21: (a) XRD patterns and (b) IPCE spectra of $FTO/ZnO/Cu_2O$ thin films containing LiOH, NaOH, and KOH.

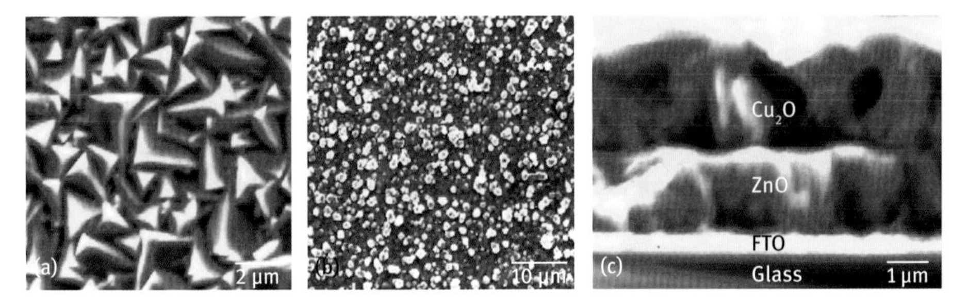


Fig. 4.22: SEM images of (a) $Cu_2O$ and (b) ZnO thin films. (c) Cross-section image of $ZnO/Cu_2O$.

upon illumination by light. The highest IPCE was obtained for the $ZnO/Cu_2O$ solar cell prepared using an electrolyte containing LiOH.

Figure 4.22(a) and (b) are SEM images of $Cu_2O$ and ZnO thin films, and Fig. 4.22(c) is a cross-section image of $FTO/ZnO/Cu_2O$ thin film, respectively. The SEM image indicated $Cu_2O$ and growth of ZnO polycrystals with sizes of 2 ~ 5 μm and 1 ~ 2 μm, respectively. The cross-section image indicated $Cu_2O$ and ZnO film thicknesses are ~ 2.5 μm and ~ 2.0 μm, respectively.

Figure 4.23(a) and (b) show a TEM image and a selected area electron diffraction pattern of a $Cu_2O$ layer prepared by electrodeposition, respectively [51]. The TEM image indicates $Cu_2O$ nanocrystal structures with sizes of 40–50 nm, which agrees with the XRD results. The comparatively larger crystallite sizes of the $Cu_2O$ resulted in unclear Debye–Scherrer rings in Fig. 4.23(b), which indicates the higher crystallinity of the

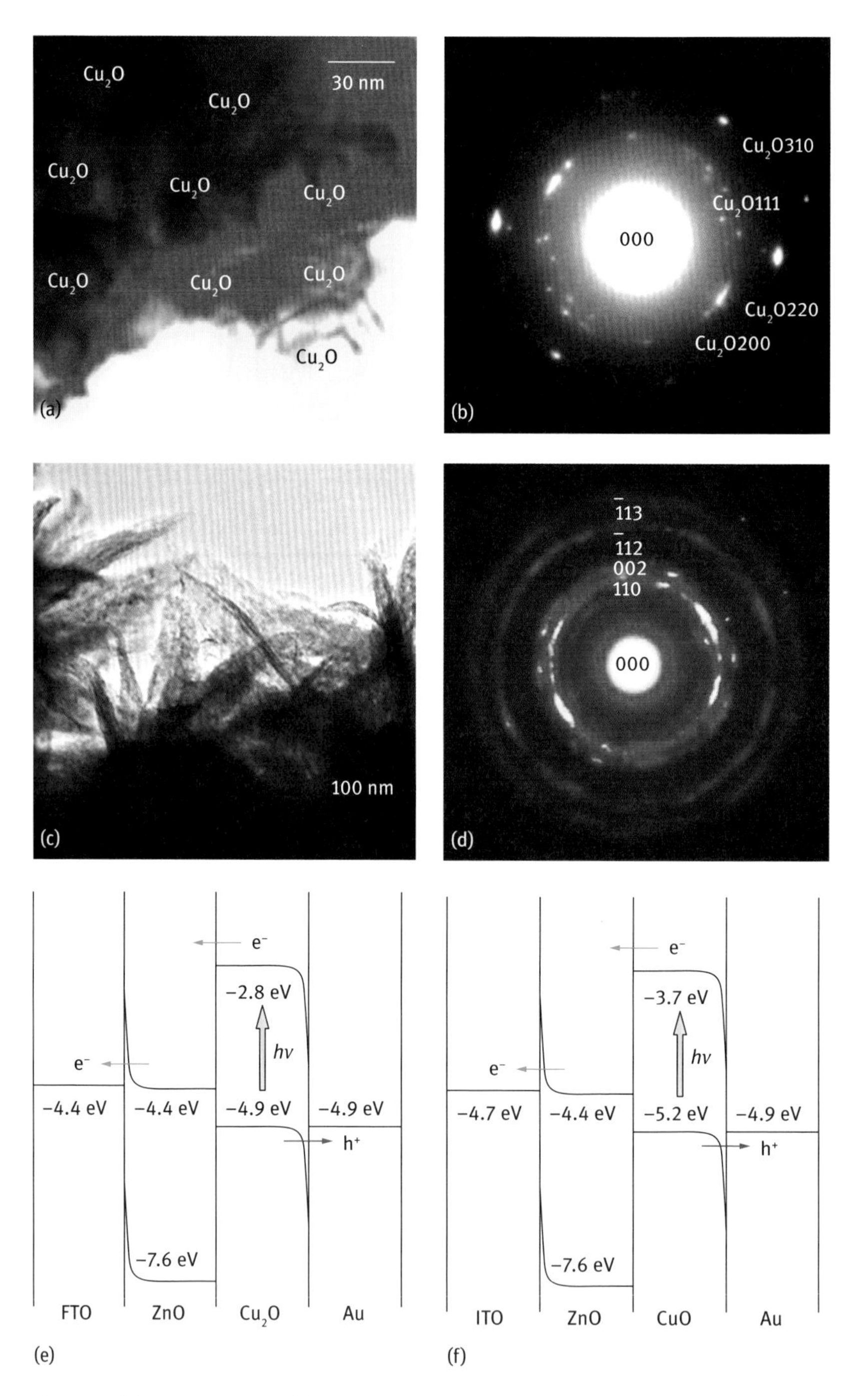


Fig. 4.23: (a) TEM image and (b) electron diffraction pattern of $Cu_2O$ thin film and (c) TEM image and (d) electron diffraction pattern of CuO thin film. Energy level diagrams of (e) $ZnO/Cu_2O$ and (f) ZnO/CuO heterojunction solar cells.

$Cu_2O$ nanoparticles. A TEM image of the CuO layer is shown in Fig. 4.23(c), which indicates the formation of CuO nanowires of ~ 50 nm diameter. Debye–Scherrer rings in the electron diffraction pattern in Fig. 4.23(d) indicate polycrystalline CuO structures within the films.

The energy level diagrams of ZnO/$Cu_2O$ and ZnO/CuO solar cells are summarized in Fig. 4.23(e) and (f). Previously reported values were used for the energy levels. It has been reported that $V_{OC}$ is nearly proportional to the band gap of the semiconductors [60], and that control of the energy levels is important to increase efficiency.
In this work, LiOH was found to be effective in improving the conversion efficiency of electrodeposited solar cells in comparison to KOH and NaOH. $Cu_2O$ prepared with LiOH provided the highest $I_{111} = I_{200}$, which indicates that the highest {111}-oriented $Cu_2O$ crystallites were formed with the addition of LiOH. This resulted in a decrease in the series resistance ($R_s$) and an increase in the shunt resistance ($R_{sh}$) of the ZnO/$Cu_2O$ solar cell, as listed in Table 4.5. As a result, carriers move more easily and this leads to a reduction in $R_s$ and an improvement in IPCE. A decrease in the leakage current and an increase in $R_{sh}$ resulted in improved photoelectric parameters.

In summary, $Cu_2O$ and ZnO layers were deposited onto precleaned FTO glass plates by electrodeposition and the fabricated FTO/ZnO/$Cu_2O$/Au solar cells were characterized. A device based on the structure of ZnO/$Cu_2O$ using LiOH provided the highest $\eta$ of 1.43 % and $V_{OC}$ of 0.535 V. XRD results indicated the presence of $Cu_2O$ crystals. The $Cu_2O$ consisted of highly oriented {111} crystallites and the crystallite sizes of $Cu_2O$ were determined to be 100 nm. The improvement in photoelectric parameters was due to the crystal orientation of $Cu_2O$ layers, which resulted in an increase in $R_{sh}$ and a decrease in $R_s$ for $Cu_2O$-based solar cells.

## 4.8 Bibliography


[1] Green MA, Emery K, Hishikawa Y, Warta W, Dunlop ED. Solar cell efficiency tables (Version 48). *Prog Photovolt.* 2016; 24: 905–913.

[2] Green MA. Third generation photovoltaics. Springer; 2003.

[3] Sai H, Matsui T, Koida T, Matsubara K, Kondo M, Sugiyama S, Katayama H, Takeuchi Y, Yoshida I. Triple-junction thin-film silicon solar cell fabricated on periodically textured substrate with a stabilized efficiency of 13.6 %. *Appl Phys Lett.* 2015; 106: 213902.

[4] Masuko K, Shigematsu M, Hashiguchi T, Fujishima D, Kai M, Yoshimura Naoki, Yamaguchi T, Ichihashi Y, Mishima T, Matsubara N, Yamanishi T, Takahama T, Taguchi M, Maruyama E, Okamoto S. Achievement of more than 25 % conversion efficiency with crystalline silicon heterojunction solar cell. *IEEE J Photovolt.* 2014; 4: 1433–1435.

[5] Rabadanov, MK, Simonov VI. Atomic structure of $Cd_{1-x}Zn_xTe$ solid solution single crystals and structural prerequisites of their ferroelectricity. *Crystallogr Rep.* 2006; 51: 778–791.

[6] Mahabaduge HP, Rance WL, Burst JM, Reese MO, Meysing DM, Wolden CA, Li J, Beach JD, Gessert TA, Metzger WK, Garner S, Barnes TM. High-efficiency, flexible CdTe solar cells on ultra-thin glass substrates. *Appl Phys Lett.* 2015; 106: 133501.

[7] Souilah M, Lafond A, Barreau N, Guillot Deudon C, Kessler J. Evidence for a modified-stannite crystal structure in wide band gap Cu-poor $CuIn_{1-x}Ga_xSe_2$: Impact on the optical properties. *Appl Phys Lett.* 2008; 92: 241923.

[8] Mitzi DB, Gunawan O, Todorov TK, Wang K, Guha S. The path towards a high-performance solution-processed kesterite solar cell. *Sol Energy Mater Sol Cells.* 2011; 95: 1421–1436.

[9] Maruyama T, Minami H. Light trapping in spherical silicon solar cell module. *Sol Energy Mater Sol Cells.* 2003; 79: 113–124.

[10] Liu Z, Masuda A, Kondo M. Investigation on the crystal growth process of spherical Si single crystals by melting. *J Cryst Growth.* 2009; 311: 4116–4122.

[11] Gharghi M and Sivoththaman S. Growth and structural characterization of spherical silicon crystals grown from polysilicon. *J Electron Mater.* 2008; 37: 1657–1664.

[12] Okamoto C, Minemoto T, Murozono M, Takakura H, Hamakawa Y. Defect evaluation of spherical silicon solar cells fabricated by dropping method. *Jpn J Appl Phys.* 2005; 44: 7805.

[13] Liu Z, Nagai T, Masuda A, Kondo M, Sakai K, Asai K. Seeding method with silicon powder for the formation of silicon spheres in the drop method. *J Appl Phys.* 2007; 101: 093505.

[14] Minemoto T, Takakura H. Fabrication of spherical silicon crystals by dropping method and their application to solar cells. *Jpn J Appl Phys.* 2007; 46: 4016.

[15] Gharghi M, Bai H, Stevens G, Sivoththaman S. Three-dimensional modeling and simulation of *p-n* junction spherical silicon solar cells. *IEEE Trans Electron Devices.* 2006; 53: 1355–1363.

[16] Liu Z, Masuda A, Kondo M. Investigating minority-carrier lifetime in small spherical Si using microwave photoconductance decay. *J Appl Phys.* 2008; 103: 104909.

[17] Hua XS, Zhang YJ, Wang HW. The effect of texture unit shape on silicon surface on the absorption properties. *Sol Energy Mater Sol Cells.* 2010; 94: 258–262.

[18] Yoo J, Yu G, Yi J. Large-area multicrystalline silicon solar cell fabrication using reactive ion etching (RIE). *Sol Energy Mater Sol Cells.* 2011; 95: 2–6.

[19] Hayashi S, Minemoto T, Takakura H, Hamakawa Y. Influence of texture feature size on spherical silicon solar cells. *Rare Met.* 2006; 25: 115–120.

[20] Minemoto T, Murozono M, Yamaguchi Y, Takakura H, Hamakawa Y. Design strategy and development of spherical silicon solar cell with semi-concentration reflector system. *Sol Energy Mater Sol Cells.* 2006; 90: 3009–3013.

[21] Liu Z, Masuda A, Nagai T, Miyazaki T, Takano M, Takano M, Yoshigahara H, Sakai K, Asai K Kondo M. A concentrator module of spherical Si solar cell. *Sol Energy Mater Sol Cells.* 2007; 91: 1805–1810.

[22] Mizuta T, Ikuta T, Minemoto T, Takakura H, Hamakawa Y, Numai T. An optimum design of antireflection coating for spherical silicon solar cells. *Sol Energy Mater Sol Cells.* 2006; 90: 46–56.

[23] Ono Y, Oku T, Akiyama T, Kanamori Y, Ohnishi Y, Ohtani Y, Murozono M. Microstructure analysis of spherical silicon solar cells coated with anti-reflection films. *J Phys Conf Ser.* 2012; 352: 012023.

[24] Huang CK, Lin HH, Chen JY, Sun KW, Chang WL. Efficiency enhancement of the poly-silicon solar cell using self-assembled dielectric nanoparticles. *Sol Energy Mater Sol Cells.* 2011; 95: 2540–2544.

[25] Kanayama M, Oku T, Akiyama T, Kanamori Y, Seo S, Takami J, Ohnishi Y, Ohtani Y, Murozono M. Microstructure analysis and properties of anti-reflection thin films for spherical silicon solar cells. *Energy Power Eng.* 2013; 5: 18–22.

[26] Shirahata Y, Zhang B, Oku T, Kanamori Y, Murozono M. Microstructures and optical properties of silicon spheres for solar cells. *Mater Trans.* 2016; 57: 1082–1087.

[27] Omae S, Minemoto T, Murozono M, Takakura H, Hamakawa Y. Crystal evaluation of spherical silicon produced by dropping method and their solar cell performance. *Sol Energy Mater Sol Cells.* 2006; 90: 3614–3623.

[28] Liu Z, Asai K, Masuda A, Nagai T, Akashi Y, Murozono M, Kondo M. Improvement of the production yield of spherical Si by optimization of the seeding technique in the dropping method. *Jpn J Appl Phys*. 2007; 46: 5695–5700.
[29] Murozono M, Akashi Y, Oshima Y. Method of producing semiconductor particle. Japan Patent. 2012: 151413.
[30] Stevens GD, Conklin HL. Process for producing semiconductor spheres. US Patent. 1995: 5431127.
[31] Nakamura T, Akashi Y, Murozono M. Method for producing spherical semiconductor particles. Japan Patent. 2010: 150106.
[32] Kanamori Y, Akashi Y, Murozono M. Method for producing crystal semiconductor particle. Japan Patent. 2012: 126592.
[33] Oku T, Kanayama M, Ono Y, Akiyama T, Kanamori Y, Murozono M. Microstructures, optical and photoelectric conversion properties of spherical silicon solar cells with anti-reflection $SnO_x$:F thin films. *Jpn J Appl Phys*. 2014; 53: 05FJ03.
[34] Oku T. Direct structure analysis of advanced nanomaterials by high-resolution electron microscopy. *Nanotechnol Rev*. 2012; 1: 389–425.
[35] Omae S, Minemoto T, Murozono M, Takakura H, and Hamakawa Y. Crystal characterization of spherical silicon solar cell by X-ray diffraction. *Jpn J Appl Phys*. 2006; 45: 3933.
[36] McCarthy GJ, Welton JM. X-Ray diffraction data for $SnO_2$. An illustration of the new powder data evaluation methods. *Powder Diffr*. 1989; 4: 156–159.
[37] Nagai T, Liu Z, Masuda A, Kondo M. Characterization of spherical Si by photoluminescence measurement. *J Appl Phys*. 2007; 101: 103530.
[38] Edwards PP, Porch A, Jones MO, Morgan DV, Perks RM. Basic materials physics of transparent conducting oxides. *Dalton Trans*. 2004; 19: 2995–3002.
[39] Minami T. Transparent conducting oxide semiconductors for transparent electrodes. *Semicond Sci Technol*. 2005; 20: S35.
[40] Han CH, Han SD, Singh I, Toupance T. Micro-bead of nano-crystalline F-doped $SnO_2$ as a sensitive hydrogen gas sensor. *Sens Actuators B*. 2005; 109: 264–269.
[41] Kumar V, Govind A, Nagarajan R. Optical and photocatalytic properties of heavily $F^-$-doped $SnO_2$ nanocrystals by a novel single-source precursor approach. *Inorg Chem*. 2011; 50: 5637–5645.
[42] Zhang B, Tian Y, Zhang JX, Cai W. Numerical simulation on aerodynamics of ramjet projectiles. *J Optoelectron Adv Mater*. 2011; 13: 89–92.
[43] Oku T, Wakimoto H, Otsuki A, Murakami M. NiGe-based ohmic contacts to n-type GaAs. I. Effects of In addition. *J Appl Phys*. 1994; 75: 2522–2529.
[44] Oku T, Furumai M, Uchibori CJ, Murakami M. Formation of WSi-based ohmic contacts to n-type GaAs. *Thin Solid Films*. 1997; 300: 218–222.
[45] Restori R, Schwarzenbach D. Charge density in cuprite, $Cu_2O$. *Acta Cryst B*. 1986, 42: 201–208
[46] Bianchi AE, Montenegro L, Viña R, Punte G. Microstructure anisotropy in CuO powders. *Powder Diffr*. 2008; 23: S81–S86.
[47] Minami T, Nishi Y, Miyata T. High-efficiency $Cu_2O$-based heterojunction solar cells fabricated using a $Ga_2O_3$ thin film as n-type layer. *Appl Phys Express*. 2013; 6: 044101.
[48] Izaki M, Shinagawa T, Mizuno K, Ida Y, Inaba M, Tasaka A. Electrochemically constructed *p*-$Cu_2O$/*n*-ZnO heterojunction diode for photovoltaic device. *J Phys D*. 2007; 40: 3326.
[49] Duan Z, Pasquier AD, Lu Y, Xu Y, Garfunkel E. Effects of Mg composition on open circuit voltage of $Cu_2O$–$Mg_xZn_{1-x}O$ heterojunction solar cells. *Sol Energy Mater Sol Cells*. 2012; 96: 292–297.
[50] Shao F, Sun J, Gao L, J. Luo J, Liu Y, Yang S. High efficiency semiconductor-liquid junction solar cells based on Cu/$Cu_2O$. *Adv Funct Mater*. 2012; 22: 3907–3913.

[51] Oku T, Takeda A, Nagata A, Kidowaki H, Kumada K, Fujimoto K, Suzuki A, Akiyama T, Yamasaki Y, Ōsawa E. Microstructures and photovoltaic properties of $C_{60}$ based solar cells with copper oxides, $CuInS_2$, phthalocyanines, porphyrin, PVK, nanodiamond, germanium and exciton diffusion blocking layers. *Mater Technol.* 2013; 28: 21–39.
[52] Oku T, Motoyoshi R, Fujimoto K, Akiyama T, Jeyadevan B, Cuya J. Structures and photovoltaic properties of copper oxides/fullerene solar cells. *J Phys Chem Solids.* 2011; 72: 1206–1211.
[53] McShane CM, Choi KS. Junction studies on electrochemically fabricated *p-n* $Cu_2O$ homojunction solar cells for efficiency enhancement. *Phys Chem Chem Phys.* 2012; 14: 6112–6118.
[54] Cui J, Gibson UJ. A simple two-step electrodeposition of $Cu_2O$/ZnO nanopillar solar cells. *J Phys Chem C.* 2010; 114: 6408–6412.
[55] Wei H, Gong H, Wang Y, Hu X, Chen L, Xu H, Liub P, Cao B. Three kinds of $Cu_2O$/ZnO heterostructure solar cells fabricated with electrochemical deposition and their structure-related photovoltaic properties. *CrystEngComm.* 2011; 13: 6065–6070.
[56] Fujimoto K, Oku T, Akiyama T, Suzuki A. Fabrication and characterization of copper oxide-zinc oxide solar cells prepared by electrodeposition. *J Phys Conf Ser.* 2013; 433: 012024.
[57] Minami T, Nishi Y, Miyata T, Nomoto J. High-efficiency oxide solar cells with ZnO/$Cu_2O$ heterojunction fabricated on thermally oxidized $Cu_2O$ sheets. *Appl Phys Express.* 2011; 4: 062301.
[58] Santos DAA, Rocha ADP, Macêdo MA. Rietveld refinement of transition metal doped ZnO. *Powder Diffr.* 2008; 23: S36–S41.
[59] Fujimoto K, Oku T, Akiyama T. Fabrication and characterization of ZnO/$Cu_2O$ solar cells prepared by electrodeposition. *Appl Phys Express.* 2013; 6: 086503.
[60] Golden TD, Shumsky MG, Zhou Y, VanderWerf RA, Van Leeuwen RA, Switzer JA. Electrochemical deposition of copper(I) oxide films. *Chem Mater.* 1996; 8: 2499–2504.

# 5 Organic-type solar cells

## 5.1 Donor-acceptor type organic solar cells

Carbon is generated by He nuclear fusion in fixed stars, and a huge amount of carbon atoms exist throughout the universe. In life-related activity such as photosynthesis, carbon compounds such as $CO_2$ and $C_6H_{12}O_6$ play an important role. Carbon has various hollow-cage nanostructures such as $C_{60}$, giant fullerenes, nanocapsules, onions, nanopolyhedra, cones, cubes, and nanotubes. These C structures show different physical properties, and there would be great potential in studying these materials at a small scale within an isolated environment. By controlling the size, number of layers, helicity, composition and included clusters, cluster-included C nanocage structures with a band-gap energy of 0–1.7 eV and nonmagnetism are expected to show various electronic, optical, and magnetic properties such as Coulomb blockade, photoluminescence, and superparamagnetism [1]. Recently, $C_{60}$-based polymer/fullerene solar cells have been investigated and reported on [2–6]. These organic solar cells can be fabricated by the use of printing methods under ordinary atmospheric conditions, and they have a potential for use in lightweight, flexible, inexpensive, and large-scale solar cells [7–12].

The photovoltaic mechanism of organic solar cells is shown in Fig. 5.1. The light is absorbed in the donor (D) layers, such as phthalocyanine and poly[3-hexylthiophene] (P3HT), and electrons are excited to form excitons from the energy levels of the highest occupied molecular orbital (HOMO) to the lowest unoccupied molecular orbital (LUMO). Then, the excitons diffuse to the donor-acceptor (DA) interface, and the charges are separated at the interface. Separated electrons are transported in the acceptor (A) such as fullerene, holes are transported in the donor to the electrodes and the current flows.

Four factors that determine the conversion efficiencies of organic solar cells are exciton generation efficiency $\eta_1$, exciton transport efficiency $\eta_2$, charge separation efficiency $\eta_3$ and carrier transport efficiency, as shown in Fig. 5.2 [8]. Since the total efficiency is calculated by multiplying the four efficiencies, all efficiencies should be high. Although the $\eta_1$ and $\eta_3$ values are high for the organic solar cells, the $\eta_2$ and $\eta_4$ values are low because of the very short diffusion length of the excitons.

A weak point of organic solar cells is their low conversion efficiency, which results from the recombination of excitons produced by light irradiation. In photovoltaic solar cells, excitons are generated by light irradiation. The excitons are separated into electrons and holes, and they are transported to each electrode to generate potential difference. However, if the electrons and holes of the excitons are recombined before their arrival at the electrode, light is emitted by their recombination, and no electric power is generated.

DOI 10.1515/9783110298505-008

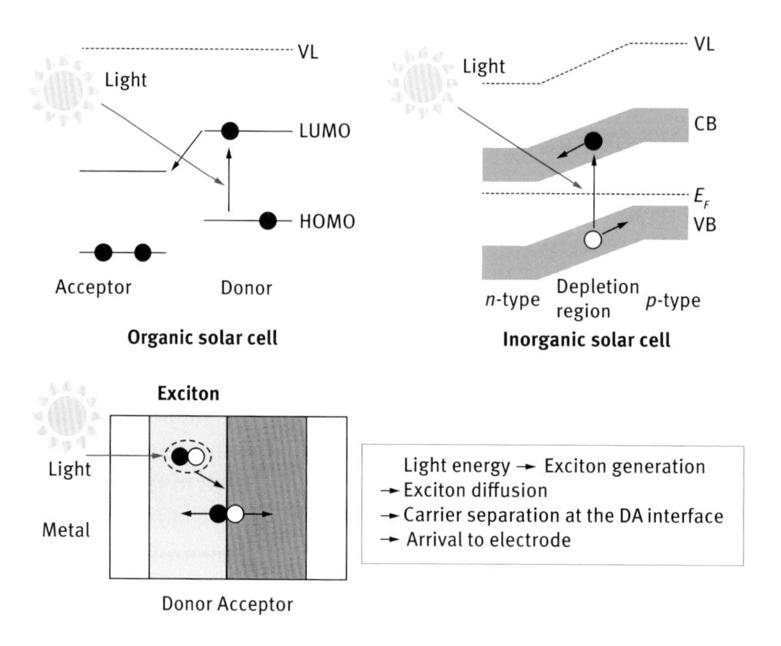


Fig. 5.1: The photovoltaic mechanism of an organic solar cell.

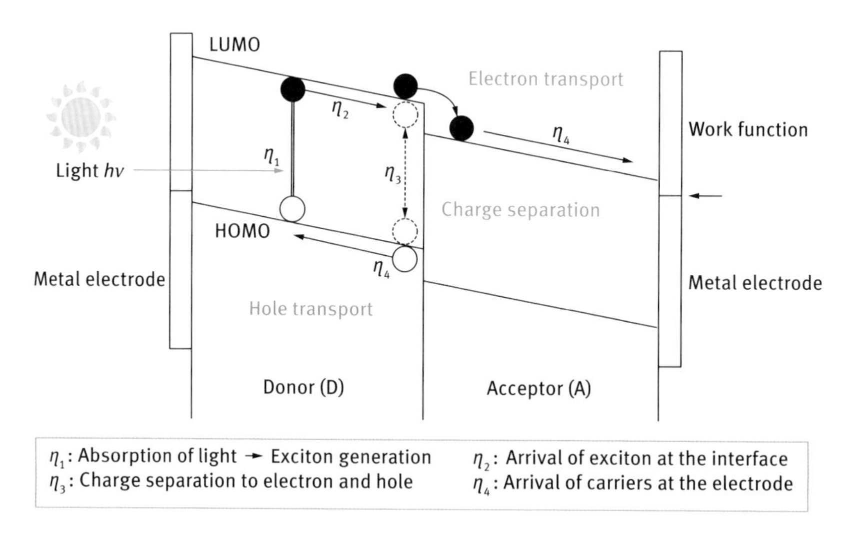


$\eta_1$: Absorption of light → Exciton generation
$\eta_2$: Arrival of exciton at the interface
$\eta_3$: Charge separation to electron and hole
$\eta_4$: Arrival of carriers at the electrode

Fig. 5.2: Carrier separation and carrier transport in organic solar cells.

One of the causes of recombination in organic solar cells is its low carrier mobility. It takes too much time for generated excitons to reach the *pn* junction for the carrier separation, and the excitons recombine on the way to the *pn* junction or metal electrodes. In addition, since the sizes of excitons in organic materials are small, they tend to recombine.

## 5.2 Exciton

An exciton is a pair of an excited electron and an excited hole restricted by the electrostatic Coulomb force, and it is an electrically neutral quasiparticle that exists in semiconductors and insulators, as shown in Fig. 5.3. The Coulomb force is expressed as follows ($\varepsilon$: relative permittivity):

$$F = \frac{1}{4\pi\varepsilon\varepsilon_0}\frac{q_1 q_2}{r^2} \tag{5.1}$$

The exciton is regarded as an elementary excitation that can transport energy without transporting electric charge. The current of solar cells flows only when the exciton is separated into an electron and a hole. Excitons are introduced physically from excited waves in the wave function of the biding state of electrons on the conduction and holes on the valence band. Frenkel excitons and Mott-Wannier excitons are limit models of the excited waves, and actual excitons have intermediate states between these excitons.

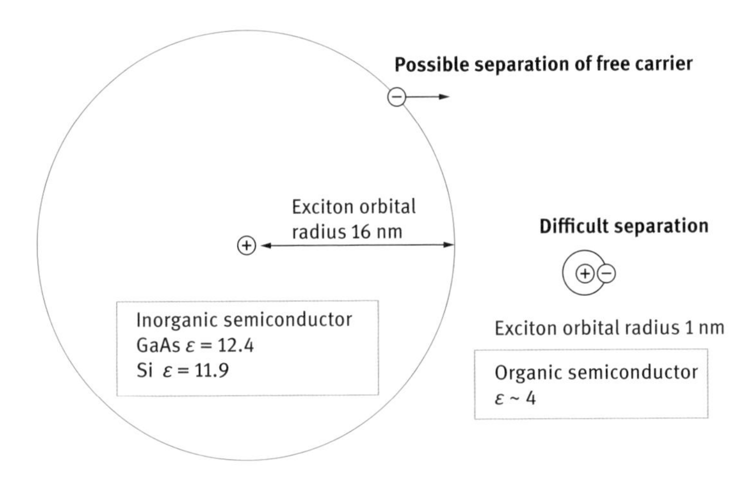


Fig. 5.3: Excitons of inorganic and organic semiconductors.

The wave function of Mott-Wannier excitons (weak binding of an electron and a hole) in the excited state is broader than the lattice constants. The excited state is a spread state at a lattice point, and an electron and a hole are in a bound state with weak restriction. In an excited state like the Mott-Wannier, excitons spread in crystals such as various ionic crystals and ionic semiconductors.

Frenkel excitons (comparatively strong binding of an electron and a hole) have a narrower wave function in the excited state compared to the lattice constants. The excited state is similar to the excited states of atoms or ions. An excited state like the

Frenkel excitons spread resonantly through lattice points with a certain wavenumber in organic-molecular crystals.

The energy required for exciton generation is lower than the bandgap energy because of the binding energy between an electron and a hole, and the exciton is in a stable state. A sharp reflection peak can be observed at lower energy compared with that of interband transition. Excitons that spread in the hard, non-deformed lattice are called free excitons, and they can be transported freely throughout the crystal. Self-restraint excitons that spread in the vibrating lattice localize at a certain position by interaction with lattice vibration.

## 5.3 Bulk heterojunction

One of the improvements presented by organic solar cells is donor-acceptor (DA) proximity in the devices by using blends of donor-like and acceptor-like molecules or polymers, which are called DA bulk-heterojunction solar cells [12–18], as shown in Fig. 5.4. Previous organic solar cells consisted of a simple *pn* heterojunction. The bulk-heterojunction is a *pin* junction which consists of a mixture intrinsic semiconductor layer (*i*-layer) between *p*- and *n*-type semiconductors. For a fullerene-based system, *p*-type molecular crystals are surrounded by an amorphous fullerene matrix.

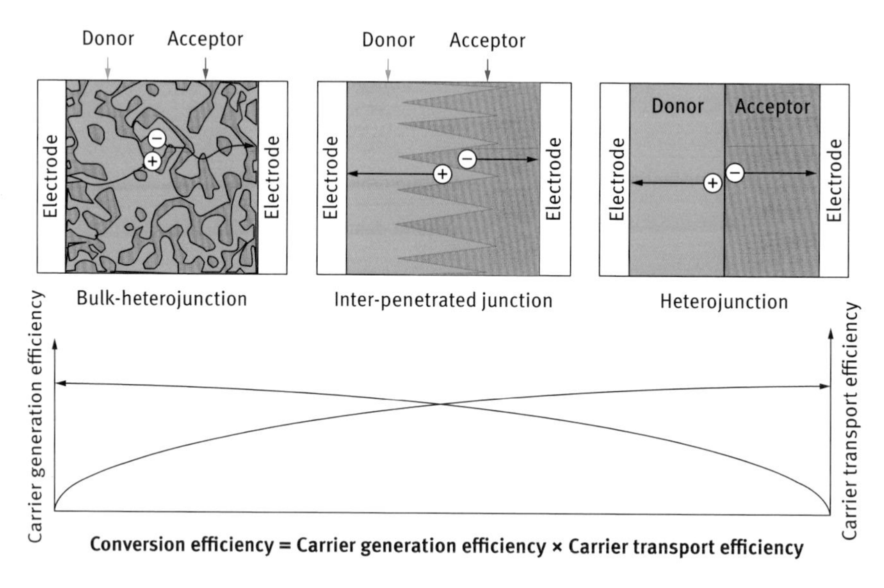


Fig. 5.4: Main device structures proposed for organic solar cells.

When light is irradiated on the junction, excitons are generated around the p-type molecular/fullerene interface, which consists of a bulk-hetero mixture layer, and the excitons can reach the DA or *pn* junction by transporting several nm. Electrons and holes are separated into an n-layer and p-layer at the interface, respectively. Each carrier transports through connected crystals and the matrix to the electrode, and current flows.

One merit of ordinary heterojunction solar cells is their high efficiency when the carrier mobility and electrical conductivity of the D and A layers are high. However, only the excitons generated near the D/A interface contribute to the photocurrent. On the other hand, the interfacial area of the bulk-heterojunction is so large that the carriers are separated effectively. However, the carrier transport pass is complicated, and the carrier could not be taken away from the cells. After all, an inter-penetrated structure would be effective for carrier generation and carrier transport, and research on new structures such as nanorods [19] and nanotubes is currently in progress.

## 5.4 P3HT:PCBM

$C_{60}$-based polymer/fullerene solar cells have been investigated, and significant improvements in photovoltaic efficiencies are mandatory for use in future solar power plants. A characterization of polymer/fullerene bulk-heterojunction solar cells using different organic polymers is presented here. Poly[3-hexylthiophene] (P3HT) and poly[2-methoxy-5-(20-ethylhexoxy)-1,4-phenylenevinylene] (MEH-PPV) were used for *p*-type semiconductors, and 6,6-phenyl $C_{61}$-butyric acid methyl ester (PCBM) was used for n-type one. Device structures were produced, and efficiencies and spectral responsivity were investigated.

A thin layer of polyethylenedioxythiophen doped with polystyrene–sulfonic acid (PEDOT:PSS) (Sigma Aldrich)was spin-coated on pre-cleaned indium tin oxide (ITO) glass plates. Then, semiconductor layers were prepared on a PEDOT layer by spin-coating using a mixed solution of P3HT, MEH-PPV and PCBM in 1,2-dichlorobenzene. The weight ratios of both P3HT:PCBM and MEH-PPV:PCBM were 1 : 8. The thickness of the blended device was approximately 150 nm. After annealing at 100 °C for 30 min in $N_2$ atmosphere, aluminum (Al) metal contacts with a thickness of 100 nm were evaporated as a top electrode. A schematic diagram of the presented solar cells is shown in Fig. 5.5(a) [4].

The typical current density–voltage ($J$–$V$) characteristics of a P3HT/PCBM structure in the dark and under illumination are shown in Fig. 5.5(a). Although no photocurrent was observed in the dark, a photocurrent over 5 mA cm$^{-2}$ is observed under illumination. The $J$–$V$ characteristics of both MEH-PPV/PCBM and P3HT/PCBM solar cell structures are shown in Fig. 5.5(a). Each structure shows characteristic curves for open-circuit voltage and short-circuit current. Measured parameters of these solar cells are summarized in Table 5.1. A solar cell with a P3HT/PCBM structure provided power convergent efficiency of 1.03 %, a fill factor of 0.53 and a short circuit current density of 5.18 mA cm$^{-2}$, which is better than that of a MEH-PPV/PCBM device. On

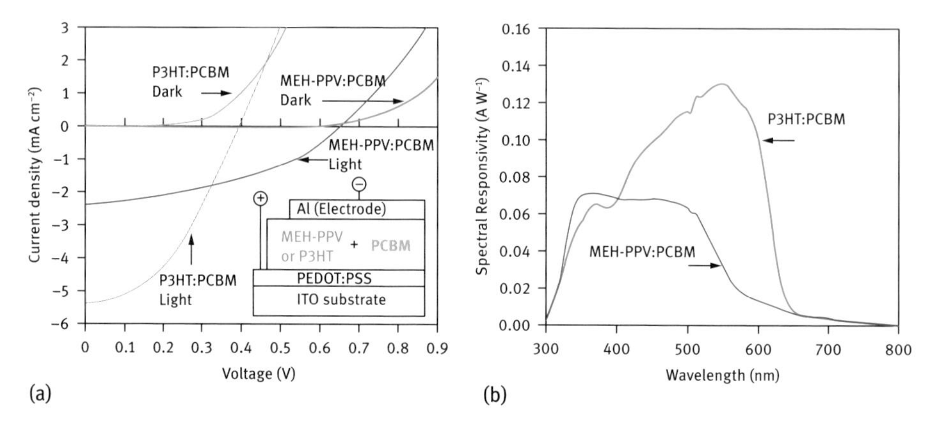


Fig. 5.5: (a) Structure of bulk-heterojunction solar cells. Measured J–V characteristic of P3HT:PCBM and MEHPPV:PCBM structure in the dark and under illumination. (b) Spectral photoresponses of the solar cells.

Table 5.1: Measured parameters of solar cells.

| ETL | $J_{SC}$ (mA cm$^{-2}$) | $V_{OC}$ (V) | *FF* | $\eta$ (%) |
|---|---|---|---|---|
| P3HT:PCBM | 5.18 | 0.37 | 0.53 | 1.03 |
| MEH-PPV:PCBM | 2.59 | 0.70 | 0.42 | 0.75 |

the other hand, the MEH-PPV/PCBM structure showed a higher open-circuit voltage of 0.70 V.

Figure 5.5(b) shows the measured spectral photoresponses of the solar cells. The MEH-PPV/PCBM structure shows a high photoresponse in the range of 300–600 nm, while the P3HT/PCBM shows higher spectral responsivity in the range of 400–650 nm, which corresponds to 3.1 and 1.9 eV, respectively. Optimization of the nanocomposite structure with P3HT and MEH-PPV would increase the efficiencies of the solar cells.

The electronic structures of the molecules were calculated, and the energy levels of HOMO of P3HT and MEH-PPV are shown in Fig. 5.6(a) and (b), respectively. HOMO levels are observed around the five and six-membered rings in the main-chain structures of the polymers, which could be due to the charge transfer from the sulfur and oxygen atoms, respectively. The energy levels of the LUMO of PCBM are also shown in Fig. 5.6(c), and the LUMO levels are observed around a $C_{60}$ molecule with high electron negativity. Effective formation and separation of excitons in the P3HT/PCBM system could be due to the nanocomposite structure, and the separated carriers could transfer from P3HT to $C_{60}$, as has been reported previously [7]. Interdiffusion of PCBM into the P3HT network would lead to the existence of $C_{60}$ molecules within the exciton diffusion radius of the P3HT network.

An energy-level diagram of P3HT/PCBM solar cells is summarized in Fig. 5.6(d). Previously reported values were used for the energy levels of the figures by adjusting them to the present work [15, 20–22]. An energy gap of 1.9 eV, which is an estimated value from Fig. 5.5(b), is used for the model. The relation between $V_{OC}$ and polymer oxidation potential has been reported as follows:

$$V_{OC} \sim e^{-1}\,|D_{HOMO} - A_{LUMO}| - 0.3\ (\mathrm{V}) \tag{5.2}$$

Where $e$ is the elementary charge [21, 22]. The value of 0.3 V is an empirical factor, and this is enough for efficient charge separation [23]. The present model agrees with this equation, and control of the energy levels is important to increase the efficiency. Combination of the present solar cells and boron nitride nanomaterials with various direct band gaps might be effective for an increase in efficiencies [24]. The performances of these solar cells could also be linked to the nanoscale structures of the polymer materials, and the control of the nanostructure should be investigated further.

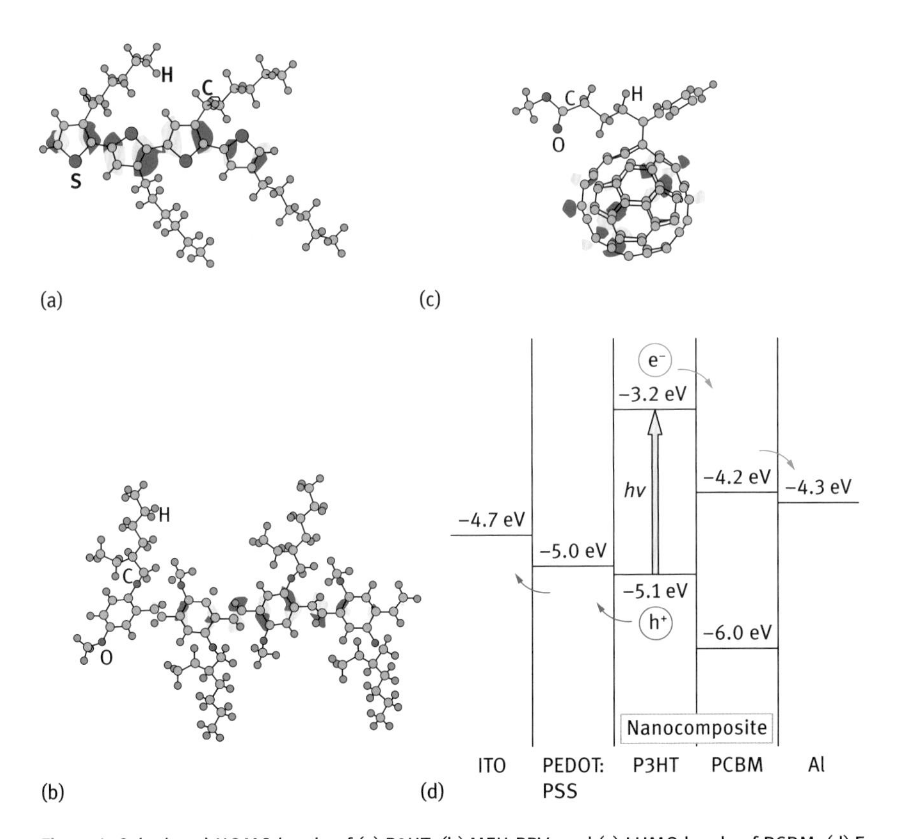


Fig. 5.6: Calculated HOMO levels of (a) P3HT, (b) MEH-PPV, and (c) LUMO levels of PCBM. (d) Energy-level diagram of P3HT/PCBM solar cells.

## 5.5 Phthalocyanine dimer

Phthalocyanines, which exhibit photovoltaic properties, heat resistance, light stability, chemical stability and a high optical absorption at visible range, are used as an oxidation catalyst, a catalyst in fuel cells and solar cells. Many studies on the metal phthalocyanine (MPc) monomers have been performed [25, 26], and the properties are strongly dependent on central metals or chemical substitutions. Organic–inorganic hybrid device structures were produced, and nanostructure, electronic property and optical absorption were investigated. When the nearest two neighboring phthalocyanines with substituents such as amino group and hydroxy group are connected by a hydrogen bridged substituent, high photoconduction has been observed [27, 28]. However, few phthalocyanine dimers have been reported, and high photoconduction can be expected for the covalently bridged phthalocyanine dimers. The purpose of the study presented here was to fabricate and characterize phthalocyanine dimer/fullerene HJ solar cells. Here, μ-oxo bridged gallium phthalocyanine (GaPc) dimer was used for p-type semiconductors, and fullerene with an excellent electron affinity was used for the n-type ones. The molecular orbital of GaPc dimer and fullerene was investigated as a solar cell material [12].

GaPc monomer with axial Cl ligand was also investigated for the comparison. The measured $J$–$V$ characteristic of ITO/ PEDOT:PSS/GaPc dimer/$C_{60}$/Al solar cell and ITO/PEDOT:PSS/GaPc/$C_{60}$/Al showed a characteristic curve for open circuit voltage and short circuit current density. All parameters were improved using GaPc dimer compared with GaPc monomer. Figure 5.7(a) shows a measured optical absorption of GaPc dimer, $C_{60}$ and GaPc dimer/$C_{60}$ cells [29]. The solar cells show a wide optical absorption ranging from 320 to 800 nm, which corresponds to 3.8 and 1.5 eV respectively. The absorption spectrum of the GaPc dimer was almost the same as that of the monomer, but a new peak was observed at ~ 450 nm. The fluorescence (FL) spectra of GaPc dimer and GaPc dimer/$C_{60}$ thin films are shown in Fig. 5.7(b), and the excitation wavelength was 300 nm. A FL peak of GaPc dimer disappeared after formation of GaPc dimer/$C_{60}$ HJ thin films. It is believed that carriers would be effectively transported from GaPc dimer to $C_{60}$.

Figure 5.7(c) shows the structure of the μ-oxo-bridged gallium phthalocyanine dimer used here. Two GaPc planes are parallel to one another, and the degree of rotation is 41.35 ° [29, 30]. The plane distance between the GaPc monomer is ~ 0.34 nm. When the nearest two neighboring phthalocyanines are arranged with hydrogen bridged substituent, high photoconduction can be expected for the covalently-bridged phthalocyanine dimer.

▸ Fig. 5.7: (a) Optical absorption and (b) photoluminescence spectra of GaPc dimer/$C_{60}$ solar cells. (c) Structure of GaPc dimer. (d) Energy level diagram of GaPc dimer/$C_{60}$ solar cells. (e) Electronic structure of GaPc dimer and $C_{60}$.

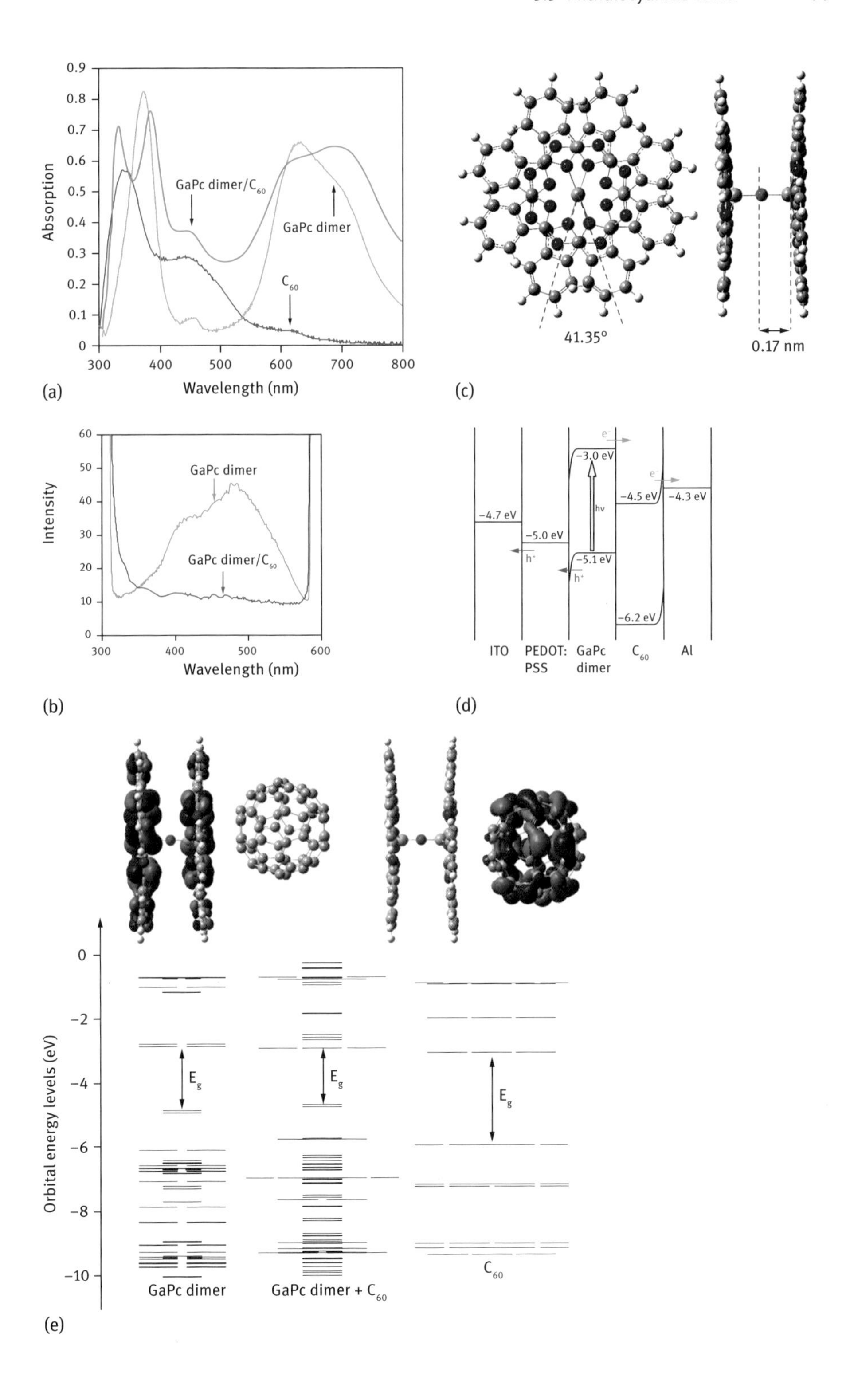

0.9
0.8
0.7
0.6
0.5
0.4
0.3
0.2
0.1
0
Absorption
GaPc dimer/$C_{60}$
GaPc dimer
$C_{60}$
300
400
500
600
700
800
Wavelength (nm)
(a)
41.35°
0.17 nm
(c)
60
50
40
30
20
10
0
Intensity
GaPc dimer
GaPc dimer/$C_{60}$
300
400
500
600
Wavelength (nm)
(b)
e⁻
−3.0 eV
e⁻
−4.5 eV
−4.3 eV
−4.7 eV
hν
−5.0 eV
h⁺
−5.1 eV
h⁺
−6.2 eV
ITO
PEDOT: PSS
GaPc dimer
$C_{60}$
Al
(d)
0
−2
−4
−6
−8
−10
Orbital energy levels (eV)
$E_g$
$E_g$
$E_g$
GaPc dimer
GaPc dimer + $C_{60}$
$C_{60}$
(e)

The energy level diagram and electronic structures of the solar cell were calculated and summarized as shown in Fig. 5.7(d) and (e). The HOMO and LUMO levels of GaPc, and HOMO and LUMO of two phthalocyanine monomers were stirred and piled up, respectively. The interaction between two phthalocyanine monomers was not able to be confirmed. Carriers could transport from −4.5 eV to −4.3 eV by hopping conduction. Figure 5.7(e) shows HOMO and LUMO energy levels of the GaPc dimer with $C_{60}$ after structural optimization using DFT/6-31G*. The electronic densities of LUMO, LUMO+1, and LUMO+2 are localized for the fullerene side, while the HOMO is localized for the GaPc-dimer side, which suggests electron transfer between the GaPc dimer and fullerene. The similar localization of frontier orbital was previously reported for the other donor-fullerene systems [31, 32]. A schematic diagram of the energy levels of GaPc dimer, $C_{60}$ and GaPc with $C_{60}$ showed that the LUMO levels of the GaPc dimmer with $C_{60}$ are comparable to the LUMO levels of fullerene, and that the HOMO levels of the GaPc dimer with $C_{60}$ are close to the HOMO levels of GaPc dimer. However, the symmetry of the GaPc dimer seems to be reduced due to a decrease in degeneracy, which could be due to the interaction with $C_{60}$.

Although the energy gap and energy level of GaPc dimer were hardly changed by dimerization here, the power conversion efficiency was significantly improved. The improvement of efficiency could be due to the decrease of career recombination in the ordered molecular orientation by dimerization. As a result, open-circuit voltage was greatly improved, which led to a high conversion efficiency.

The X-ray diffraction pattern of the GaPc dimer layer showed a peak of lattice spacing of 1.27 nm. Various crystallizations of μ-oxo bridged GaPc dimer have been reported [27, 28]. When the crystallographic structure is different, the initial surface potential, photosensitivity and residual surface potential are also different. Further investigation of the crystallographic structure should be carried out in the future.

## 5.6 ZnTPP:$C_{60}$

The fabrication and characterization of porphyrin:$C_{60}$ BHJ solar cells are presented here. 5,10,15,20-tetraphenyl-21,23H-porphin zinc (ZnTPP) was used for p-type semiconductors [5, 33] and $C_{60}$ was used for n-type semiconductors. Porphyrin has high optical absorption in the visible spectrum and high hole mobility [34–36] and was expected to form cocrystallites [37, 38] with $C_{60}$ that would be suitable for the BHJ structure [39, 40]. The second purpose of the investigation was to look at the effects of the electron transport layer (ETL). 3,4,9,10-perylenetetracarboxylic dianhydride (PTCDA) is a perylene derivative with a simple structure, which was reported to be used for solar cells [41]. In the present work, PTCDA was used as the ETL for porphyrin/$C_{60}$ BHJ solar cells. The ETL prevents hole transfer between an active layer and an electrode, and an improvement in conversion efficiency was expected through the introduction of the ETL. A thin layer of PEDOT:PSS was spin coated on pre-cleaned ITO glass plates. The PEDOT:PSS serves as a hole transport layer (HTL) for an electron blocking layer. Then, semiconductor layers were prepared on a PEDOT layer by spin coating using a mixed solution of $C_{60}$, ZnTPP in 1 mL odichlorobenzene. The total weight of ZnTPP:$C_{60}$ was 18 mg, and the weight ratio of ZnTPP to $C_{60}$ was in the range of 1 : 9 ~ 5 : 5. The thickness of the blended device was ~ 150 nm. To increase efficiencies, PTCDA with a thickness of ~ 20 nm was also added over the active layers. After annealing at 100 °C for 30 min in $N_2$ atmosphere, PTCDA was evaporated between an active layer and a metal layer. Finally, Al metal contacts were evaporated as a top electrode.

Figure 5.8(a) and (b) shows the optical absorption of $C_{60}$, ZnTPP, ZnTPP:$C_{60}$ and ZnTPP:$C_{60}$/PTCDA BHJ solar cells, respectively. The ZnTPP:$C_{60}$/PTCDA structure provided absorption in the range of 300–800 nm (which corresponds to 4.0 and 1.5 eV respectively), which was higher than that in the ZnTPP:$C_{60}$ structure.

The *J–V* characteristics of ZnTPP:$C_{60}$ BHJ solar cells measured under illumination are shown in Fig. 5.8(c). The BHJ indicates one-layered composite structures with p and n-type semiconductors, denoted as ZnTPP:$C_{60}$. The effects of the addition of PTCDA to the ZnTPP:$C_{60}$ BHJ solar cells was also investigated and this is denoted as ZnTPP:$C_{60}$/PTCDA. Each structure shows a characteristic curve for open circuit voltage and short circuit current. The current density of ZnTPP:$C_{60}$ increased through the addition of PTCDA, and the best efficiency was obtained for the ZnTPP:$C_{60}$/PTCDA sample. Exciton migration of $C_{60}$ can be efficiently suppressed using PTCDA, and excitons would be generated for both ZnTPP/$C_{60}$ and $C_{60}$/PTCDA interfaces, which results in an increase in conversion efficiency, as shown in Fig. 5.8(c). A schematic illustration of electron and hole transport is shown in Fig. 5.8(d) and (e).

The X-ray diffraction patterns of the ZnTPP and ZnTPP:$C_{60}$ BHJ layers are shown in Fig. 5.9(a) and (b) respectively. In Fig. 5.9(a), diffraction peaks corresponding to ZnTPP crystal are observed. After the formation of the ZnTPP:$C_{60}$ BHJ layer, the diffraction peaks corresponding to ZnTPP disappeared, and $C_{60}$ peaks are observed as shown in Fig. 5.9(b). In addition, a new diffraction peak is observed as indicated by an arrow, which is believed to be porphyrin/$C_{60}$ cocrystallites [33, 34]. Figure 5.9(c) is an electron diffraction pattern of the ZnTPP:$C_{60}$ BHJ layer, taken along the $[\bar{1}23]$ direction of $C_{60}$. A twin structure with a (112) twin plane is observed in Fig. 5.9(c), as indicated by a dotted line. Diffraction spots, which could correspond to cocrystallites of ZnTPP:$C_{60}$, are also observed as indicated by arrows.

Since the microstructure of the ZnTPP and $C_{60}$ BHJ layer is strongly dependent on their weight ratio, it is necessary to control the microstructures to form cocrystallites of ZnTPP:$C_{60}$. In the present work, higher efficiencies were obtained for the ZnTPP:$C_{60}$ sample with a weight ratio of 3 : 7, which would be suitable for the formation of cocrystallites, as observed for weak reflections in X-ray and electron diffraction patterns. Recombination of electrons of $C_{60}$ and holes of ZnTPP could occur in the BHJ layer with intermittent cocrystallite structure. If continuous cocrystallite structures form perpendicular to the thin films, it is believed that the recombination of

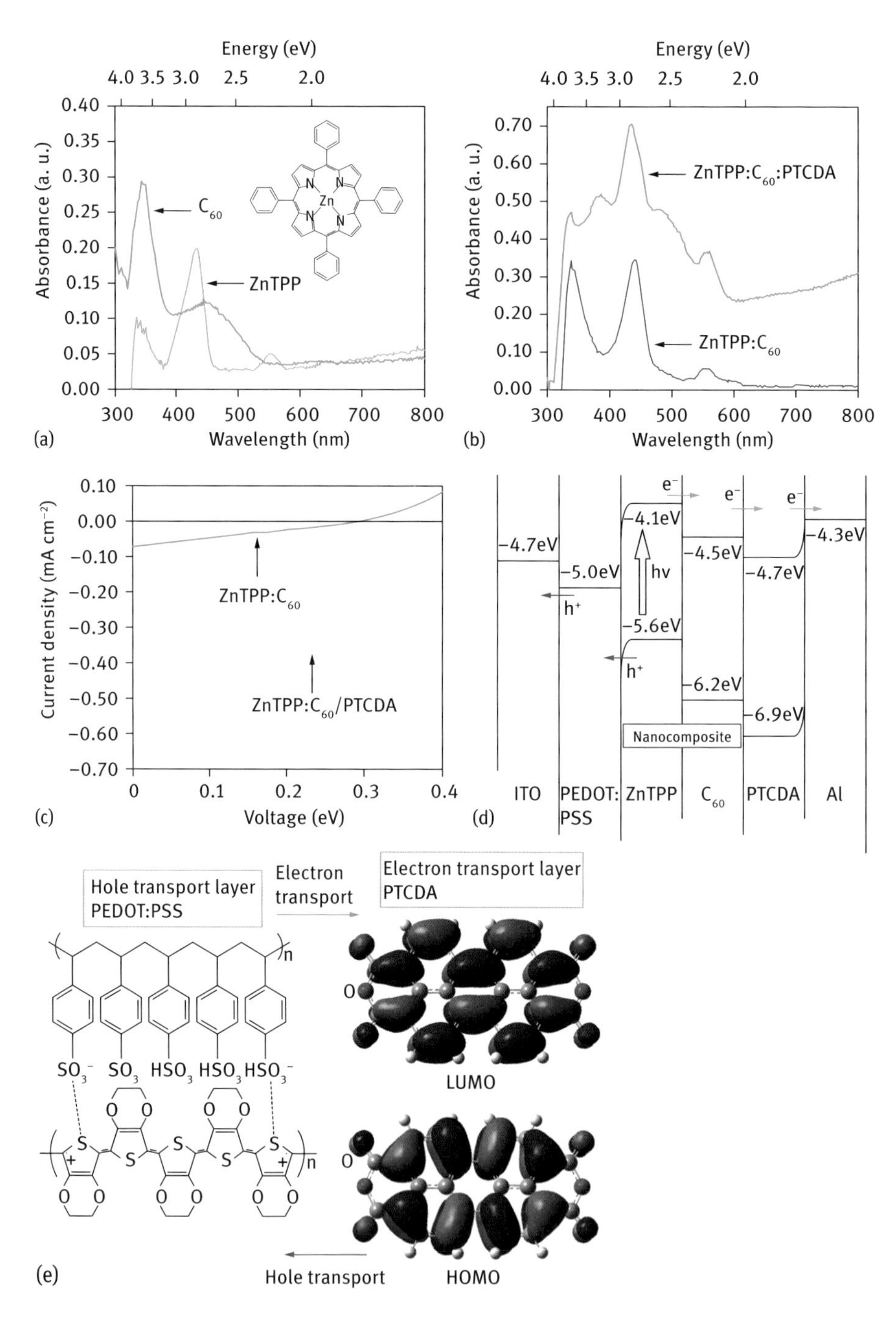


Fig. 5.8: Absorbance spectra of (a) $C_0$, ZnTPP and (b) ZnTPP:$C_{60}$ BHJ solar cells. (c) *J–V* characteristics measured for ZnTPP:$C_{60}$ BHJ solar cells under illumination. (d) Energy level diagram of ZnTPP/$C_{60}$ solar cell. (e) Carrier transfer of electron transport layer (ETL) and hole transport layer (HTL).

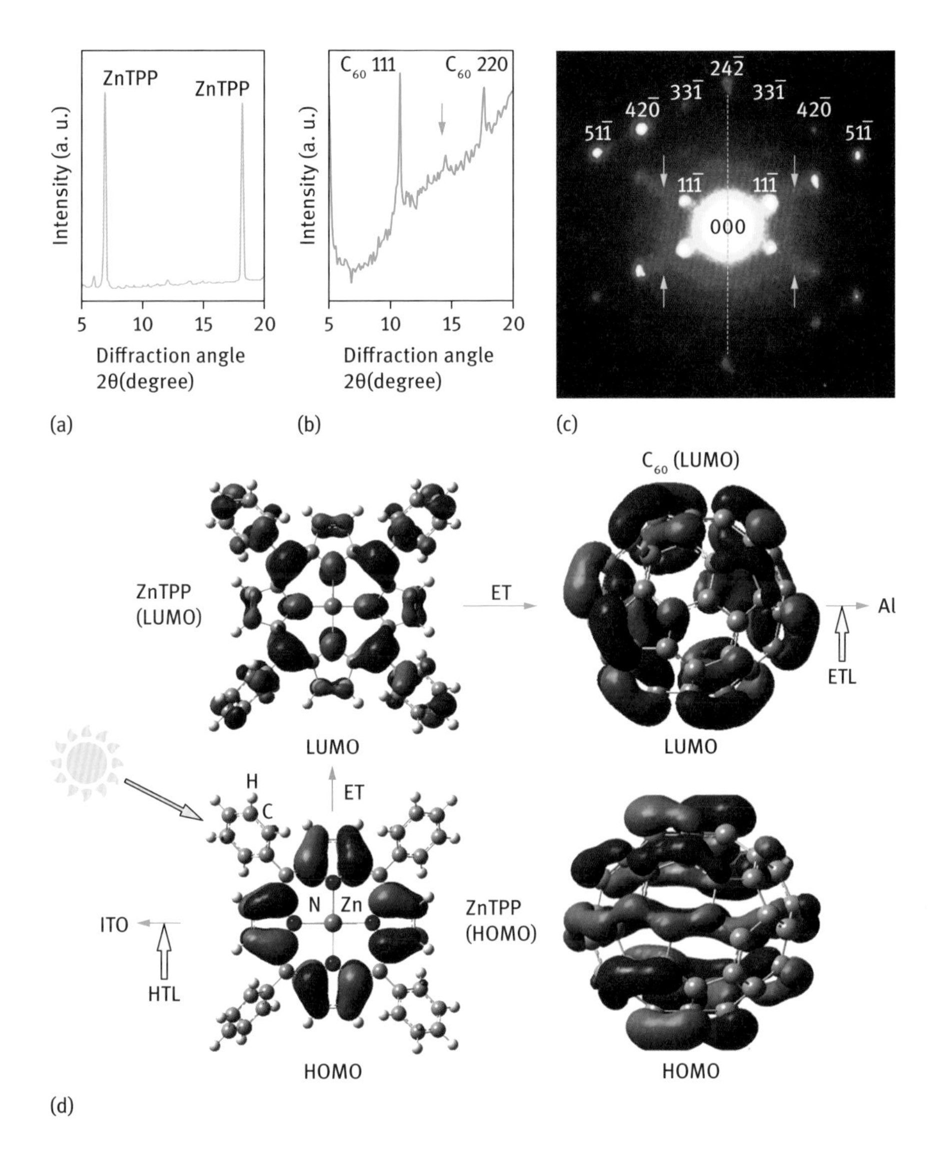


Fig. 5.9: (a) X-ray diffraction pattern of (a) ZnTPP and (b) ZnTPP:$C_{60}$ BHJ layer. (c) Electron diffraction pattern of ZnTPP:$C_{60}$ BHJ layer. (d) Electron transport (ET) of ZnTPP/$C_{60}$ system.

electrons and holes could be suppressed, which would lead to an improvement in conversion efficiency.

An energy level diagram, and the carrier transfer of electron transport layer (ETL) and hole transport layer (HTL) of a ZnTPP/$C_{60}$/PTCDA solar cell are summarized as shown in Fig. 5.9(d) and (e), respectively. The incident direction of light is from the ITO side. An energy barrier could exist near the semiconductor/metal interface [42, 43]. Electronic charge-transfer separation was caused by light irradiation from the ITO substrate side. Electrons are transported to an Al electrode, and holes are transported to an ITO substrate. The $V_{OC}$ of organic solar cells is reported to be determined by the energy gap, as indicated by Eq. (5.2). The present experimental data of $V_{OC}$ indicated smaller values compared to the ones calculated from the equation, which might be due to the voltage descent at the metal/semiconductor interface. Control of the energy levels is also important to increase the efficiency.

## 5.7 Diamond:$C_{60}$

The fabrication and to characterization of $C_{60}$/phthalocyanine based BHJ and HJ solar cells are presented here. $C_{60}$ and fullerenol [$C_{60}(OH)_{10\text{-}12}$] were used for n-type semiconductors, and nanodiamoond (ND) and MPc derivatives were used for p-type semiconductors. A schematic diagram of the present $C_{60}$/phthalocyanine based BHJ and HJ solar cells is shown in Fig. 5.10(a). A thin layer of PEDOT:PSS was spin coated on precleaned ITO glass plates. The PEDOT:PSS plays a role as an electron blocking layer for hole transport. Two types of solution for p-type semiconductors were produced [44, 45]. The first was produced by ND and tetra carboxy phthalocyaninate cobalt (Tc-CoPc) in deionized water. The solution for n-type semiconductors was prepared by dissolving $C_{60}$ in 1,2-dichlorobenzene. On the thin layer of PEDOT:PSS, p-type semiconductor layers were prepared by spin coating a mixed solution of Tc-CoPc and ND in deionized water. The NDs were dispersed in the Tc-CoPc thin film. The n-type semiconductor layers were deposited on top of the p-type semiconductor layer by spin coating a $C_{60}$ solution in 1,2-dichlorobenzene.

The second solution was also produced using tetra carboxy phthalocyaninate copper (Tc-CuPc), fullerenol [$C_{60}(OH)_{10\text{-}12}$] and ND in deionized water. On the thin layer of PEDOT:PSS, semiconductor layers were prepared by spin coating using a mixed solution of Tc-CuPc, $C_{60}(OH)_{10\text{-}12}$ and ND in deionized water. The NDs were obtained using the bead milling method in water and were dispersed in the active layer [46, 47].

The parameters measured for diamond based thin films indicated that the thin film structure with ND provided a higher cell performance on the $J_{SC}$ values than that of thin film structure without ND. Figure 5.10(b) and (c) shows the optical absorption spectra of the ND based thin films, and a solid line and a dashed line show thin film structure with ND and thin film structure without ND respectively. These thin films provided photo absorption in the range of 300–800 nm, and thin film structure with ND indicates a higher optical absorption compared with that of thin film structure without ND. The optical absorption properties of the thin film were improved by adding ND to the active layer.

Figure 5.10(d) and (e) shows the X-ray diffraction patterns of diamond powder and the thin films. In Fig. 5.10(d), the diffraction peaks of the diamond powder were confirmed as 111, 220 and 311 of the diamond structure. In Fig. 5.10(d) and (e), diffraction peaks corresponding to diamond are observed for the Tc-CoPc:ND/$C_{60}$ and Tc-

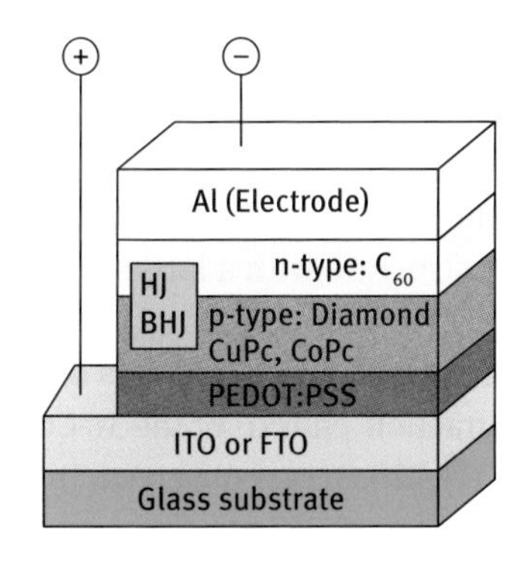


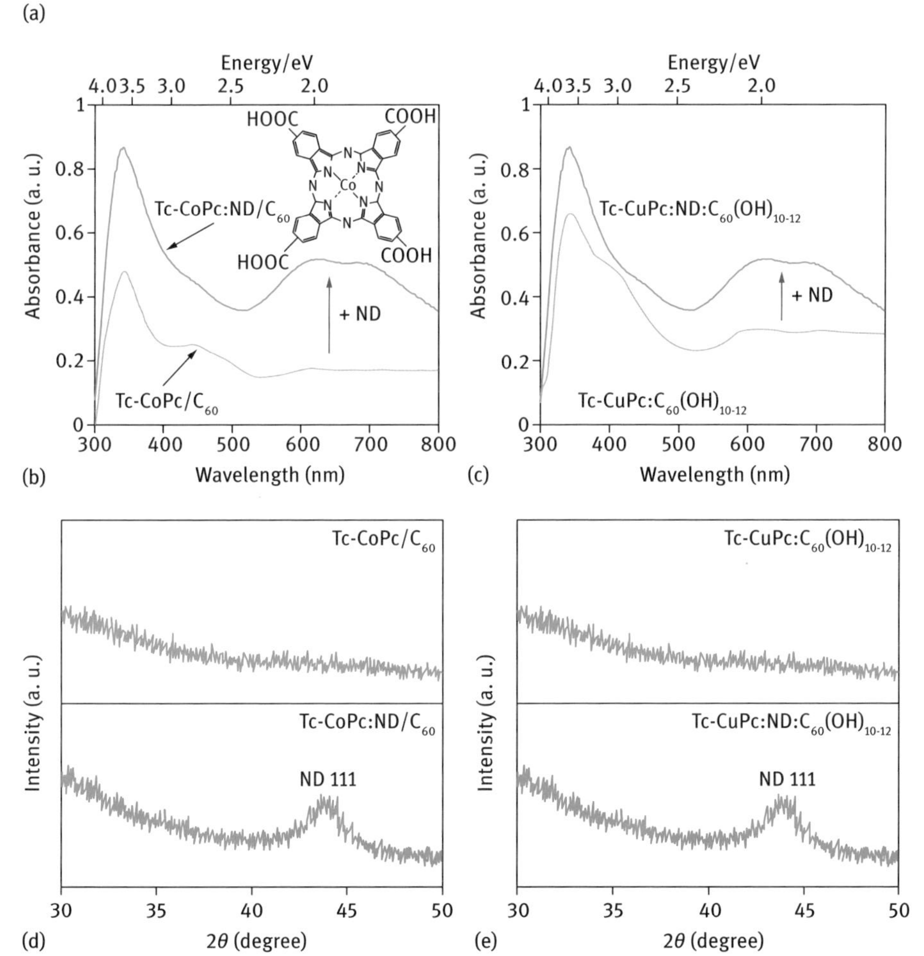


Fig. 5.10: (a) Device structure of MPc:ND/$C_{60}$ cells. Optical absorption spectra of (b) Tc-CoPc:ND/$C_{60}$ and Tc-CoPc/$C_{60}$ layers and (c) Tc-CuPc:ND:$C_{60}(OH)_{10-12}$ and Tc-CuPc:$C_{60}(OH)_{10-12}$ layers and X-ray diffraction patterns of (d) Tc-CoPc:ND/$C_{60}$ and Tc-CuPc/$C_{60}$ layers and (e) Tc-CuPc:ND:$C_{60}(OH)_{10-12}$ and Tc-CuPc:$C_{60}(OH)_{10-12}$ layers.

CuPc:ND:$C_{60}(OH)_{10\text{-}12}$ sample. The average particle sizes of the ND were calculated to be 4.5 and 5.5 nm from Scherrer's formula.

Figure 5.11(a) is a TEM image of a $C_{60}$ layer, and the lattice image of $C_{60}$ {111} is observed. Figure 5.11(b) is an electron diffraction pattern of a $C_{60}$ layer, and the diffraction peaks of $C_{60}$ are observed. $C_{60}$ also has an fcc structure with a lattice parameter of $a = 1.42$ nm. Figure 5.11(c) is an HREM image of the Tc-CoPc:ND composite layer. In Fig. 5.11(c), the lattice image of diamond {111} is observed. Tc-CoPc shows dark contrast in the image. Figure 5.11(d) is an electron diffraction pattern of the Tc-CoPc:ND composite layer, and diffraction peaks of diamond 111, 220, 311 are observed. Diamond powder has an fcc structure with a lattice parameter of $a = 0.357$ nm. Since no diffraction peak of Tc-CoPc was observed, Tc-CoPc could have an amorphous structure. Figure 5.11(e) is a TEM image of the Tc-CuPc:ND:$C_{60}$ $(OH)_{10\text{-}12}$ composite layer. The TEM image indicated ND of 4–6 nm as indicated by arrows, which agrees with the XRD results. Figure 5.11(f) is an electron diffraction pattern of the active layer, and the diffraction peaks of diamond 111, 220, 311 are observed. Since no diffraction peaks of Tc-CuPc and $C_{60}(OH)_{10\text{-}12}$ were observed, Tc-CuPc and $C_{60}(OH)_{10\text{-}12}$ could have amorphous structures. The increase in $p/n$ HJ interface presents an advantage for the nanocomposite structure. However, due to the disarray of the donor/acceptor microstructure, electrons and holes could not transport smoothly through carrier recombination at the electronic acceptor/Al interface and at the PEDOT:PSS/electronic donor interface, re-

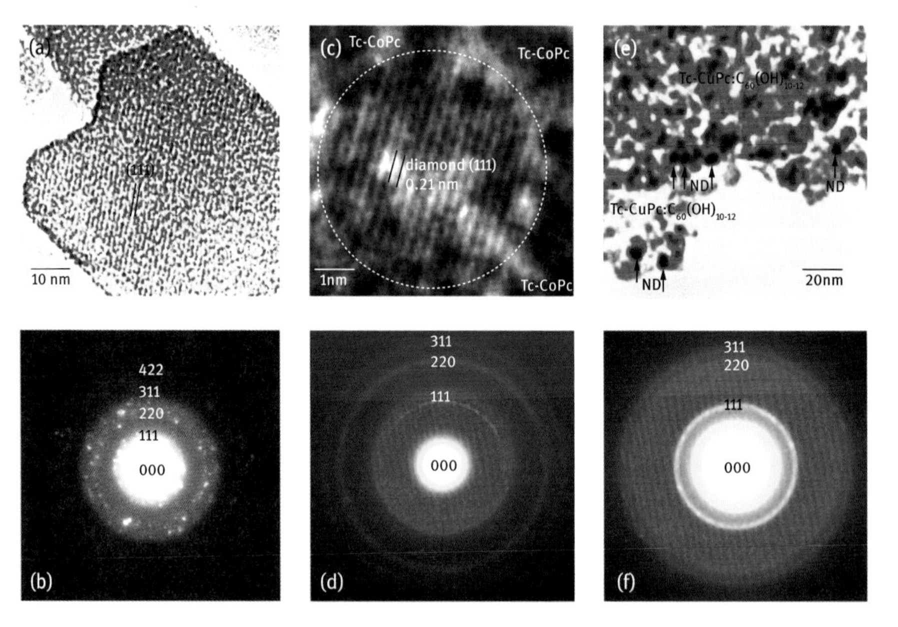


Fig. 5.11: (a) TEM image and (b) electron diffraction pattern of $C_{60}$ layer. (c) HREM image and (d) electron diffraction pattern of Tc-CoPc:ND layer. (e) TEM image and (f) electron diffraction pattern of Tc-CuPc:ND:$C_{60}$ $(OH)_{10\text{-}12}$ layer.

spectively. To solve these problems, it will be necessary to introduce a layer preventing carrier recombination and improve the crystalline structure to have fewer defects. In the study presented in this chapter, ND based solar cells were fabricated and characterized. In the carbon-based solar cells presented earlier, thin films are fabricated by a chemical vapor deposition method [48, 49]. In the work presented here, solar cells with $C_{60}$, $C_{60}(OH)_{10\text{-}12}$ and MPc as an organic semiconductor, and diamond particles and ND as an inorganic semiconductor were fabricated using a spin coating method, which is a low cost method.

The $J_{SC}$ values of the cells with ND increased compared to those without ND. In addition, the optical absorption spectra of the cells with ND were higher than those without ND in the range of 600–800 nm. The bandgap energy of diamond is typically ~ 5.5 eV and its carrier mobility is low. However, the ND could have a core shell structure as shown in Fig. 5.12(a), which indicates that the surface of the ND is covered by graphene sheets with $sp^2$ hybridized orbitals, and there is an intermediate layer between the ND core and the graphene sheets. If the ND has such a three layered structure, the ND have various bandgap energies as shown in Fig. 5.12(b), and light with various wavelengths can be absorbed by the ND [12]. The XRD results also showed lattice distances of ~ 3.2 Å, which could be related to the intermediated layers. The energy level diagram of the cell with ND is shown in Fig. 5.12(c), and the carrier can be transported by hopping mechanism.

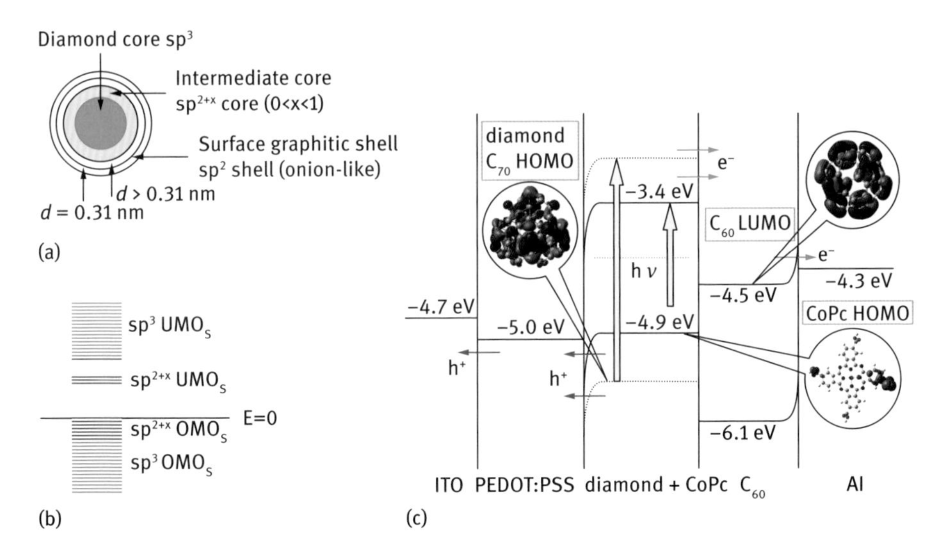


Fig. 5.12: (a) Structure of nanodiamond cluster. (b) Schematic occupied molecular orbitals (OMOs) and unoccupied molecular orbitals (UMOs) levels of diamond cluster. (c) Energy level diagram of diamond:CoPc/$C_{60}$ solar cells.

## 5.8 Ge nanoparticles

The fabrication and characterization of fullerene based solar cells with Ge nanoparticles are presented here. Copper tetrakis (4-cumylphenoxy) phthalocyanine (Tc-CuPc) was used for p-type semiconductors as shown in Fig. 5.13(a), and $C_{60}$ was used for n-type semiconductors. In addition, Ge(IV) bromide ($GeBr_4$) was added to the solar cells in order to form Ge based quantum dots to increase their photovoltaic efficiencies [50]. Device structures were produced, and efficiencies, optical absorption and nanostructures were investigated.

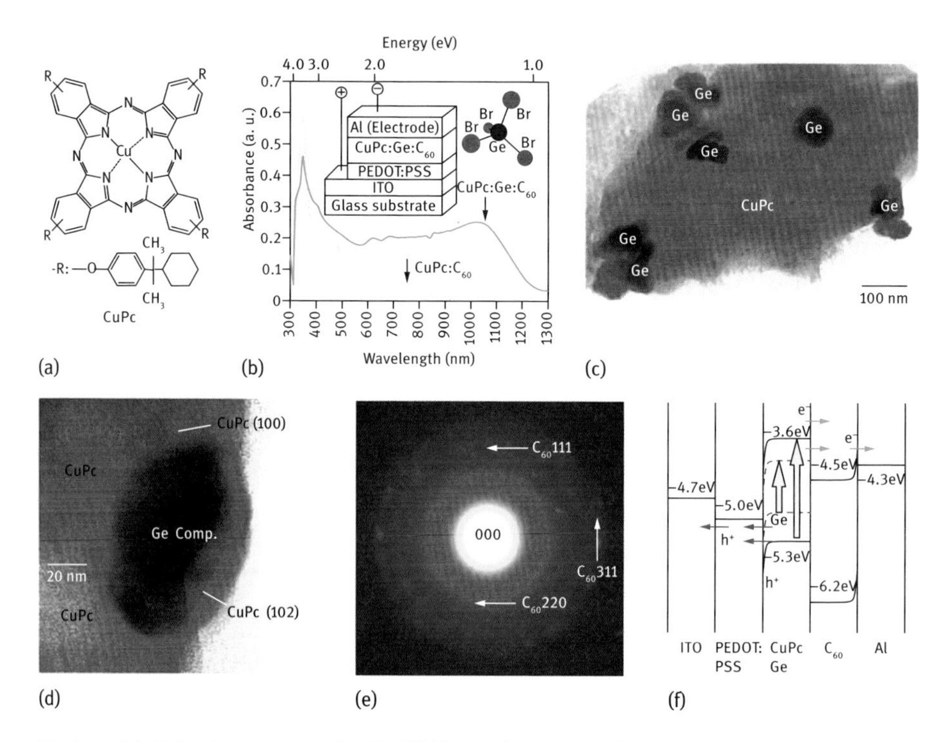


Fig. 5.13: (a) Molecular structure of CuPc. (b) Absorption spectra of Tc-CuPc:Ge:$C_{60}$ and Tc-CuPc:$C_{60}$ bulk heterojunction structure. (c) TEM image, (d) enlarged image and (e) electron diffraction pattern of Tc-CuPc:Ge:$C_{60}$ bulk heterojunction layer. (f) Energy level diagram of Tc-CuPc:Ge/$C_{60}$ solar cell.

A thin layer of PEDOT:PSS (Sigma Aldrich Corp.) was spin coated onto precleaned ITO glass plates. Then, semiconductor layers were prepared on a PEDOT layer by spin coating using a mixed solution of $C_{60}$, Tc-CuPc and $GeBr_4$ in 1 mL o-dichlorobenzene. The weight ratio of Tc-CuPc:$C_{60}$ was 1 : 8 (2 mg : 16 mg), and 0.03 mL of $GeBr_4$ was added into the solution [50]. The thickness of the blended device was ~ 150 nm. A schematic diagram of the Tc-CuPc:$C_{60}$ BHJ and HJ solar cells with a Tc-CuPc/$C_{60}$ structure is shown in Fig. 5.13(b). To increase efficiencies, $GeBr_4$ was also added in the Tc-CuPc layers for both structures. After annealing at 100 °C for 30 min in $N_2$ atmosphere, Al metal contacts with a thickness of 100 nm were evaporated as a top electrode.

The BHJ indicates one layered composite structures with p and n type semiconductors, which is denoted as Tc- CuPc:$C_{60}$. The common HJ solar cell that has separated two layers was also investigated for comparison, which is denoted as Tc-CuPc/$C_{60}$. The open circuit voltages measured for Tc-CuPc:$C_{60}$ and Tc-CuPc/$C_{60}$ were increased by several times through the addition of $GeBr_4$, and slight increases were also observed in the short circuit current density of both structures. Figure 5.13(b) shows the optical absorption of Tc-CuPc:Ge:$C_{60}$ and Tc-CuPc:$C_{60}$ BHJ solar cells. The Tc-CuPc:Ge:$C_{60}$ structure provided photo-absorption in the range of 500 to 1200 nm (which corresponds to 2.5 and 1.0 eV. respectively), which was higher than that of the Tc-CuPc:$C_{60}$ structure. The energy gap between HOMO and LUMO for $C_{60}$ is 1.7 eV, which corresponds to an absorbance of 730 nm [51].

A TEM image of the Tc-CuPc:Ge:$C_{60}$ BHJ layer is shown in Fig. 5.13(c). Nanoparticles containing Ge (the element with the largest atomic number in this cell) are observed in the Tc-CuPc layer. An enlarged TEM image is shown in Fig. 5.13(d), and the lattice fringes of Tc-CuPc are observed. The nanoparticle with Ge compounds is denoted as Ge comp. The electron diffraction pattern of the Tc-CuPc:Ge:$C_{60}$ BHJ layer is shown in Fig. 5.13(e), and many diffraction spots and rings corresponding to $C_{60}$ 111, 220 and 311 were observed, which indicates microcrystalline structures in $C_{60}$. The dispersion of Ge based nanoparticles is effective for optical absorption in the range of 500 to 1200 nm. Although Ge has a band gap energy of 0.7 eV, optical absorption was observed in the range of 2.5 and 1.0 eV in the present work, which could be due to Ge compound formation and the nanodispersion effect of the nanoparticles, which I have reported on previously [52]. The interpenetrating DA network has a large interfacial area, which could be effective for charge generation. Since the microstructures of Tc-CuPc and $C_{60}$ were disordered, recombination of electrons of $C_{60}$ and holes of Tc-CuPc could occur. Therefore, the ordered column-like structure would be suitable for carrier transport. If continuous nanocomposite structures are perpendicular to the thin films, it is believed that the recombination of electrons and holes could be avoided, and the conversion efficiency of the solar cells would increase.

The quantum dot solar cell including intermediate band structures is an important candidates for future high efficiency solar cells [53–58]. In the present work, the efficiencies of the solar cells were increased through the formation of Ge based nanoparticles. The technique presented here as a solution is a very simple and cost effective method for the formation of nanoparticles. To improve the efficiencies, the arrangement of the quantum dots and the control of size distribution are necessary. Combination of solar cells presented here and copper oxide nanomaterials with various direct band gaps or other organic materials might also be effective increasing efficiency. [59, 60] The performances of the solar cells presented here could also be due to the nanoscale structures.

## 5.9 Dye-sensitized solar cells

Although silicon solar cells have high conversion efficiencies and a long lifetime, their production processes are complicated and expensive. Dye-sensitized solar cells, based on the concept of photo-sensitization of wide band-gap mesoporous oxide semiconductors [61], are now in a state of advanced development. This technology has been established as a promising low-cost method [62], and the cells are lightweight and can be colorized. However, dye-sensitized solar cells have a short lifetime due to the leakage and vaporization of electrolytes. Therefore, studies of solidification of dye-sensitized solar cells have been performed [63, 64], and organic dyes without noble metals are expected to have potential as low-cost dyes.

An investigation into the electrical and optical properties of dye-sensitized solar cells (DSSC) with an amorphous $TiO_2$ layer to introduce electrons at trap levels in acceptor and donor levels is presented here. The effects caused by the addition of organic dyes such as protoporphyrin IX (PPIX), xylenol orange (XO) and rose Bengal (RB) to dye-sensitized solar cells were investigated through the fabrication and characterization of such cells. The schematic illustrations of the solar cells presented here are shown in Fig. 5.14(a) [65].

Nanocrystalline $TiO_2$ photoelectrodes were prepared, beginning by dispersing $TiO_2$ powder in an aqueous solution (1 mL) in a mixture of acetylacetone (0.02 mL) with Triton X–100 (0.01 mL) with polyethylene glycol (0.2 g) [66, 67]. The $TiO_2$ paste was then coated on pre-cleaned FTO using the squeegee method. After the $TiO_2$ coating, the FTO substrate was sintered for 30 min at 450 °C, and the prepared titanium isopropoxide (TTIP) solution was dropped onto the substrate. After that, the substrate was sintered for 60 min at 180 °C. The TTIP solution was prepared by mixing TTIP (0.3 mL), acetilaceton (0.08 mL), ethanol (0.64 mL) and PEG# 20 000 (0.2 g). The $TiO_2$ electrodes were dissolved into the solved organic dyes, which included protoporphyrin, xylenol orange and rose Bengal solutions in distilled water, methanol or ethanol [65, 68]. Bellfine (0.1 g) and Denka black (0.02 g) as carbon was dispersed in the distilled water (0.8 mL) and ethanol (0.4 mL) with sodium carboxymethyl cellulose (0.012 g). The carbon paste was applied to the indium tin oxide (ITO) as an opposite electrode using the squeegee method. A heat treatment of the carbon on the ITO substrate was carried out at 180 °C for 30 min. The electrolyte fabricated in a mixture of iodine (0.05 g), lithium iodide (0.09 g), ethylene carbonate (0.41 g), propylene carbonate (0.5 mL) and polyacrylonitrile (0.17 g) was agitated and heated at 80 °C. The dye-sensitized solar cells were assembled by putting the electrolyte between the adsorbed dye materials at the $TiO_2$ layer on the FTO glass substrate and the carbon film on the ITO substrate.

The measured $J$–$V$ curves of DSSC with or without an amorphous $TiO_2$ layer are shown in Fig. 5.14(b). It can be seen that the current density of DSSC with an amorphous $TiO_2$ layer was higher than that of DSSC without an amorphous $TiO_2$ layer. The current density improved from 0.60 to 0.83 mA cm$^{-2}$. The other parameters

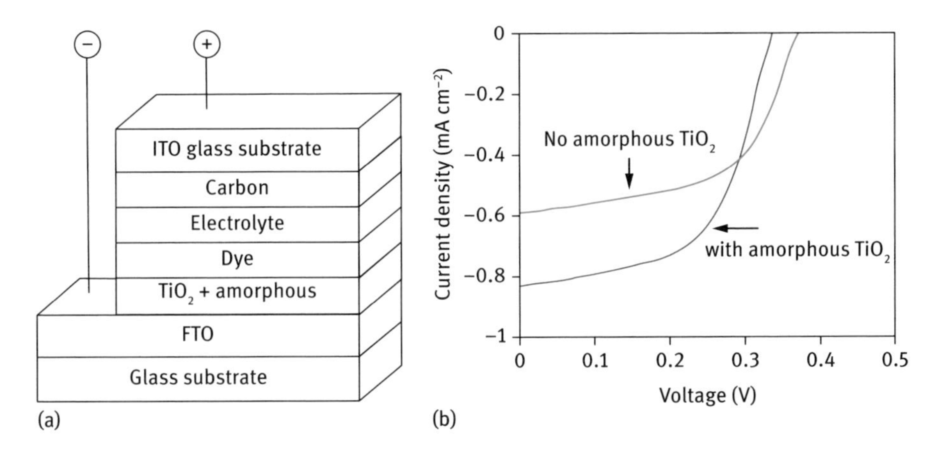


Fig. 5.14: (a) Structures of dye-sensitized solar cells with dyes. (b) *J*–*V* curves of DSSCs with or without amorphous$TiO_2$ layer.

remained almost the same, and the conversion efficiency improved from 0.12 % to 0.16 %.

Figure 5.15(a) shows XRD patterns of $TiO_2$ layers prepared from the TTIP solution as a function of temperature. No peak of an anatase phase is observed when annealing at 250 °C. The small peaks of the anatase phase are observed after annealing at 350 °C. The diffraction peaks of $TiO_2$ (P25) are shown for comparison with the amorphous $TiO_2$, which indicates the existence of rutile and anatase structures of $TiO_2$, as shown in Fig. 5.15(b) and (c), respectively. The optical absorption of DSSC with an amorphous $TiO_2$ layer was compared to that without an amorphous $TiO_2$ layer, indicating that the optical absorption peak for $TiO_2$ of around 360 nm was increased through the introduction of an amorphous $TiO_2$ layer.

A TEM image and an electron diffraction pattern of the DSSC with PPIX are shown in Figs. 5.15(d) and (e), respectively. The TEM image indicates the presence of $TiO_2$ nanoparticles with sizes of 20 ~ 60 nm, and the electron diffraction pattern shows 101, 103 and 200 reflections of a tetragonal $TiO_2$ anatase phase. A high-resolution electron microscopy (HREM) image of an interface of $TiO_2$ nanoparticles is shown in Fig. 5.15(f). The interface is directly connected at the atomic scale, which could results in good carrier transport between $TiO_2$ nanoparticles. A HREM image of surface of the $TiO_2$ nanoparticle is shown in Fig. 5.13(g), and the $TiO_2$ nanoparticle is covered by a 2 nm thick PPIX layer with an amorphous structure. The TEM image and an ED pattern also showed the amorphous $TiO_2$ layer [65]. The amorphous $TiO_2$ layer around the $TiO_2$ nanoparticle was observed by TEM, and the ED pattern indicated 101 and 103 reflections of the polycrystalline coagulation of $TiO_2$ with a tetragonal anatase structure. The diffuse ring of an amorphous phase was also observed, which indicates that the structure of the $TiO_2$ layer is a mixture of anatase and amorphous phases.

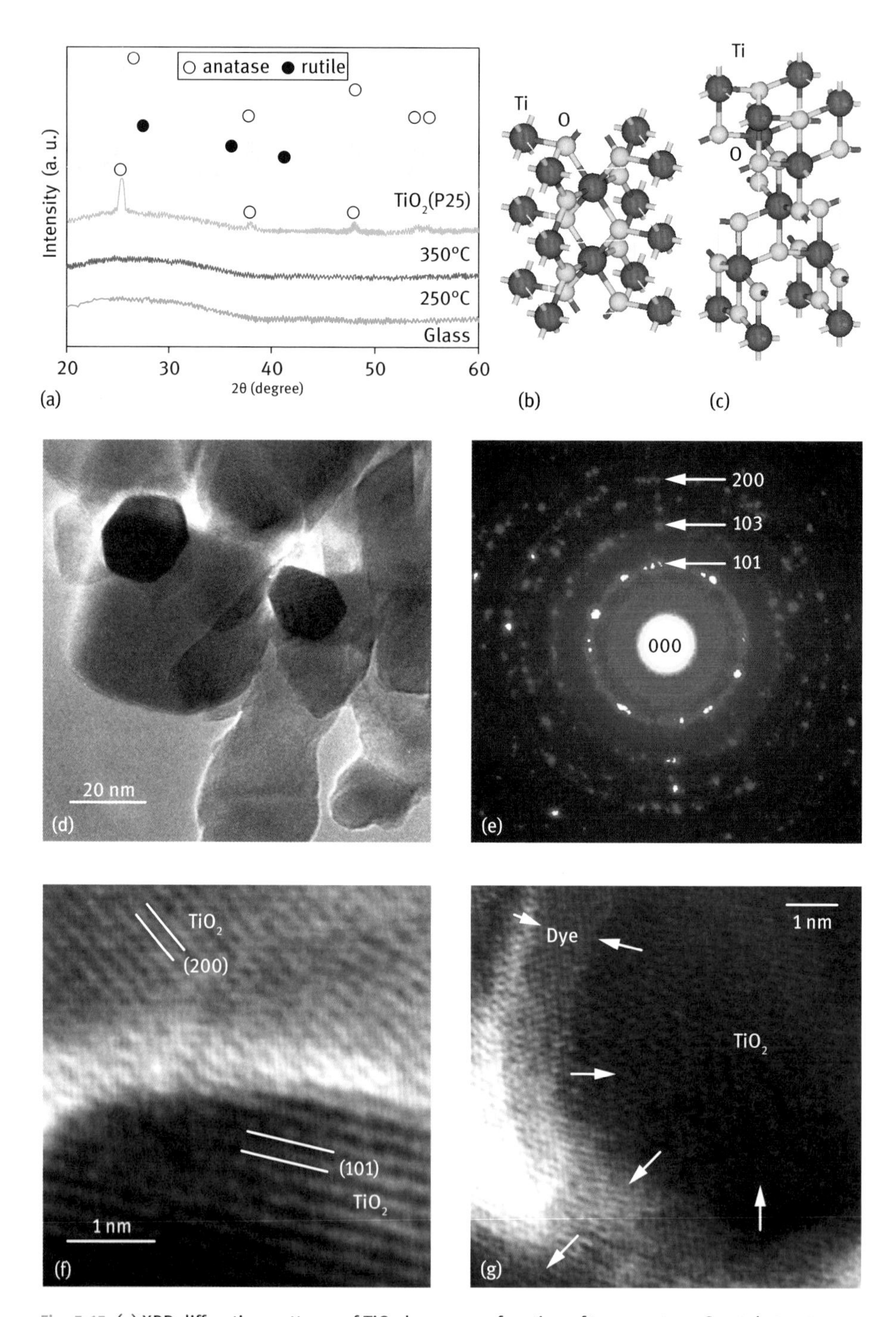


Fig. 5.15: (a) XRD diffraction patterns of $TiO_2$ layers as a function of temperature. Crystal structures of (b) rutile and (c) anatase $TiO_2$. (d) TEM image and (e) electron diffraction pattern of $TiO_2$ DSSC with protoporphyrin. HREM images of interface (c) and surface (d) of $TiO_2$ nanoparticles.

A schematic diagram of a $TiO_2$ electrode with amorphous $TiO_2$ layer is shown in Fig. 5.16(a). An energy level diagram of DSSC with an amorphous $TiO_2$ layer is shown in Fig. 5.16(b). The values of HOMO and LUMO were calculated by B3LYP/6–31G*. An energy barrier could exist at the semiconductor metal interface. The generation of electronic charge is caused by light irradiation from the FTO substrate side. The $TiO_2$ layer or amorphous $TiO_2$ layer receives the electrons from the dye, and the electrons are trapped by several trap levels of the amorphous $TiO_2$. The electrons are transported to an FTO electrode through the $TiO_2$ layer, and electrons are transferred to the outside circuit, and flow through the carbon electrode. Then, electrons return to the electrolyte through an oxidation-reduction reaction. The improvements seen in the solar cells presented here resulted from the introduction of the amorphous $TiO_2$ electrode, and this structure should be investigated further. Carrier recombination could be a main cause for the low conversion efficiency. It is expected that the conversion efficiency of the cells presented here can be improved through charge separation and electronic charge transfer.

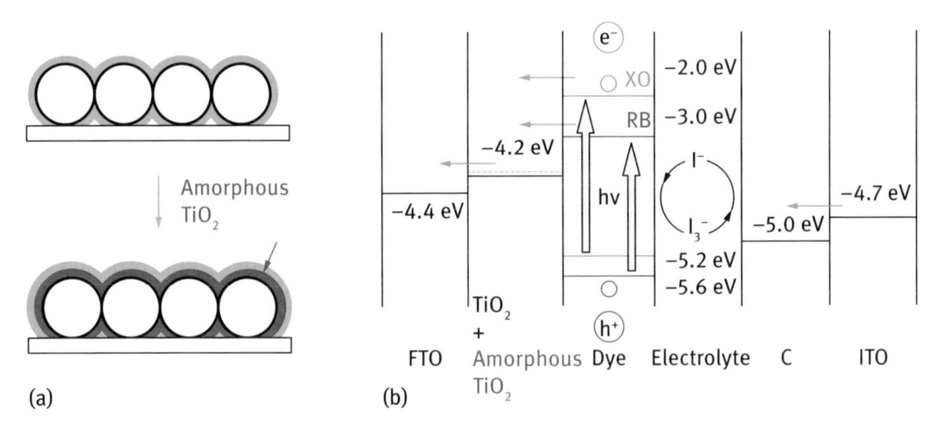


Fig. 5.16: (a) Schematic diagram of $TiO_2$ electrode with amorphous $TiO_2$ layer. (b) Energy level diagram of present DSSC with amorphous $TiO_2$ layer.

Improvements in conversion efficiency have been reported as an effect of coating materials on $TiO_2$. Insulator oxides such as $Al_2O_3$ or $SiO_2$ on $TiO_2$ have been introduced for DSSCs [69]. These oxides suppress carrier recombination by reverse transfer of electrons in the cells. In the present study, DSSCs with amorphous $TiO_2$ layers on $TiO_2$ were fabricated, and the amorphous $TiO_2$ could attract electrons because of the many trap levels in the acceptor and donor levels. In addition, the carrier separation effectiveness is high as a result of the large contact area. Further improvement would be possible by introducing nanoparticles as light-harvesting materials.

In summary, amorphous $TiO_2$ was introduced into DSSC with mixed xylenol orange and rose Bengal. The optical absorption peak of $TiO_2$, around 360 nm, was increased through the introduction of an amorphous $TiO_2$ layer. Diffraction spots of the anatase phase and a diffuse ring from the amorphous phase were observed. The energy diagram shows that the amorphous $TiO_2$ can attract electrons to several trap levels in the $TiO_2$ layers. In addition, the carrier separation is high, which could be due to the large contact area. As a result, the current density and the conversion efficiency were improved.

## 5.10 Polysilane system

Polysilane is a p-type semiconductor, and has been used as an electrically conductive material and in photovoltaic systems [70–76]. Polysilanes are known as $\sigma$-conjugate polymers, and their hole mobility is $10^{-4}$ $cm^2$ $V^{-1}$ $s^{-1}$ [70]. Although polysilanes could be applied to p-type semiconductors on organic thin-film solar cells, few studies have previously been carried out on polysilane solar cells [72, 77–79].

The purpose of the study presented here is twofold. First, the aim was to fabricate and characterize bulk heterojunction solar cells with polysilane and fullerenes of $C_{60}$ and [6,6]-phenyl-$C_{61}$-butyric acid methyl ester (PCBM). $C_{60}$ was selected as a good electronic acceptor material for devices. The second purpose of the study way to fabricate heterojunction solar cells by using a mixture solution of polysilane doped with phosphorus and boron, and to investigate the effects of annealing temperature and doping of boron (B) and phosphorous (P) on their electronic properties and microstructures. It is expected that amorphous silicon doped with boron would function as a p-type semiconductor, and amorphous silicon doped with phosphorus would function as an n-type semiconductor. Spin-coating is a low-cost method, and is essential for the mass production of solar cells. Four types of polysilane were used in the present work: dimethyl-polysilane (DMPS), poly(methyl phenyl silane) (PMPS), poly(phenyl silane) (PPSi) and decaphenyl cyclopentasilane (DPPS). Figure 5.17 shows the solar cell structures and molecular structures of the DMPS, PMPS, PPSi, and DPPS used to fabricate bulk-heterojunction and heterojunction solar cells [79]. Indium tin oxide (ITO) grass plates were cleaned in an ultrasonic bath with acetone and methanol, and then were dried with nitrogen gas. A thin layer of polyethylendioxythiophen doped with poly(3,4-ethylene dioxythiophene):poly(styrene sulfonate) (PEDOT:PSS) was spin-coated onto the ITO substrates. Then, semiconductor layers were prepared on a PEDOT:PSS layer by spin-coating using mixture solutions of DMPS, PMPS, PPSi or DPPS, and $C_{60}$ or PCBM in 1 mL o-dichlorobenzene. The total weight of $C_{60}$:PMPS, $C_{60}$:PPSi, $C_{60}$:DPPS, or PCBM:DPPS was 10 mg, the weight ratio of $C_{60}$:PMPS, $C_{60}$:PPSi, or $C_{60}$:DPPS was 8 : 2 and the weight ratio of PCBM:DPPS was 7 : 3. The total weight of $C_{60}$:DMPS was 19 mg, and the weight ratio of $C_{60}$:DMPS was 16 : 3. Another type of thin-film solar cells was prepared as described below. Semiconductor layers were prepared on the PEDOT:PSS layer by spin-coating using a mixture solution of DPPS (7 mg) and diethyl-methoxyborane (0.2 mL) in 1mL o-dichlorobenzene. After annealing at 300 °C

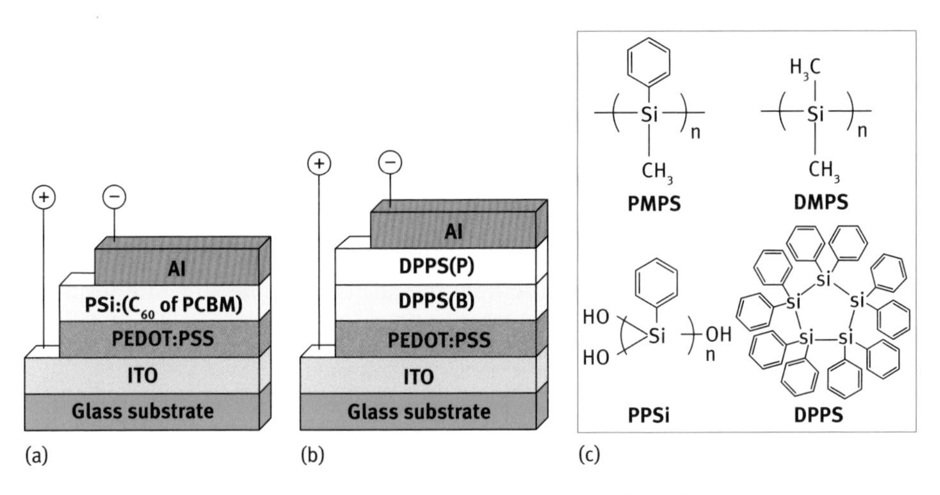


Fig. 5.17: Structures of (a) ITO/PEDOT:PSS/polysilane:($C_{60}$ or PCBM)/Al bulk heterojunction and (b) ITO/PEDOT:PSS/DPPS(B)/DPPS(P)/Al heterojunction solar cells. (c) Molecular structures of polysilanes.

for 10 min in $N_2$ atmosphere, a subsequent layer was prepared by spin-coating using the mixture solution of DPPS and phosphoric acid (0.2 mL) or phosphorus bromine (0.02 mL) in 1 mL of o-dichlorobenzene, and then the samples were annealed at 300 °C for 10 min in $N_2$ atmosphere. Aluminum (Al) metal contacts were evaporated as a top electrode. Finally, the devices were annealed at 140 °C for 30 min in $N_2$ atmosphere.

The *J*–*V* characteristics of the solar cells showed open-circuit voltage and short-circuit current. The measured parameters of the present solar cells are summarized in Table 5.2. A solar cell with the DMPS/$C_{60}$ structure provided a power conversion efficiency of 0.020 %, which was better than in other polysilane-based devices containing $C_{60}$. A solar cell with the DPPS/PCBM structure also provided a conversion efficiency of 0.032 %. The DPPS(B)/DPPS(P) solar cells prepared at 300 °C provided a conversion efficiency of 0.028 % and the highest open-circuit voltage of 0.81 V.

Table 5.2: Measured parameters of the present solar cells.

| Active layer | $V_{OC}$ (V) | $J_{SC}$ (mA cm$^{-2}$) | *FF* | $\eta$ (%) |
|---|---|---|---|---|
| DMPS:$C_{60}$ | 0.15 | 0.42 | 0.31 | 0.020 |
| PMPS:$C_{60}$ | 0.10 | 0.045 | 0.25 | $1.2\times 10^{-4}$ |
| PPSi:$C_{60}$ | $2.5\times 10^{-3}$ | $7.0\times 10^{-5}$ | 0.26 | $1.2\times 10^{-8}$ |
| DPPS:$C_{60}$ | 0.050 | $5.0\times 10^{-3}$ | 0.24 | $6.0\times 10^{-5}$ |
| DPPS:PCBM | 0.40 | 0.24 | 0.34 | 0.032 |
| DPPS(B)/DPPS(P) | 0.81 | 0.11 | 0.28 | 0.028 |

XRD patterns for DMPS:$C_{60}$ and DPPS:PCBM thin films are shown in Figs. 5.18(a) and (b), respectively. The diffraction patterns showed several diffraction peaks, which corresponded to DMPS, $C_{60}$, DPPS, and PCBM, as indicated in the figures. The XRD patterns of the annealed DPPS thin films are also shown in Fig. 5.18(c). Some diffraction peaks were observed for DPPS after annealing at 250 °C, which indicates the presence of microcrystalline structures. After annealing at 300 °C, most of the diffraction peaks decreased and disappeared for DPPS(B) and DPPS(P) thin films, which indicates the formation of an amorphous structure from the microcrystalline structure resulting from the addition of doping elements. The addition of phosphoric acid and diethyl-

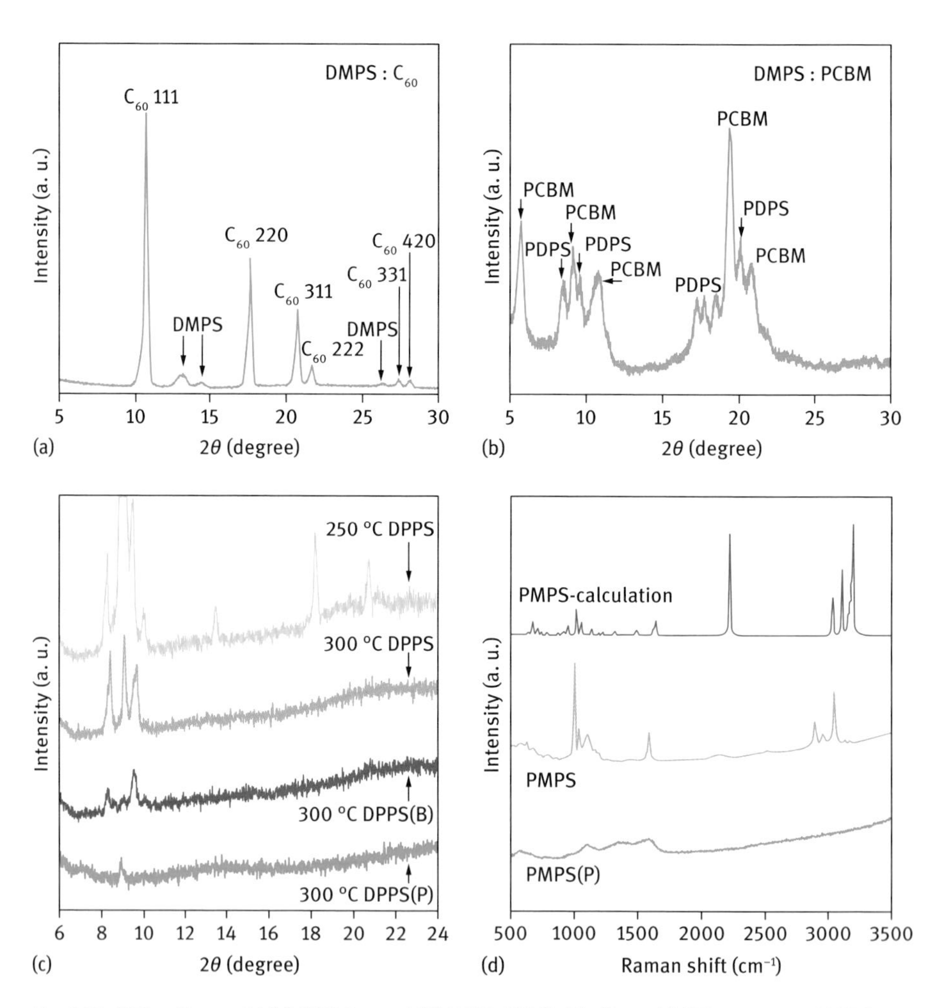


Fig. 5.18: XRD patterns of (a) DMPS:$C_{60}$ and (b) DPPS:PCBM thin films. (c) XRD patterns of DPPS thin films with and without phosphorus or boron after annealing. (d) Calculated Raman scattering spectra of PMPS, and Raman scattering spectra of PMPS and PMPS(P) thin films.

methoxyborane to DPPS could enhance the decomposition of the DPPS to form an amorphous structure.

Figure 5.19(a)–(f) show TEM images and electron diffraction patterns for DMPS:$C_{60}$, DPPS:PCBM, and DPPS thin films. A TEM image and electron diffraction pattern of the DMPS:$C_{60}$ composite film are shown in Fig. 5.19(a) and (d), respectively. In Fig. 5.19(a), nanoparticles consisting of the Si element in the $C_{60}$ matrix were observed. The electron diffraction pattern in Fig. 5.19(d) shows many diffraction spots and Debye-Scherrer rings, which indicates the microcrystalline structures of $C_{60}$ and DMPS.

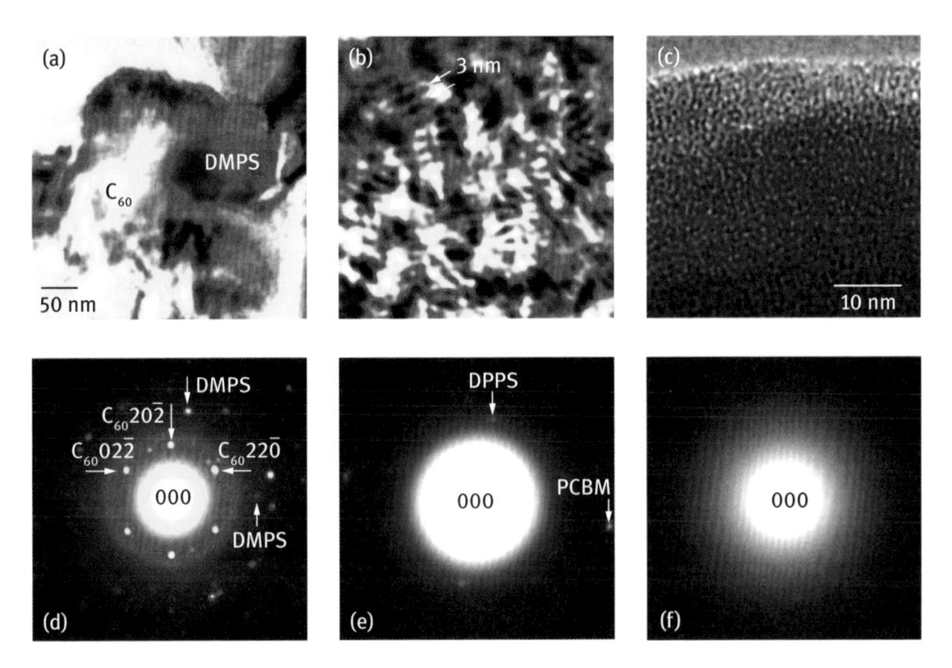


Fig. 5.19: (a–c) TEM images and (d–f) electron diffraction patterns of DMPS:$C_{60}$, DPPS:PCBM, and DPPS thin films, respectively.

Figure 5.19(b) and (e) show a HREM image and an electron diffraction pattern of DPPS:PCBM bulk heterojunction thin films, respectively. In Fig. 5.19(e), the diffraction spots corresponding to DPPS and PCBM were observed. The HREM image in Fig. 5.19(e) indicates that the DPPS:PCBM thin films have a nanocomposite structure, which has a lamella structure with a periodicity of ~ 3 nm. The optimization of the nanocomposite structure of DPPS:PCBM would increase the conversion efficiency of the solar cells.

Figure 5.19(c) shows a HREM image of DPPS annealed at 300 °C, and Fig. 5.19(f) shows an electron diffraction pattern of the DPPS. Contrast in the image corresponding to an amorphous structure is observed in Fig. 5.19(c), and (f) shows a halo-like intensity signal, which indicates that the DPPS thin films that annealed at 300 °C had an amorphous structure.

The Raman scattering spectra of PMPS were calculated to identify active modes, as shown in Fig. 5.18(d). The Raman scattering spectra of PMPS and doped PMPS thin films were obtained, as also shown in Fig. 5.18(d). The Raman active modes of the film were obtained: 668, 1012, 1644, 2228, 3044, 3116, and 3196 $cm^{-1}$. The peaks at 668, 1012, 1644, and 2228 $cm^{-1}$ were identified as the phenyl group vibration modes in PMPS. The strong peaks at 3044, 3116, and 3196 $cm^{-1}$ were identified as the methyl group vibration modes in PMPS. The Raman peaks observed in non-doped PMPS conformed to the vibration mode of the carbon bonds of the phenyl and methyl groups in PMPS, as shown in Fig. 5.18(d). When phosphorus was doped in the PMPS thin film that was then annealed at 300 °C, the peaks at 2228, 3044, 3116, and 3196 $cm^{-1}$ disappeared.

The effect of doping phosphorus on polysilane was investigated by Hall effect measurements. The resistivity, carrier concentration, and mobility of PMPS thin films doped with phosphorus are listed in Table 5.3. Although non-doped PMPS films showed a high electrical resistivity (~ $10^8$ Ω cm), the P-doped PMPS thin films showed *n*-type semiconducting behavior, low resistivity, and high electron mobility. After annealing at 500 °C, electronic properties deteriorated. The phosphorus doping to PMPS provided good transport phenomena after annealing at 300 °C.

Table 5.3: Resistivity, carrier concentration and mobility of PMPS(P) thin films.

| Temperature (°C) | Resistivity (Ω cm) | Carrier concentration ($cm^{-3}$) | Mobility ($cm^2\ V^{-1}\ s^{-1}$) |
|---|---|---|---|
| As-prepared | 0.15 | $2.1\times 10^{19}$ | 2.0 |
| 300 | 0.26 | $1.0\times 10^{17}$ | 2.4 |
| 500 | 48 | $1.9\times 10^{17}$ | 0.69 |

The energy level diagrams of the polysilane-based solar cells discussed here are summarized in Fig. 5.20. In this study, the $V_{OC}$ of DPPS was higher than that of PMPS, which could indicate that the HOMO–LUMO levels are different, as indicated in Fig. 5.20. Mixing $C_{60}$ with polysilanes would suppress the recombination of photo-separated charge carriers by promoting electron transfer from polysilanes. Amorphous silicon doped with boron or phosphorus could function as a *p*- or *n*-type semiconductor, respectively, and energy levels would be different between the DPPS with B and DPPS with P.

The DPPS-based solar cells presented here were compared with other silicon-based solar cells such as amorphous silicon solar cells, prepared by inductively coupled plasma chemical vapor deposition (ICP-CVD), and a spin-coating method, as listed in Table 5.4. Solar cells fabricated using a spin-coating method had a simpler fabrication process and better cost performance. The low conversion efficiency of the

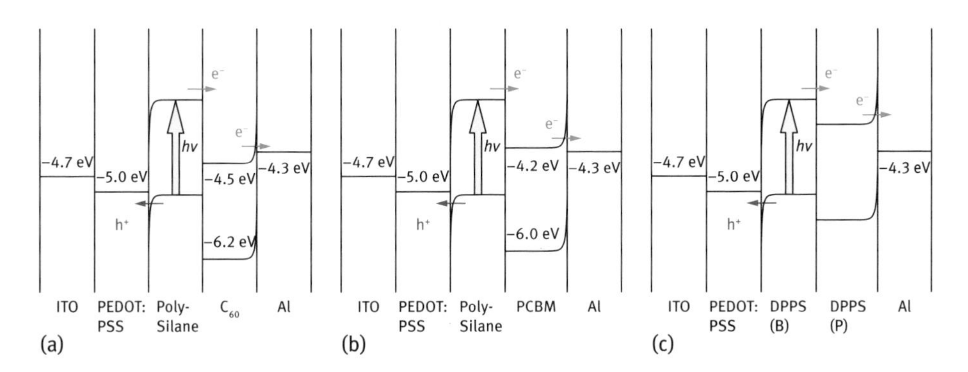


Fig. 5.20: Energy level diagrams of polysilane:$C_{60}$, DPPS:PCBM, and DPPS(B)/DPPS(P) solar cells.

Table 5.4: Comparison of silicon-based solar cells.

| Material | $\eta$ (%) | Fabrication process | Reference |
|---|---|---|---|
| amorphous Si | 9.6 | ICP-CVD | [80] |
| PMPS:$C_{60}$ | 0.33 | Spin-coating | [78] |
| PSi-Phth:$C_{60}$ | 0.013 | Spin-coating | [72] |
| DPPS | 0.028 | Spin-coating | [79] |

solar cells presented here could be due to the high electrical resistance and carrier recombination caused by defects, and further improvement of these solar cells will be necessary.

In summary, polysilane-based solar cells were fabricated using a mixture solution of DMPS, PMPS, PMSi, DPPS, $C_{60}$, PCBM, phosphorus, and boron, and characterized using electrical measurements, Raman scattering and microstructural analyses. Bulk heterojunction devices of DMPS:$C_{60}$ and DPPS:PCBM and heterojunction devices of DPPS(B)/ DPPS(P) exhibited photovoltaic properties. XRD and TEM results indicated that the DMPS:$C_{60}$ and DPPS:PCBM layers had nanocomposite structures, and that doped DPPS thin films formed an amorphous structure following their annealing at 300 °C. Raman scattering results indicated a desorption of the methyl and phenyl groups owing to phosphorus doping. Energy level diagrams, and carrier transport mechanisms were discussed.

## 5.11 PCBM:P3HT with SiPc or SiNc

Metal phthalocyanines (MPc) and metal naphthalocyanines (MNc) are groups of small molecular materials with Q-band absorption in the red to near-infrared range, and they have high optical and chemical stabilities, and photovoltaic properties. Therefore, they are used as donor materials in organic solar cells. Heterojunction solar cells using copper phthalocyanine and fullerene have been fabricated by an evaporation method, and their power conversion efficiency was ~ 3 % [81]. The characteristics of such solar cells with various MPc and MNc, including electronic conductivity, crystalline structure, and absorption range, were investigated by several research groups [82–85]. Organic solar cells such as those using P3HT and PCBM exhibit good incident photon-to-current conversion efficiency and fill factor. The device performance of such polymer solar cells can be affected by the preparation conditions such as the annealing temperature, the concentration of the starting material, and film thickness. The addition of third components such as phthalocyanines, naphthalocyanines, and low-band-gap polymers is expected to help absorb light that the P3HT and PCBM cannot collect. In particular, phthalocyanines absorb near-infrared light, and the effects of adding silicon phthalocyanine, silicon naphthalocyanine, or germanium phthalocyanine to the P3HT:PCBM system have been investigated in previous studies [86–90]. MPc and MNc can be dissolved in organic solvents, and application to the device process using a spin-coating method is possible with solubilization.

The fabrication and characterization of bulk heterojunction polymer solar cells with an inverted structureusing PCBM, P3HT, soluble tetrakis(tert-butyl)[bis(trihexylsiloxy)sillicon phthalocyanine] (SiPc) and tetrakis(tert-butyl)[bis(trihexylsiloxy)sillicon naphthalocyanine] (SiNc) were presented here. SiPc or SiNc was added as the third component for PCBM:P3HT solar cells, as shown in Fig. 5.21(a) [91, 92]. The direction of electron flow in the inverted device structure is opposite to that in conventional devices [93–95], and the inverted structure also provided stable device performances in air. Layered structures of bulk heterojunction solar cells with inverted structures were denoted as ITO/TiOx/PCBM:P3HT(SiPc or SiNc)/PEDOT:PSS/Au, as shown schematically in Fig. 5.21.

The device performance of the solar cell doped with SiPc was better than that of the solar cell with PCBM:P3HT structure. A solar cell with a PCBM:P3HT(SiPc, 3wt.%) structure provided $\eta$ of 0.768 %, which is the best $\eta$ value of any of the devices presented in this work [91]. The maximum $\eta$ occured for SiPc concentrations between 3 to 7 wt.%. The stability of inverted-structure solar cells with a PCBM:P3HT(SiPc) active layer was also investigated. $J$–$V$ characteristics of the device were measured after 1 week of exposure to ambient atmosphere, as listed in Table 5.5, and the devices exhibited stability in air. The conversion efficiency slightly increased upon exposure to air for 1 week, and similar increases in efficiencies have been observed in previous works [96, 97]. The morphology of the solar cells after one week would be improved in comparison to the as-prepared device, leading to improved carrier mobility and a decreased energy barrier of the metal/semiconductor interface. Devices maintained similar conversion efficiencies after 2 weeks of exposure to air.

▸ Fig. 5.21: (a) Molecular structures of P3HT, PCBM, SiPc and SiNc. (b) EQE and (c) IQE spectra of PCBM:P3HT(SiPc) solar cells. (d) EQE and (e) IQE spectra of PCBM:P3HT(SiPc) solar cells.

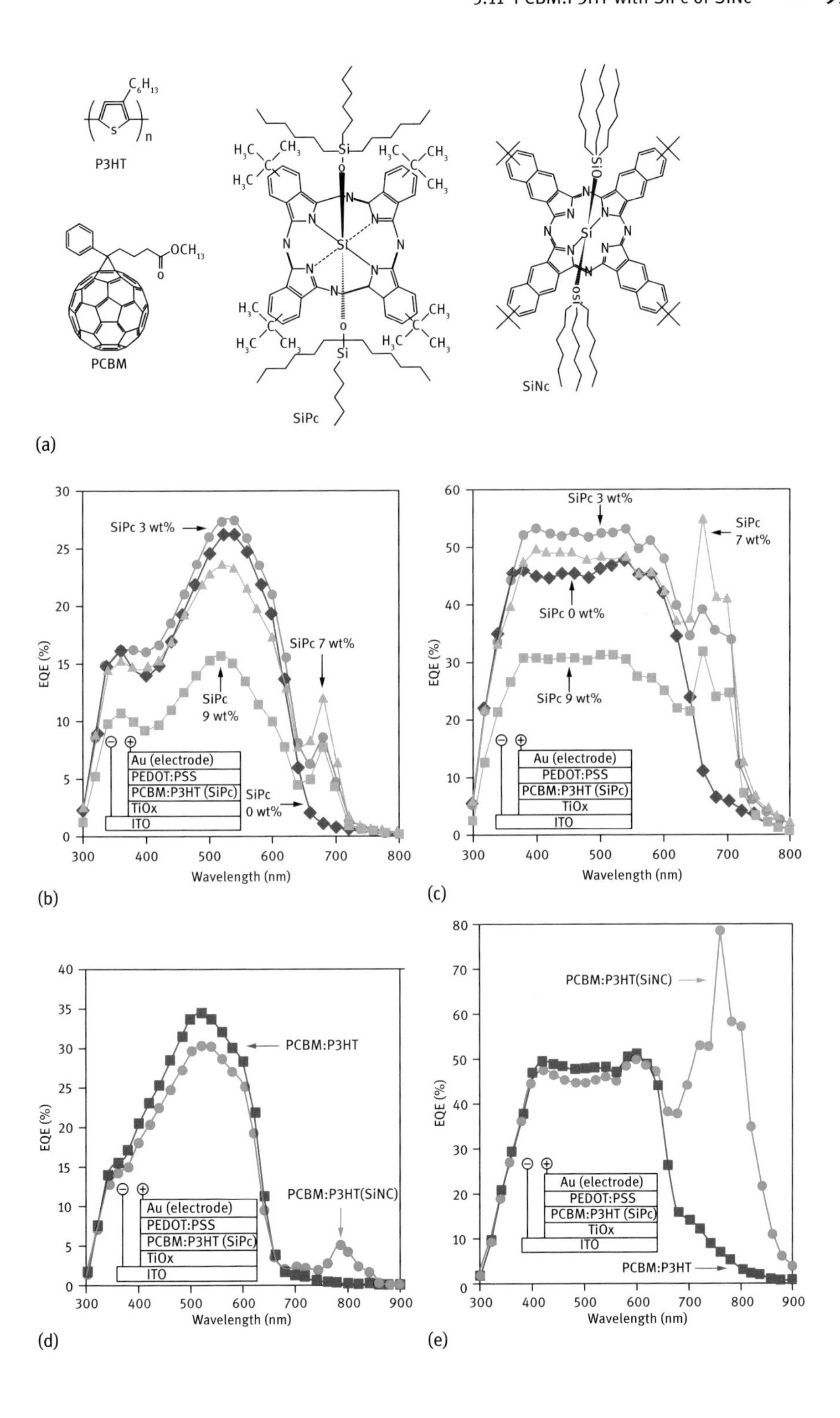

P3HT
PCBM
SiPc
SiNc
(a)
SiPc 3 wt%
SiPc 7 wt%
SiPc 9 wt%
SiPc 0 wt%
Au (electrode)
PEDOT:PSS
PCBM:P3HT (SiPc)
TiOx
ITO
EQE (%)
Wavelength (nm)
(b)
(c)
PCBM:P3HT
PCBM:P3HT(SiNC)
(d)
(e)

Table 5.5: Measured parameters of PCBM:P3HT(SiPc, 1wt.%) solar cell.

| Sample | $J_{SC}$ (mA cm$^{-2}$) | $V_{OC}$ (V) | *FF* | $\eta$ (%) |
|---|---|---|---|---|
| As-prepared | 3.4 | 0.50 | 0.41 | 0.69 |
| After 1 week | 3.4 | 0.59 | 0.52 | 1.1 |
| After 2 week | 3.1 | 0.57 | 0.53 | 0.95 |

The EQE spectra of PCBM:P3HT(SiPc) solar cells are shown in Fig. 5.21(b). EQE peaks were observed at ~ 680 nm for the solar cell with SiPc. If SiPc aggregates exist in the P3HT domain, no charge separation occurs, and if the SiPc exists in the PCBM domain, charge transfer does not occur. Therefore, the power conversion efficiency would be improved only when SiPc exists at the PCBM:P3HT interface. Since the peaks of EQE for SiPc were observed at ~ 680 nm, SiPc could exist at the PCBM:P3HT interface. The internal quantum efficiencies of PCBM:P3HT(SiPc) solar cells were calculated from the EQE, as shown in Fig. 5.21(c). A relatively higher IQE is observed for the PCBM:P3HT(SiPc) solar cells in the range of 640–700 nm. EQE spectra of PCBM:P3HT(SiNc) solar cells are also shown in Fig. 5.21(d), which indicates an EQE peak ~ 800 nm. The IQE spectra of the PCBM:P3HT(SiNc) solar cell is shown in Fig. 5.21(e), which also indicates higher IQE in the range of 700-900 nm, which is an effective range for photovoltaic devices.

The XRD patterns of PCBM, P3HT, SiPc, PCBM:P3HT, and PCBM:P3HT(SiPc) thin films are shown in Fig. 5.22(a). Diffraction peaks are observed for P3HT and SiPc, which indicate that they have crystalline structures. No sharp diffraction peak is observed for PCBM, which indicates that the PCBM has an amorphous structure. For the PCBM:P3HT(SiPc) and PCBM:P3HT thin films, diffraction peaks due to P3HT are weaker and broader than that of the single P3HT phase, and no sharp peak due to SiPc is observed. This indicates that P3HT has nanocrystalline structures dispersed in the amorphous PCBM, and that SiPc molecules without a crystalline structure could exist at the PCBM/P3HT interface. No sharp diffraction peak is observed for SiNc, which also indicates the PCBM:P3HT(SiNc) thin films have a similar structure.

A TEM image and an electron diffraction pattern of the PCBM:P3HT(SiPc) thin film are shown in Figs. 5.22(b) and (c), respectively [91]. In Fig. 5.22(b), P3HT nanowires are observed in the amorphous PCBM matrix. The electron diffraction pattern in Fig. 5.22(c) shows a halo-like intensity, which indicates the presence of amorphous and nanocrystalline structures, which agrees well with the XRD results shown in Fig. 6.15(a).

An interfacial structural model and an energy level diagram of the PCBM: P3HT(SiPc) solar cells are summarized and illustrated in Figs. 5.22(d) and 5.23(a), respectively. In Fig. 5.22(d), P3HT nanowires are dispersed in the PCBM amorphous matrix, which was confirmed by XRD and TEM analyses. Since an increase in photocurrent originating from the SiPc was observed, it is believed that the SiPc molecules

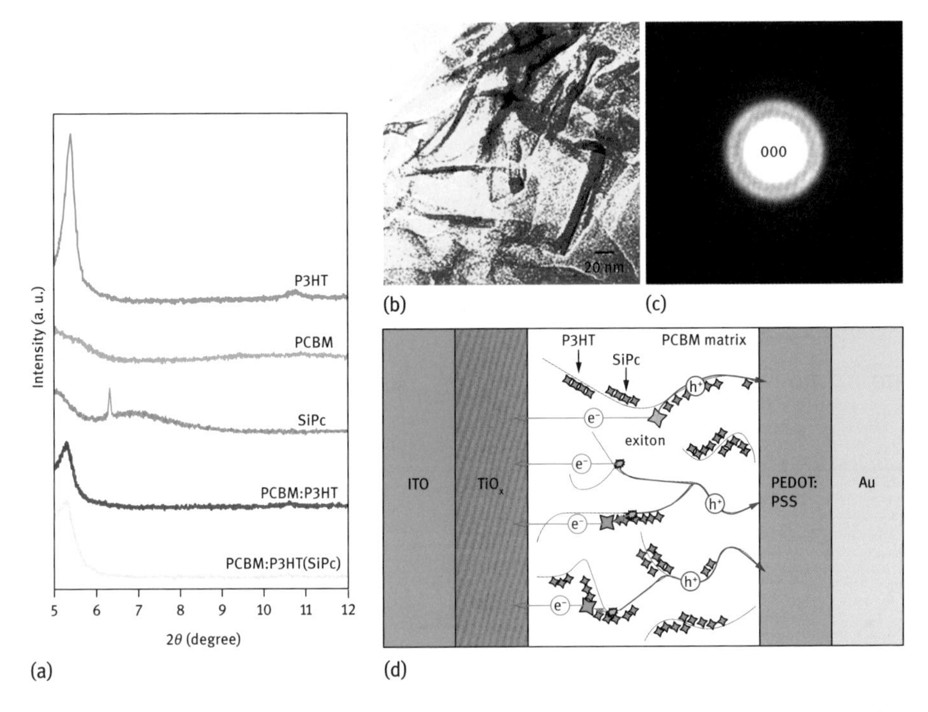


Fig. 5.22: (a) XRD patterns of P3HT, PCBM, SiPc, PCBM:P3HT and PCBM:P3HT(SiPc) thin films. (b) TEM image and (c) electron diffraction pattern of PCBM:P3HT(SiPc) thin film. (d) Schematic interfacial structure of a PCBM:P3HT(SiPc) solar cell.

were located at the PCBM/P3HT interface. The localization of SiPc molecules at the PCBM/P3HT interface could be caused by the insolubility of SiPc both in P3HT nanowires and the PCBM matrix. When the amount of SiPc added increased, the efficiencies decreased, which indicates that the single-layer localization of SiPc at the PCBM/P3HT interface could be effective for the conversion efficiencies. There could be two mechanisms that cause an increase in efficiency upon the addition of SiPc. Additional absorption of SiPc directly contributes to the increase in $J_{SC}$, which was observed in EQE and IQE spectra at 680 nm in Figs. 5.21(b) and (c), respectively. In addition, EQE and IQE intensities in the range of 400–600 nm were increased by the addition of SiPc, which could be explained by the Förster energy transfer from P3HT to SiPc [98, 99], as shown in Fig. 5.23(b). In the cell with an inverted structure, electrons are transported to an ITO substrate, and holes are transported to the Au electrode. Electrons could be transported only when the value of the LUMO for SiPc (X eV) is $-4.2 \le X \le -3.2$ eV, and holes could be transported only when the value of the HOMO for SiPc (Y eV) is $Y \le -5.2$ eV. Control of HOMO and LUMO of the phthalocyanine is required for polymer solar cells. Since a molecule exists in monomeric molecule, a bulky molecular structure is required, and the present SiPc has a suitable structure. The addition of silicon naphthalocyanine to PCBM:P3HT bulk heterojunction solar

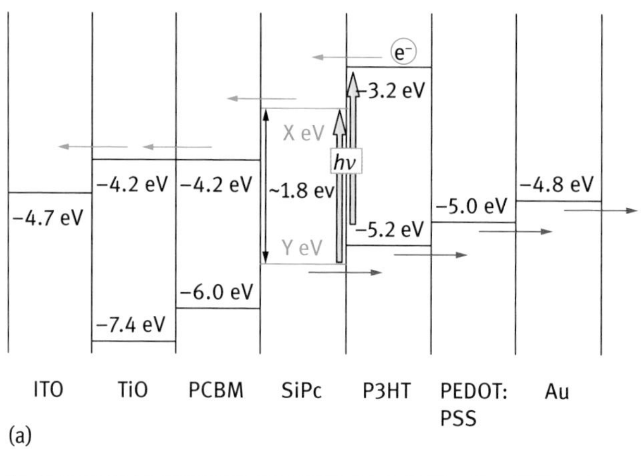


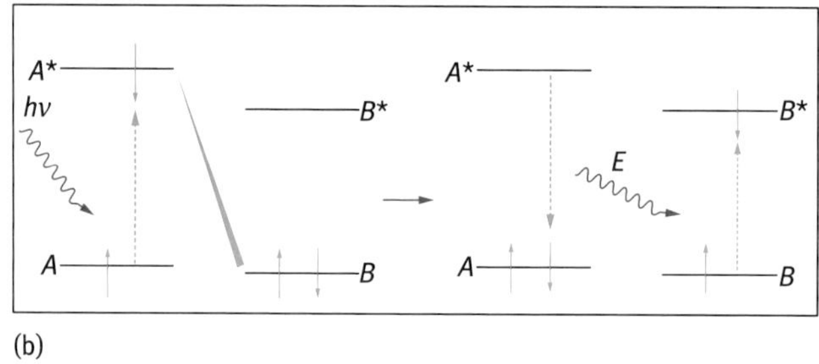


Fig. 5.23: (a) Energy level diagram of a PCBM:P3HT(SiPc) solar cell. (b) Schematic illustration of Förster resonance energy transfer mechanism.

cells also resulted in a higher IQE, as observed in Fig. 5.21(e), which is due to the different compositions of PCBM:P3HT. [92]

In summary, PCBM:P3HT(SiPc) and PCBM:P3HT(SiNc) bulk heterojunction solar cells with inverted structures were fabricated and characterized. The photovoltaic properties of the solar cells were improved by the addition of SiPc, and they were almost stable after two weeks of exposure in air. Microstructural analysis showed that P3HT nanowires are dispersed in the amorphous PCBM matrix, and it is believed that the SiPc or SiNc molecules are located at the PCBM/P3HT interface. The EQE and IQE spectra of the PCBM:P3HT(SiPc) solar cell showed peaks at ~ 680 nm, and the intensities of EQE and IQE in the range of 400–600 nm were increased by the addition of SiPc. The increase in conversion efficiency caused by the addition of SiPc could be explained in terms of the direct charge transfer from SiPc to PCBM and Förster energy transfer from P3HT to SiPc. PCBM:P3HT(SiNc) solar cells also showed higher values at the near-infrared region (~ 800 nm), which indicated that solar cell performance can be enhanced by the addition of SiNc.

## 5.12 Bibliography


[1] Oku T, Narita I, Nishiwaki A, Koi N, Suganuma K, Hatakeyama R, Hirata T, Tokoro H, Fujii S. Formation, atomic structures and properties of carbon nanocage materials. *Topics Appl Phys*. 2006; 100: 187–216.

[2] Sariciftci NS, Smilowitz L, Heeger AJ, Wudl F. Photoinduced electron transfer from a conducting polymer to buckminsterfullerene. *Science*. 1992; 258: 1474–1476.

[3] Hayashi Y, Yamada I, Takagi S, Takasu A, Soga T, Jimbo T. Influence of structure and $C_{60}$ composition on properties of blends and bilayers of organic donor-acceptor polymer/$C_{60}$ photovoltaic devices. *Jpn J Appl Phys*. 2005; 44: 1296–1300.

[4] Oku T, Nagaoka S, Suzuki A, Kikuchi K, Hayashi Y, Inukai H, Sakuragi H, Soga T. Formation and characterization of polymer/fullerene bulk heterojunction solar cells. *J Phys Chem Solids*. 2008; 69: 1276–1279.

[5] Oku T, Noma T, Suzuki A, Kikuchi K, Kikuchi S. The effects of exciton-diffusion blocking layers on pentacene/$C_{60}$ bulk heterojunction solar cells. *J Phys Chem Solids*. 2010; 71: 551–555.

[6] Lin YH, Yang PC, Huang JS, Huang GD, Wang IJ, Wu WH, Lin MY, Su WF, Lin CF. High-efficiency inverted polymer solar cells with solution-processed metal oxides. *Sol Energy Mater Sol Cells*. 2011; 95: 2511–2515.

[7] Ma W, Yang C, Gong X, Lee K, Heeger AJ. Thermally Stable, Efficient Polymer Solar Cells with Nanoscale Control of the Interpenetrating Network Morphology. *Adv* Funct Mater. 2005; 15: 1617–1622.

[8] Nikkei Microdevices and Nikkei Electronics Edit. Solar cells 2008/2009 (in Japanese). Nikkei BP. 2008.

[9] Oku T, Takeda A, Nagata A, Noma T, Suzuki A, Kikuchi K. Fabrication and characterization of fullerene-based bulk heterojunction solar cells with porphyrin, $CuInS_2$, diamond and exciton-diffusion blocking layer. *Energies*. 2010; 3: 671–685.

[10] Manceau M, Angmo D, Jørgensen M, Krebs FC. ITO-free flexible polymer solar cells: From small model devices to roll-to-roll processed large modules. *Org Electron*. 2011; 12: 566–574.

[11] Dou L, You J, Yang J, Chen CC, He Y, Murase S, Moriarty T, Emery K, Li G, Yang Y. Tandem polymer solar cells featuring a spectrally matched low-bandgap polymer. *Nat Photonics*. 2012; 6: 180–185.

[12] Oku T, Takeda A, Nagata A, Kidowaki H, Kumada K, Fujimoto K, Suzuki A, Akiyama T, Yamasaki Y, Ōsawa E. Microstructures and photovoltaic properties of $C_{60}$ based solar cells with copper oxides, $CuInS_2$, phthalocyanines, porphyrin, PVK, nanodiamond, germanium and exciton diffusion blocking layers. *Mater Technol*. 2013; 28: 21–39.

[13] Yu G, Heeger AJ. Charge separation and photovoltaic conversion in polymer composites with internal donor/acceptor heterojunctions. *J Appl Phys*. 1995; 78: 4510–4515.

[14] Padinger F, Rittberger RS, Sariciftci NS. Effects of Postproduction Treatment on Plastic Solar Cells. *Adv Funct Mater*. 2003; 13: 85–88.

[15] Oku T, Kakuta N, Kawashima A, Nomura K, Motoyoshi R, Suzuki A, Kikuchi K, Kinoshita G. Formation and characterization of bulk hetero-junction solar cells using $C_{60}$ and perylene. *Mater Trans*. 2008; 49: 2457–2460.

[16] Nomura K, Oku T, Suzuki A, Kikuchi K, Kinoshita G. The effects of exciton-diffusion blocking layers on pentacene/$C_{60}$ bulk heterojunction solar cells. *J Phys Chem Solids*. 2010; 71: 210–213.

[17] Liang Y, Xu Z, Xia J, Tsai ST, Wu Y, Li G, Ray C, Yu L. For the Bright Future—Bulk Heterojunction Polymer Solar Cells with Power Conversion Efficiency of 7.4 %. *Adv Mater*. 2010; 22: E135–138.

[18] Yang L, Zhou H, Price SC, You W. Parallel-like Bulk Heterojunction Polymer Solar Cells. *J Am Chem Soc*. 2012; 134: 5432–5435.

[19] Shirahata Y, Tanaike K, Akiyama T, Fujimoto K, Suzuki A, Balachandran J, Oku T. Fabrication and photovoltaic properties of ZnO nanorods/perovskite solar cells. *AIP Conf Proc*. 2016; 1709: 020018-1-9.
[20] Koster LJA, Mihailetchi VD, Blom PWM. Ultimate efficiency of polymer/fullerene bulk heterojunction solar cells. *Appl Phys Lett*. 2006; 88: 093511–093513.
[21] Scharber MC, Mühlbacher D, Koppe M, Denk P, Waldauf C, Heeger AJ, Brabec CJ. Design rules for donors in bulk-heterojunction solar cells – towards 10 % energy-conversion efficiency. *Adv Mater*. 2006; 18: 789–794.
[22] Brédas JL, Beljonne D, Coropceanu V. Charge-transfer and energy-transfer processes in pi-conjugated oligomers and polymers: a molecular picture. *Chem Rev*. 2004; 104: 4971–5004.
[23] Ulbricht R, Lee SB, Jiang X, Inoue K, Zhang M, Fang S, Baughman RH, Zakhid AA. Transparent carbon nanotube sheets as 3-D charge collectors in organic solar cells. *Sol Energy Mater Sol Cells*. 2007; 91: 416–419.
[24] Oku T, Narita I, Koi N, Nishiwaki A, Suganuma K, Inoue M, Hiraga K, Matsuda T, Hirabayashi M, Tokoro H, Fujii S, Gonda M, Nishijima M, Hirai T, Belosludov RV, Kawazoe Y. Boron nitride nanocage clusters, nanotubes, nanohorns, nanoparticles, and nanocapsules. In: Yap YK ed. B-C-N nanotubes and related nanostructures. Springer. 2009: 149–194.
[25] Yook KS, Chin BD, Lee JY, Lassiter BE, Forrest SR. Vertical orientation of copper phthalocyanine in organic solar cells using a small molecular weight organic templating layer. *Appl Phys Lett*. 2011; 99: 043308.
[26] Mori S, Nagata M, Nakahata Y, Yasuta K, Goto R, Kimura M, Taya M. Enhancement of incident photon-to-current conversion efficiency for phthalocyanine-sensitized solar cells by 3D molecular structuralization. *J Am Chem Soc*. 2010; 132: 4054–4055.
[27] Yamasaki Y, Takaki K. Synthesis of μ-oxo-bridged hetero-metal phthalocyanine dimer analogues and application for charge generating material in photoreceptor. *Dyes Pigm*. 2006; 70: 105–109.
[28] Yamasaki Y, Kuroda K, Yakaki K. Synthesis of new polymorphs of μ-oxo-meta(III) phthalocyanine dimers and their photoconductive properties. *J Chem Soc Jpn Chem Indust Chem*. 1997; 12: 887–898.
[29] Takeda A, Minowa A, Oku T, Suzuki A, Kikuchi K, Yamasaki Y. Formation and characterization of phthalocyanine dimer/$C_{60}$ solar cells. *Prog Nat Sci*. 2011; 21: 27–30.
[30] Takeda A, Oku T, Suzuki A, Yamasaki Y. Theoretical study of gallium phthalocyanine dimer-fullerene complex for photovoltaic device. *J Mod Phys*. 2011; 2: 966–969.
[31] Mizuseki H, Igarashi N, Belosludov RV, Farajian AA, Kawazoe Y. Theoretical study of phthalocyanine–fullerene complex for a high efficiency photovoltaic device using ab initio electronic structure calculation. *Synth Met*. 2003; 138: 281–283.
[32] Hameed AJ. Theoretical investigation of a phthalocyanine–fulleropyrrolidine adduct and some of its metallic complexes. *J Mol Struct Theochem*. 2006; 764: 195–199.
[33] Takahashi K, Kuraya N, Yamaguchi T, Komura T, Murata K. Three-layer organic solar cell with high-power conversion efficiency of 3.5 %. *Sol Energy Mater Sol Cells*. 2000; 61: 403–416.
[34] Hasobe T, Imahori H, Kamat PV, Ahn TK, Kim SK, Kim D, Fujimoto A, Hirakawa T, Fukuzumi S. Photovoltaic cells using composite nanoclusters of porphyrins and fullerenes with gold nanoparticles. *J Am Chem Soc*. 2005; 127: 1216–1228.
[35] Sun Q, Dai L, Zhou X, Li L, Li Q. Bilayer- and bulk-heterojunction solar cells using liquid crystalline porphyrins as donors by solution processing. *Appl Phys Lett*. 2007; 91: 253505.
[36] Hasobe T, Sandanayaka ASD, Wada T, Araki Y. Fullerene-encapsulated porphyrin hexagonal nanorods. An anisotropic donor–acceptor composite for efficient photoinduced electron transfer and light energy conversion. *Chem Commun*. 2008: 3372–3374.

[37] Ishii T, Aizawa N, Kanehama R, Yamashita M, Sugiura K, Miyasaka H. Cocrystallites consisting of metal macrocycles with fullerenes. *Coord Chem Rev.* 2002; 226: 113–124.
[38] Konarev DV, Kovalevsky AY, Li X, Neretin IS, Litvinov AL, Drichko NV, Slovokhotov YL, Coppens P, and Lyubovskaya RN. Synthesis and Structure of Multicomponent Crystals of Fullerenes and Metal Tetraarylporphyrins. *Inorg Chem.* 2002; 41: 3638–3646.
[39] Belcher WJ, Wagner KI, Dastoor PC. The effect of porphyrin inclusion on the spectral response of ternary P3HT:porphyrin:PCBM bulk heterojunction solar cells. *Sol Energy Mater Sol Cells.* 2007; 91: 447–452.
[40] Dastoor PC, McNeill CR, Frohne H, Foster CJ, Dean B, Fell CJ, Belcher WJ, Campbell WM, Officer DL, Blake IM, Thordarson P, Crossley MJ, Hush NS, Reimers JR. Understanding and improving solid-state polymer/$C_{60}$-fullerene bulk-heterojunction solar cells using ternary porphyrin blends. *J Phys Chem C.* 2007; 111C: 15415–15426.
[41] Berredjem Karst YN, Cattin L, Lakhdar-Toumi A, Godoy A, Soto G, Diaz F, Del Valle MA, Morsli M, Drici A, Boulmokh A, Gheid AH, Khelil A, Bernède JC. The open circuit voltage of encapsulated plastic photovoltaic cells. *Dyes Pigm.* 2008; 78: 148–156.
[42] Oku T, Wakimoto H, Otsuki A, Murakami M. NiGe-based ohmic contacts to *n*-type GaAs. I. Effects of In addition. *J Appl Phys.* 1994; 75: 2522–2529.
[43] Oku T, Furumai M, Uchibori CJ, Murakami M. Formation of WSi-based ohmic contacts to *n*-type GaAs. *Thin Solid Films.* 1997; 300: 218–222.
[44] Nagata A, Oku T, Kikuchi K, Suzuki A, Yamasaki Y, Ōsawa E. Fabrication, nanostructures and electronic properties of nanodiamond-based solar cells. *Prog Nat Sci.* 2010; 20: 38–43.
[45] Nagata A, Oku T, Suzuki A, Kikuchi K, Kikuchi S. Fabrication and photovoltaic property of diamond:fullerene nanocomposite thin films. *J Ceram Soc Jpn.* 2010; 118: 1006–1008.
[46] Ōsawa E. Recent progress and perspectives in single-digit nanodiamond. *Diamond Relat Mater.* 2007; 16: 2018–2022.
[47] Korobova MV, Avramenko NV, Bogachev AG, Rozhkova NV, Osawa E. Nanophase of water in nano-diamond gel. *J Phys Chem C.* 2007; 111: 7330–7334.
[48] Soga T, Kokubu T, Hayashi Y, Jimbo T. Effect of rf power on the photovoltaic properties of boron-doped amorphous carbon/*n*-type silicon junction fabricated by plasma enhanced chemical vapor deposition. *Thin Solid Films.* 2005; 82: 86–89.
[49] Krishna KM, Umeno M, Nukaya Y, Soga T, Jimbo T. Photovoltaic and spectral photoresponse characteristics of *n*-C/*p*-C solar cell on a *p*-silicon substrate. *Appl Phys Lett.* 2000; 77: 1472–1474.
[50] Oku T, Kumada K, Suzuki A, Kikuchi K. Effects of germanium addition to copper phthalocyanine/fullerene-based solar cells. *Cent Eur J Eng.* 2012; 2: 248–252.
[51] Yang J, Nguyen TQ. Effects of thin film processing on pentacene/$C_{60}$ bilayer solar cell performance. *Org Electron.* 2007; 8: 566–574.
[52] Oku T, Nakayama T, Kuno M, Nozue Y, Wallenberg LR, Niihara K and Suganuma K. Formation and photoluminescence of Ge and Si nanoparticles with oxide layers. *Mater Sci Eng B.* 2000; B74: 242–247.
[53] Conibeer G, Green M, Corkish R, Cho Y, Fangsuwannarak T, Pink E, Huang Y, Puzzer T, Trupke T, Richards B, Shalav A, Lin KL. Silicon nanostructures for third generation photovoltaic solar cells. *Thin Solid Films.* 2006; 511–512: 654–662.
[54] Martí A, López N, Antolín E, Cánovas E, Stanley C, Farmer C, Cuadra L, Luque A. Novel semiconductor solar cell structures: The quantum dot intermediate band solar cell. *Thin Solid Films.* 2006; 511–512: 638–644.
[55] Choi SH, Song H, Park IK, Yum JH. Synthesis of size-controlled CdSe quantum dots and characterization of CdSe–conjugated polymer blends for hybrid solar cells. *J Photochem Photobiol A: Chem.* 2006; 179: 135–141.

[56] Nozik AJ. Quantum dot solar cells. *Physica E*. 2002; 14E: 115–120.
[57] Brown P, Kamat PV. Quantum dot solar cells. Electrophoretic deposition of CdSe–$C_{60}$ composite films and capture of photogenerated electrons with $nC_{60}$ cluster shell. *J Am Chem Soc*. 2008; 130: 8890–8891.
[58] Emin S, Singh SP, Han L, Satoh N, Islam A. Colloidal quantum dot solar cells. *Sol Energy*. 2011; 85: 1264–1282.
[59] Oku T, Motoyoshi R, Fujimoto K, Akiyama T, Jeyadevan B, Cuya J. Structures and photovoltaic properties of copper oxides / fullerene solar cells. *J Phys Chem Solids*. 2011; 72: 1206–1211.
[60] Oku T, Nomura K, Suzuki A, Kikuchi K. Effect of perylenetetracarboxylic dianhydride layer as a hole blocking layer on photovoltaic performance of poly-vinylcarbazole:$C_{60}$ bulk heterojunction thin films. *Thin Solid Films*. 2012; 520: 2545–2548.
[61] Grätzel M. Dye-sensitized solar cells. *J Photochem Photobiol C*. 2003; 4: 145–153.
[62] Hagfeldt A, Boschloo G, Sun L, Kloo L, Pettersson H. Dye-sensitized solar cells. *Chem Rev*. 2010; 110: 6595–6663.
[63] Murai S, Mikoshiba S, Hayase S. Influence of alkyl dihalide gelators on solidification of dye-sensitized solar cells. *Sol Energy Mater Sol Cells*. 2007; 91: 1707–1712.
[64] Schmidt-Mende L, Bach U., Humphry-Baker R, Horiuchi T., Miura H, Ito S, Uchida S, Grätzel M. Organic dye for highly efficient solid-state dye-sensitized solar cells. *Adv Mater*. 2005; 17: 813–815.
[65] Oku T, Kakuta N, Kobayashi K, Suzuki A, Kikuchi K. Fabrication and characterization of $TiO_2$-based dye-sensitized solar cells. *Prog Nat Sci*. 2011: 21; 122–126.
[66] Lewis LN, Spivack JL, Gasaway S, Williams ED, Gui JY, Manivannan V, Siclovan OP. A novel UV-mediated low-temperature sintering of $TiO_2$ for dye-sensitized solar cells. *Sol Energy Mater Sol Cells*. 2006; 90: 1041–1051.
[67] Nazeeruddin MK, Kay A, Rodicio I, Humphry-Baker R, Mueller E, Liska P, Vlachopoulos N, Graetzel M. Conversion of light to electricity by cis-X2bis(2,2'-bipyridyl-4,4'-dicarboxylate) ruthenium (II) charge-transfer sensitizers (X = Cl-, Br-, I-, CN-, and SCN-) on nanocrystalline titanium dioxide electrodes. *J Am Chem Soc*. 1993; 115: 6382–6390.
[68] Kakuta N, Oku T, Suzuki A, Kikuchi K, Kikuchi S. Fabrication and characterization of mixture type dye-sensitized solar cells with organic dyes. *J Ceram Soc Jpn*. 2009; 117: 964–966.
[69] Palomares E, Clifford JN, Haque SA, Lutz T, Durrant JR. Control of charge recombination dynamics in dye sensitized solar cells by the use of conformally deposited metal oxide blocking layers. *J Am Chem Soc*. 2003; 125: 475–482.
[70] Silence S, Scott J, Hache F, Ginsbrug E, Jenkner P, Miller R, Twieg R, W. Moerner. Poly(silane)-based high-mobility photorefractive polymers. *J Opt Soc Am*. 1993; B10: 2306–2312.
[71] Silence S, Scott J, Hache F, Ginsbrug E, Jenkner P, Miller R, Twieg R, Moerner W. Poly(silane)-based high-mobility photorefractive polymers. *J Opt Soc Am*. 1993; B10: 2306–2312.
[72] Rybak A, Jung J, Ciesielski W, Ulanski J. Photovoltaic effect in novel polysilane with phenothiazine rings and its blends with fullerene. *Mater Sci Pol*. 2006; 24: 527–534.
[73] Kakimoto M, Kashihara H, Kashiwagi T, Takiguchi T. Visible light photoconduction of poly(disilanyleneoligothienylene)s and doping effect of $C_{60}$. *Macromolecules*. 1997, 30, 7816–7820.
[74] Haga Y, Harada Y. Photovoltaic characteristics of phthalocyanine-polysilane composite films. *Jpn J Appl Phys*. 2001; 40: 855.
[75] Watanabe A, Ito O. Photoinduced electron transfer between $C_{60}$ and polysilane studied by laser flash photolysis in the near-IR region. *J Phys Chem*. 1994; 98: 7736–7740.
[76] Kim S, Lee C, Jin MH. Fourier-transform infrared spectroscopic studies of pristine polysilanes as precursor molecules for the solution deposition of amorphous silicon thin-films. *Sol Energy Mater Sol Cells*. 2012; 100: 61–64.

[77] Shimoda T, Matsuki Y, Furusawa M, Aoki T, Yudasaka I, Tanaka H, Iwasaki H, Wang D, Miyasaka M, Takeuchi Y. Solution-processed silicon films and transistors. *Nature*. 2006; 440: 783–786.
[78] Lee J, Seoul C, Park J, Youk J. Fullerene/poly(methylphenylsilane) (PMPS) organic photovoltaic cells. *Synth Met*. 2004; 145: 11–14.
[79] Oku T, Nakagawa J, Iwase M, Kawashima A, Yoshida K, Suzuki A, Akiyama T, Tokumitsu K, Yamada M, Nakamura M. Microstructures and photovoltaic properties of polysilane-based solar cells. *Jpn J Appl Phys*. 2013; 52: 04CR07.
[80] Huang JY, Lin CY, Shen CH, Shieh JM, Dai BT. Low cost high-efficiency amorphous silicon solar cells with improved light-soaking stability. *Sol Energy Mater Sol Cells*. 2012; 98: 277–282.
[81] Peumans P, Forrest SR. Very-high-efficiency double-heterostructure copper phthalocyanine/$C_{60}$ photovoltaic cells. *Appl Phys Lett*. 2001; 79: 126.
[82] Li L, Tang Q, Li H, Hu W, Yang X, Shuai Z, Liu Y, Zhu D. Organic thin-film transistors of phthalocyanines. *Pure Appl Chem*. 2008; 80: 2231–2240.
[83] Bamsey NM, Yuen AP, Hor AM, Klenkler R, Preston JS, Loutfy RO. Integration of an M-phthalocyanine layer into solution-processed organic photovoltaic cells for improved spectral coverage. *Sol Energy Mater Sol Cells*. 2011; 95: 1970–1973.
[84] Bamsey NM, Yuen AP, Hor AM, Klenkler R, Preston JS, Loutfy RO. Heteromorphic chloroindium phthalocyanine films for improved photovoltaic performance. *Sol Energy Mater Sol Cells*. 2011; 95: 2937–2940.
[85] Kim DY, So F, Gao Y. Aluminum phthalocyanine chloride/$C_{60}$ organic photovoltaic cells with high open-circuit voltages. *Sol Energy Mater Sol Cells*. 2009; 93: 1688–1691.
[86] Honda S, Yokoya S, Ohkita H, Benten H, Ito S. Light-harvesting mechanism in polymer/-fullerene/dye ternary blends studied by transient absorption spectroscopy. *J Phys Chem C*. 2011; 115: 11306–11317.
[87] Honda S, Ohkita H, Benten H, Ito S. Selective dye loading at the heterojunction in polymer/-fullerene solar cells. *Adv Energy Mater*. 2011; 1: 588–598.
[88] Honda S, Ohkita H, Benten H, Ito S. Multi-colored dye sensitization of polymer/fullerene bulk heterojunction solar cells. *Chem Commun*. 2010; 46: 6596–6598.
[89] Oku T, Nose S, Yoshida K, Suzuki A, Akiyama T, Yamasaki Y. Fabrication and characterization of silicon naphthalocyanine, gallium phthalocyanine and fullerene-based organic solar cells with inverted structures. *J Phys: Conf Ser*. 2013; 433: 012025.
[90] Yoshida K, Oku T, Suzuki A, Akiyama T, Yamasaki Y. Fabrication and characterization of PCBM:P3HT bulk heterojunction solar cells doped with germanium phthalocyanine or germanium naphthalocyanine. *Mater Sci Appl*. 2013; 4: 1–5.
[91] Oku T, Hori S, Suzuki A, Akiyama T, Yamasaki Y. Fabrication and characterization of PCBM:P3HT:silicon phthalocyanine bulk heterojunction solar cells with inverted structures. *Jpn J Appl Phys*. 2014; 53: 05FJ08.
[92] Oku T, Yoshida K, Suzuki A, Akiyama T, Yamasaki Y. Fabrication and characterization of PCBM:P3HT bulk heterojunction solar cells doped with silicon naphthalocyanine. *Phys Status Solidi C*. 2013; 10: 1836–1839.
[93] Krebs FC. Air stable polymer photovoltaics based on a process free from vacuum steps and fullerenes. *Sol Energy Mater Sol Cells*. 2008; 92: 715–726.
[94] Walker B, Kim C, Nguyen TQ. Small molecule solution-processed bulk heterojunction solar cells. *Chem Mater*. 2011; 23: 470–482.
[95] Lin HY, Huang WC, Chen YC, Chou HH, Hsu CY, Lin JT, Lin HW. BODIPY dyes with $\beta$-conjugation and their applications for high-efficiency inverted small molecule solar cells. *Chem Commun*. 2012; 48: 8913–8915.

[96] Kuwabara T, Nakayama T, Uozumi K, Yamaguchi T, Takahashi K. Highly durable inverted-type organic solar cell using amorphous titanium oxide as electron collection electrode inserted between ITO and organic layer. *Sol Energy Mater Sol Cells*. 2008; 92: 1476–1482.
[97] Takeda A, Oku T, Suzuki A, Akiyama T, Yamasaki Y. Fabrication and characterization of fullerene-based solar cells containing phthalocyanine and naphthalocyanine dimers. *Synth Met*. 2013; 177: 48–51.
[98] Scully SR, Armstrong PB, Edder C, Fréchet JM, McGehee MD. Long-range resonant energy transfer for enhanced exciton harvesting for organic solar cells. *Adv Mater*. 2007; 19: 2961–2966.
[99] Honda S, Nogami T, Ohkita H, Benten H, Ito S. Improvement of the light-harvesting efficiency in polymer/fullerene bulk heterojunction solar cells by interfacial dye modification. *ACS Appl Mater Interfaces*. 2009; 1: 804–810.

# 6 Perovskite-type solar cells

## 6.1 Perovskite structures and synthesis

Recently, organic-inorganic hybrid solar cells with perovskite-type pigments have been widely fabricated and rapidly studied [1–4]. Solar cells with a perovskite structure have high conversion efficiencies and stability as the organic solar cells [5–7]. Since a photoconversion efficiency of 15 % was achieved [8], higher efficiencies have been reported for various device structures and processes [9–12], and more recently the photoconversion efficiency has been increased to ~ 20 % [13–19]. The photovoltaic properties of solar cells are strongly dependent on the fabrication process, hole transport layers, electron transport layers, nanoporous layers, interfacial microstructures, and the crystal structures of the perovskite compounds [20]. The crystal structures of the perovskite-type compounds affect electronic structures such as energy band gaps and carrier transport particularly strongly [21], and a detailed analysis of them is mandatory.

In this section, the crystal structures of perovskite-type compounds such as $CH_3NH_3PbI_3$, $CH_3NH_3PbCl_3$, $CH_3NH_3PbBr_3$, $CsSnI_3$, $CH_3NH_3GeCl_3$, and $CH_3NH_3SnCl_3$, which are expected to be usable as solar cell materials, are reviewed and summarized. Since these perovskite-type materials often have nanostructures in the solar cell devices, summarized information on the crystal structures could be useful for the structural analysis of perovskite-type crystals. The nanostructures of solar cell devices are often analyzed by using X-ray diffraction (XRD) and transmission electron microscopy (TEM), and the diffraction conditions are investigated and summarized. Transmission electron microscopy, electron diffraction, and high-resolution electron microscopy are powerful tools for the structural analysis of solar cells [22] and perovskite-type structures at the atomic scale [23–25].

There are various fabrication processes for the methylammonium trihalogenoplumbates (II) ($CH_3NH_3PbI_3$) compound with perovskite structures. Two typical synthesis methods for the $CH_3NH_3PbI_3$ ($MAPbI_3$) were reported [26]. $MAPbI_3$ could be synthesized from an equimolar mixture of $CH_3NH_3I$ and $PbI_2$ using the reported method [2]. $CH_3NH_3I$ was synthesized at first by reacting a concentrated aqueous solution of hydroiodic acid with methylamine, and the cleaned precipitant was mixed with $PbI_2$ in gamma-butyrolactone to obtain the $MAPbI_3$ product. Crystalline $MAPbI_3$ was obtained by drop-casting the solutions on glass substrates, and annealed at 100 °C. Polycrystalline $MAPbI_3$ could also be prepared by precipitation from a hydroiodic acid solution [27]. Lead(II) acetate was dissolved in concentrated aqueous HI and heated. An HI solution with $CH_3NH_2$ was added to the solution, and black precipitates were formed upon cooling from 100 °C.

DOI 10.1515/9783110298505-009

## 6.2 Crystal structures of $CH_3NH_3PbX_3$ (X = Cl, Br, or I)

The crystals of methylammonium trihalogenoplumbates(II) ($CH_3NH_3PbX_3$, X = Cl, Br, or I) have perovskite structures and undergo structural transitions upon heating [27–29]. The crystal systems and transition temperatures are summarized in Table 6.1, as reported in previous works [27, 29]. Atomic sites were indicated from the space group table [30]. Although the $CH_3NH_3PbX_3$ perovskite crystals have cubic symmetry for the highest temperature phase, the $CH_3NH_3$ ion is polar and has $C_{3v}$ symmetry, which should result in a disordered cubic phase [31]. In addition to the disordering of the $CH_3NH_3$ ion, the halogen ions were also disordered in the cubic phase, as shown in Fig. 6.1(a) and Table 6.2 [31]. Site occupancies were set to 1/4 for I and 1/12 for C and N. The $CH_3NH_3$ ion occupies 12 equivalent orientations of the $C_2$ axis, and hydrogen atoms have two kinds of configurations on the $C_2$ axis. The total degree of freedom is 24 [29].

Table 6.1: Crystal systems and transition temperatures of $CH_3NH_3PbX_3$ (X = Cl, Br, or I).

| Material | $CH_3NH_3PbCl_3$ | $CH_3NH_3PbBr_3$ | $CH_3NH_3PbI_3$ |
|---|---|---|---|
| Crystal system | Cubic | Cubic | Cubic |
| Transition temperature (K) | 177 | 236 | 330 |
| Crystal system | Tetragonal | Tetragonal | Tetragonal |
| Transition temperature (K) | 172 | 149 ~ 154 | 161 |
| Crystal system | Orthorhombic | Orthorhombic | Orthorhombic |

Table 6.2: Structural parameters of cubic $CH_3NH_3PbI_3$. Space group $Pm\bar{3}m$ ($Z = 1$), $a = 6.391$ Å at 330 K. $B$ is isotropic displacement parameter.

| Atom | site | $x$ | $y$ | $z$ | $B$ (Å$^2$) |
|---|---|---|---|---|---|
| Pb | 1*a* | 0 | 0 | 0 | 3.32 |
| I | 12*h* | 0 | 0.0435 | 0.5 | 8.68 |
| N | 12*j* | 0.413 | 0.413 | 0.5 | 5.82 |
| C | 12*j* | 0.578 | 0.578 | 0.5 | 7.05 |

As the temperature decreases, a transition takes place from the cubic phase to the tetragonal phase, as shown in Fig. 6.1(b) and Table 6.3 [32]. The isotropic displacement parameters $B$ were calculated as $8\pi^2\,U_{iso}$. In the tetragonal phase, I ions are ordered, which results in lower symmetry than in the cubic phase. Site occupancies were set as 1/4 for C and N for the tetragonal $CH_3NH_3PbI_3$. As the temperature decreases, the tetragonal phase transitions to an orthorhombic system due to the ordering of $CH_3NH_3$ ions in the unit cell, as shown in Fig. 6.1(c) and Table 6.4 [26]. The energy gaps of $CH_3NH_3PbI_3$ were also measured and calculated [26], as summarized in Table 6.5. The

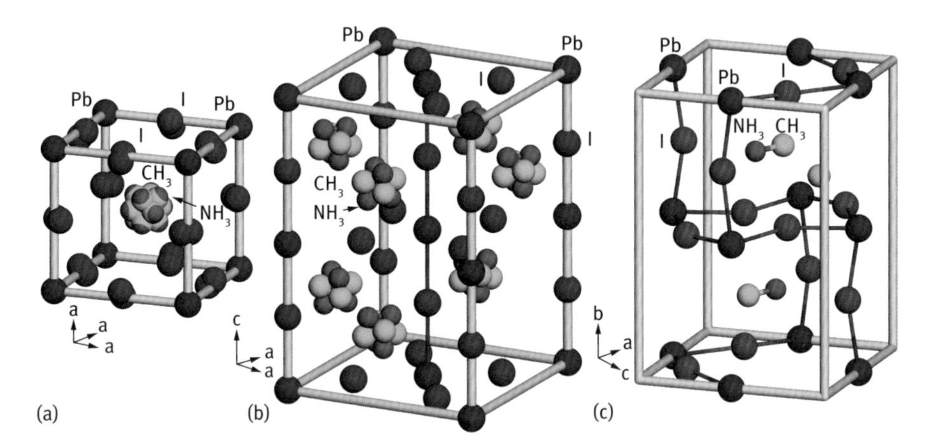


Fig. 6.1: Structural models of $CH_3NH_3PbI_3$ with (a) cubic, (b) tetragonal and (c) orthorhombic structures.

Table 6.3: Structural parameters of tetragonal $CH_3NH_3PbI_3$ at 220 K. Space group $I4/mcm$ ($Z = 4$), $a = 8.800$ Å, $c = 12.685$ Å. $B$ is the isotropic displacement parameter.

| Atom | site | $x$ | $y$ | $z$ | $B$ (Å$^2$) |
|---|---|---|---|---|---|
| Pb | 4*c* | 0 | 0 | 0 | 1.63 |
| I(1) | 8*h* | 0.2039 | 0.2961 | 0 | 4.38 |
| I(2) | 4*a* | 0 | 0 | 0.25 | 4.11 |
| N | 16*l* | 0.459 | 0.041 | 0.202 | 4.60 |
| C | 16*l* | 0.555 | −0.055 | 0.264 | 3.19 |

Table 6.4: Structural parameters of orthorhombic $CH_3NH_3PbI_3$ at 100 K. Space group $Pnma$ ($Z = 4$), $a = 8.8362$ Å, $b = 12.5804$ Å, $c = 8.5551$ Å. All occupancy factors 1.0. $B$ is the isotropic displacement parameter.

| Atom | site | $x$ | $y$ | $z$ | $B$ (Å$^2$) |
|---|---|---|---|---|---|
| Pb | 4*b* | 0.5 | 0 | 0 | 4.80 |
| I(1) | 4*c* | 0.48572 | 0.25 | −0.05291 | 1.03 |
| I(2) | 8*d* | 0.19020 | 0.01719 | 0.18615 | 1.33 |
| N | 4*c* | 0.932 | 0.75 | 0.029 | 2.37 |
| C | 4*c* | 0.913 | 0.25 | 0.061 | 1.50 |

Table 6.5: Energy band gaps of $CH_3NH_3PbI_3$.

| Material | $CH_3NH_3PbI_3$ | $CH_3NH_3PbI_3$ | $CH_3NH_3PbI_3$ |
|---|---|---|---|
| Crystal system | Cubic | Tetragonal | Orthorhombic |
| Measured energy gap (eV) | | 1.51 | |
| Calculated energy gap (eV) | 1.3 | 1.43 | 1.61 |

energy gap increases with increasing temperature from the *ab-initio* calculation, and the measured energy gap of ~ 1.5 eV is suitable for solar cell materials.

The structural parameters of cubic $CH_3NH_3PbCl_3$ and $CH_3NH_3PbBr_3$ are summarized in Table 6.6 and 6.7, respectively [31, 33]. They have similar structural parameters compared with the cubic $CH_3NH_3PbI_3$, except for their lattice constants. The lattice parameters of these compounds are strongly dependent on the size of halogen ions, as shown in Fig. 6.2. As summarized in Table 6.8, the ion radii of halogen elements increase with increasing atomic numbers, which affects the lattice constants of $CH_3NH_3PbX_3$, as observed in Fig. 6.2.

Table 6.6: Structural parameters of cubic $CH_3NH_3PbCl_3$. Space group $Pm\bar{3}m$ ($Z = 1$), $a = 5.666$ Å at 200 K. $B$ is the isotropic displacement parameter.

| Atom | $x$ | $y$ | $z$ | $B$ (Å$^2$) |
|---|---|---|---|---|
| Pb | 0 | 0 | 0 | 1.13 |
| Cl | 0 | 0.0413 | 0.5 | 6.73 |
| N | 0.413 | 0.409 | 0.5 | 8.1 |
| C | 0.578 | 0.583 | 0.5 | 5.8 |

Table 6.7: Structural parameters of cubic $CH_3NH_3PbBr_3$. Space group $Pm\bar{3}m$ ($Z = 1$), $a = 5.933$ nm at 298 K. $B$ is the isotropic displacement parameter.

| Atom | $x$ | $y$ | $z$ | $B$ (Å$^2$) |
|---|---|---|---|---|
| Pb | 0 | 0 | 0 | 1.61 |
| Br | 0 | 0.0413 | 0.5 | 5.41 |
| N | 0.413 | 0.417 | 0.5 | 6.02 |
| C | 0.578 | 0.582 | 0.5 | 6.05 |

Table 6.8: Ion radii of halogen and 14 group elements.

| Hologen element | F- | $Cl^-$ | $Br^-$ | $I^-$ |
|---|---|---|---|---|
| Ion radius (Å) | 1.33 | 1.81 | 1.96 | 2.20 |
| 14 group element | | $Ge^{2+}$ | $Sn^{2+}$ | $Pb^{2+}$ |
| Lattice parameters | | 0.73 | 0.93 | 1.18 |

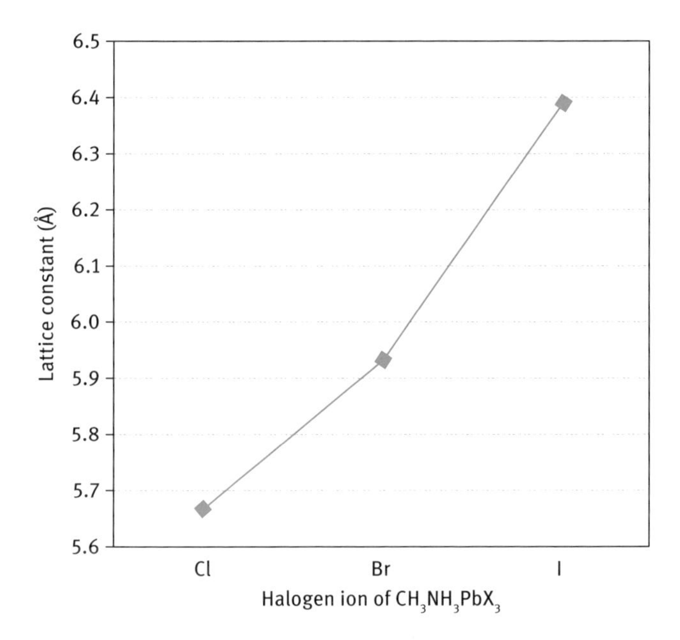


Fig. 6.2: Lattice constants of $CH_3NH_3PbX_3$ (X = Cl, Br, or I).

## 6.3 X-ray diffraction of $CH_3NH_3PbI_3$

The microstructure of the perovskite phases can be investigated by XRD. XRD will indicate whether the sample is single phase or mixed phase. If the sample consists of nanoparticles or nanocrystals, the crystallite size can be estimated from the full width at half maximum (FWHM). Through the use of XRD data, analyses of high-resolution TEM image and electron diffraction becomes easier. If the sample is a known material, plane distances ($d$) and indices can be clarified from the diffraction peaks of XRD. When the sample has an unknown structure, the values of the plane distances can be obtained by XRD, which will effectively stimulate the structural analysis.

Calculated X-ray diffraction patterns on the $CH_3NH_3PbI_3$ with cubic, tetragonal and orthorhombic structures is shown in Fig. 6.3(a), and calculated X-ray diffraction parameters of cubic, tetragonal and orthorhombic $CH_3NH_3PbI_3$ are listed in Table 6.9, 6.10, and 6.11, respectively. For the cubic phase, site occupancies were set as 1/4 for I and 1/12 for C and N. Structure factors were averaged for each index. Site occupancies were set as 1/4 for C and N for the tetragonal $CH_3NH_3PbI_3$. Fig. 6.3(b) is an enlarged calculated X-ray diffraction patterns of $CH_3NH_3PbI_3$. Reflection positions of 211 and 213 inconsistent with cubic symmetry for tetragonal structure are indicated by asterisks, which is helpful in distinguishing between the cubic and tetragonal phase [26].

Calculated X-ray diffraction patterns of $CH_3NH_3PbI_3$ with various FWHM values are shown in Fig. 6.4(a). When the crystallite sizes decrease, the FWHM values

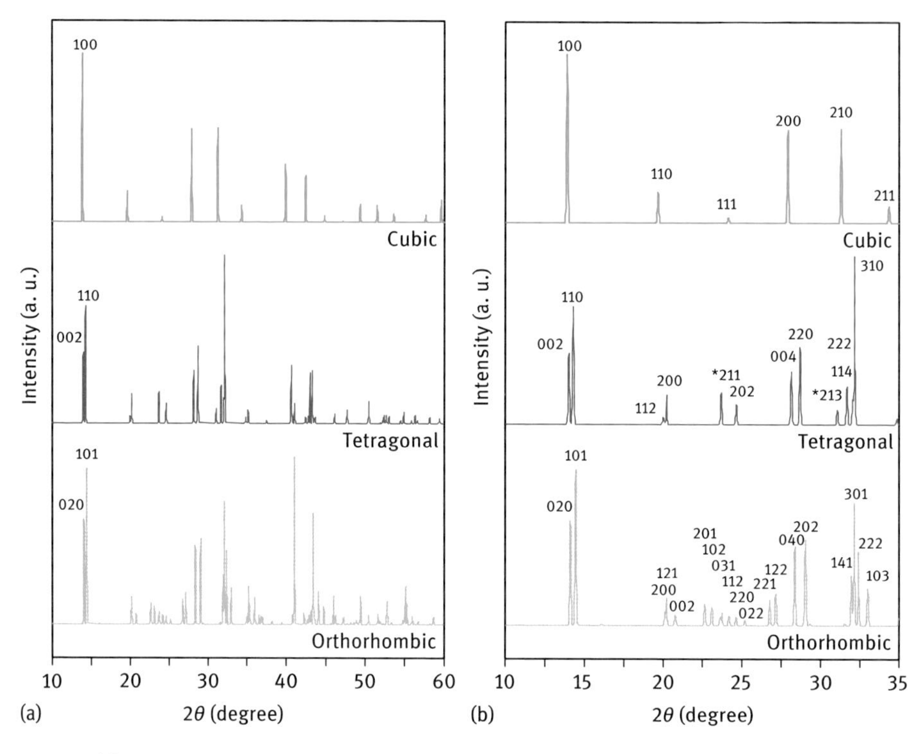


Fig. 6.3: (a) Calculated X-ray diffraction patterns of $CH_3NH_3PbI_3$ with cubic, tetragonal and orthorhombic structures. (b) Enlarged diffraction patterns of (a).

Table 6.9: Calculated X-ray diffraction parameters of cubic $CH_3NH_3PbI_3$. Equivalent indices were combined. Space group $Pm\bar{3}m$ ($Z = 1$), $a = 6.391$ Å at 330 K. $F$ is the structure factor.

| Index | 2θ (°) | d-spacing (Å) | \|F\| | Relative intensity (%) | Multiplicity |
|---|---|---|---|---|---|
| 1 0 0 | 13.8449 | 6.3910 | 107.1 | 100 | 6 |
| 1 1 0 | 19.6279 | 4.5191 | 46.3 | 18 | 12 |
| 1 1 1 | 24.0990 | 3.6898 | 29.4 | 3 | 8 |
| 2 0 0 | 27.8973 | 3.1955 | 164.3 | 55 | 6 |
| 2 1 0 | 31.2695 | 2.8581 | 93.4 | 56 | 24 |
| 2 1 1 | 34.3423 | 2.6091 | 44.4 | 10 | 24 |
| 2 2 0 | 39.8633 | 2.2596 | 136.0 | 35 | 12 |
| 2 2 1 | 42.3942 | 2.1303 | 84.0 | 23 | 24 |
| 3 0 0 | 42.3942 | 2.1303 | 76.0 | 5 | 6 |
| 3 1 0 | 44.8082 | 2.0210 | 35.9 | 4 | 24 |
| 3 1 1 | 47.1237 | 1.9270 | 8.6 | 0.2 | 24 |
| 2 2 2 | 49.3555 | 1.8449 | 116.1 | 10 | 8 |
| 3 2 0 | 51.5149 | 1.7725 | 69.5 | 10 | 24 |
| 3 2 1 | 53.6114 | 1.7081 | 35.9 | 5 | 48 |
| 4 0 0 | 57.6458 | 1.5978 | 100.9 | 4 | 6 |
| 4 1 0 | 59.5956 | 1.5500 | 66.8 | 13 | 48 |

Table 6.10: Calculated X-ray diffraction parameters of tetragonal $CH_3NH_3PbI_3$. Space group $I4/mcm$ ($Z = 4$), $a$ = 8.800 Å, c = 12.685 Å at 220 K.

| Index | $2\theta$ (°) | $d$-spacing (Å) | $\|F\|$ | Relative intensity (%) | Multiplicity |
|---|---|---|---|---|---|
| 0 0 2 | 13.9513 | 6.3425 | 477.0 | 60 | 2 |
| 1 1 0 | 14.2216 | 6.2225 | 442.5 | 100 | 4 |
| 1 1 2 | 19.9730 | 4.4418 | 116.8 | 7 | 8 |
| 2 0 0 | 20.1647 | 4.4000 | 211.6 | 11 | 4 |
| 2 1 1 | 23.6509 | 3.7587 | 195.7 | 27 | 16 |
| 2 0 2 | 24.6041 | 3.6152 | 227.6 | 17 | 8 |
| 0 0 4 | 28.1149 | 3.1713 | 852.4 | 45 | 2 |
| 2 2 0 | 28.6684 | 3.1113 | 744.0 | 66 | 4 |
| 2 1 3 | 31.0176 | 2.8808 | 184.5 | 14 | 16 |
| 1 1 4 | 31.6405 | 2.8255 | 410.3 | 33 | 8 |
| 2 2 2 | 32.0148 | 2.7933 | 331.7 | 21 | 8 |
| 3 1 0 | 32.1387 | 2.7828 | 511.0 | 49 | 8 |
| 2 0 4 | 34.8441 | 2.5727 | 180.5 | 5 | 8 |
| 3 1 2 | 35.1881 | 2.5483 | 199.6 | 12 | 16 |
| 3 2 1 | 37.4940 | 2.3967 | 117.9 | 4 | 16 |
| 2 2 4 | 40.5874 | 2.2209 | 665.6 | 50 | 8 |
| 4 0 0 | 40.9903 | 2.2000 | 566.6 | 18 | 4 |
| 2 1 5 | 42.3526 | 2.1323 | 165.5 | 6 | 16 |
| 0 0 6 | 42.7343 | 2.1142 | 415.0 | 4 | 2 |
| 3 2 3 | 42.7418 | 2.1138 | 109.2 | 2 | 16 |
| 4 1 1 | 42.9354 | 2.1047 | 277.8 | 15 | 16 |
| 3 1 4 | 43.2169 | 2.0917 | 479.9 | 45 | 16 |
| 4 0 2 | 43.5043 | 2.0785 | 222.4 | 5 | 8 |
| 3 3 0 | 43.5998 | 2.0742 | 225.3 | 2 | 4 |
| 4 2 0 | 46.0901 | 1.9677 | 317.2 | 8 | 8 |
| 2 0 6 | 47.6844 | 1.9056 | 155.2 | 2 | 8 |
| 4 1 3 | 47.6913 | 1.9053 | 265.2 | 11 | 16 |
| 4 0 4 | 50.4445 | 1.8076 | 523.0 | 19 | 8 |
| 3 2 5 | 51.9446 | 1.7589 | 106.0 | 1 | 16 |
| 2 2 6 | 52.2716 | 1.7487 | 303.6 | 6 | 8 |
| 4 3 1 | 52.4442 | 1.7433 | 171.5 | 4 | 16 |
| 3 3 4 | 52.6864 | 1.7359 | 218.2 | 3 | 8 |
| 5 1 0 | 53.0165 | 1.7258 | 307.9 | 6 | 8 |
| 3 1 6 | 54.4599 | 1.6834 | 165.0 | 3 | 16 |
| 4 2 4 | 54.8632 | 1.6720 | 294.6 | 10 | 16 |
| 2 1 7 | 55.8045 | 1.6460 | 149.8 | 2 | 16 |
| 4 1 5 | 56.2804 | 1.6332 | 250.9 | 7 | 16 |
| 4 3 3 | 56.5963 | 1.6249 | 171.4 | 3 | 16 |
| 0 0 8 | 58.1285 | 1.5856 | 657.6 | 5 | 2 |
| 4 4 0 | 59.3600 | 1.5556 | 423.9 | 4 | 4 |
| 1 1 8 | 60.1739 | 1.5365 | 353.3 | 6 | 8 |

Table 6.11: Calculated X-ray diffraction parameters of orthorhombic $CH_3NH_3PbI_3$. Space group *Pnma* ($Z = 4$), $a = 8.8362$ Å, $b = 12.5804$ Å, $c = 8.5551$ Å at 100 K.

| Index | $2\theta$ (°) | *d*-spacing (Å) | \|*F*\| | Relative intensity (%) | Multiplicity |
|---|---|---|---|---|---|
| 0 2 0 | 14.0679 | 6.2902 | 462.7 | 67 | 2 |
| 1 0 1 | 14.3989 | 6.1463 | 408.5 | 100 | 4 |
| 1 1 1 | 16.0356 | 5.5225 | 32.8 | 1 | 8 |
| 2 0 0 | 20.0813 | 4.4181 | 238.7 | 9 | 2 |
| 1 2 1 | 20.1828 | 4.3961 | 106.9 | 7 | 8 |
| 0 0 2 | 20.7483 | 4.2775 | 225.2 | 7 | 2 |
| 2 0 1 | 22.6324 | 3.9255 | 239.3 | 14 | 4 |
| 1 0 2 | 23.0816 | 3.8501 | 231.4 | 12 | 4 |
| 0 3 1 | 23.6082 | 3.7654 | 168.3 | 6 | 4 |
| 2 1 1 | 23.7239 | 3.7473 | 86.8 | 3 | 8 |
| 1 1 2 | 24.1539 | 3.6816 | 128.7 | 7 | 8 |
| 2 2 0 | 24.6029 | 3.6154 | 172.8 | 6 | 4 |
| 0 2 2 | 25.1559 | 3.5372 | 132.9 | 3 | 4 |
| 2 2 1 | 26.7471 | 3.3302 | 221.7 | 16 | 8 |
| 1 2 2 | 27.1323 | 3.2838 | 256.6 | 21 | 8 |
| 0 4 0 | 28.3536 | 3.1451 | 834.3 | 51 | 2 |
| 2 0 2 | 29.0316 | 3.0732 | 627.4 | 55 | 4 |
| 2 3 0 | 29.3402 | 3.0415 | 108.2 | 2 | 4 |
| 1 3 2 | 31.5191 | 2.8361 | 80.9 | 2 | 8 |
| 1 4 1 | 31.9379 | 2.7998 | 373.2 | 32 | 8 |
| 3 0 1 | 32.1130 | 2.7850 | 560.5 | 35 | 4 |
| 0 1 3 | 32.1584 | 2.7811 | 155.6 | 3 | 4 |
| 2 2 2 | 32.3965 | 2.7612 | 304.2 | 20 | 8 |
| 1 0 3 | 32.9780 | 2.7139 | 473.6 | 24 | 4 |
| 2 4 0 | 34.9912 | 2.5622 | 239.4 | 5 | 4 |
| 3 2 1 | 35.2135 | 2.5465 | 244.8 | 11 | 8 |
| 0 4 2 | 35.3949 | 2.5339 | 225.2 | 5 | 4 |
| 1 2 3 | 36.0126 | 2.4918 | 317.4 | 18 | 8 |
| 2 4 1 | 36.5799 | 2.4545 | 193.6 | 6 | 8 |
| 1 4 2 | 36.8717 | 2.4357 | 185.8 | 6 | 8 |
| 3 0 2 | 37.0262 | 2.4259 | 145.8 | 2 | 4 |
| 2 1 3 | 38.2065 | 2.3536 | 127.4 | 3 | 8 |
| 1 3 3 | 39.5204 | 2.2784 | 101.4 | 1 | 8 |
| 4 0 0 | 40.8148 | 2.2091 | 432.0 | 6 | 2 |
| 2 4 2 | 41.0283 | 2.1980 | 555.9 | 41 | 8 |
| 4 0 1 | 42.2164 | 2.1389 | 281.7 | 5 | 4 |
| 0 0 4 | 42.2189 | 2.1388 | 305.7 | 3 | 2 |

Table 6.11: continued.

| Index | 2θ (°) | d-spacing (Å) | \|F\| | Relative intensity (%) | Multiplicity |
|---|---|---|---|---|---|
| 2 5 1 | 42.6467 | 2.1183 | 161.0 | 3 | 8 |
| 1 5 2 | 42.9036 | 2.1062 | 142.9 | 2 | 8 |
| 3 3 2 | 43.0398 | 2.0999 | 157.9 | 3 | 8 |
| 0 6 0 | 43.1073 | 2.0967 | 330.6 | 3 | 2 |
| 3 4 1 | 43.3616 | 2.085 | 506.1 | 30 | 8 |
| 4 2 0 | 43.3783 | 2.0843 | 145.8 | 1 | 4 |
| 2 3 3 | 43.4643 | 2.0803 | 157.5 | 3 | 8 |
| 1 0 4 | 43.4991 | 2.0787 | 318.3 | 6 | 4 |
| 1 4 3 | 44.0352 | 2.0547 | 429.7 | 21 | 8 |
| 1 1 4 | 44.1197 | 2.0509 | 140.7 | 2 | 8 |
| 4 2 1 | 44.7146 | 2.0250 | 292.5 | 9 | 8 |
| 0 2 4 | 44.7169 | 2.0249 | 202.1 | 2 | 4 |
| 1 6 1 | 45.6801 | 1.9844 | 100.8 | 1 | 8 |
| 1 2 4 | 45.9414 | 1.9738 | 258.0 | 7 | 8 |
| 4 0 2 | 46.2136 | 1.9628 | 352.1 | 6 | 4 |
| 3 2 3 | 46.5832 | 1.9481 | 73.0 | 1 | 8 |
| 2 5 2 | 46.6144 | 1.9468 | 93.9 | 1 | 8 |
| 2 0 4 | 47.1728 | 1.9251 | 308.8 | 5 | 4 |
| 3 4 2 | 47.2817 | 1.9209 | 124.0 | 1 | 8 |
| 0 5 3 | 48.1925 | 1.8867 | 181.8 | 2 | 4 |
| 4 2 2 | 48.5489 | 1.8737 | 135.7 | 2 | 8 |
| 1 3 4 | 48.8598 | 1.8625 | 183.1 | 3 | 8 |
| 2 6 1 | 49.2266 | 1.8494 | 148.3 | 2 | 8 |
| 1 5 3 | 49.3503 | 1.8451 | 124.9 | 1 | 8 |
| 1 6 2 | 49.4567 | 1.8414 | 175.3 | 3 | 8 |
| 2 2 4 | 49.4736 | 1.8408 | 218.7 | 4 | 8 |
| 4 4 0 | 50.442 | 1.8077 | 403.4 | 7 | 4 |
| 4 4 1 | 51.6367 | 1.7686 | 236.3 | 4 | 8 |
| 0 4 4 | 51.6388 | 1.7686 | 287.3 | 3 | 4 |
| 0 7 1 | 51.9470 | 1.7588 | 234.2 | 2 | 4 |
| 4 0 3 | 52.3451 | 1.7464 | 135.0 | 1 | 4 |
| 2 5 3 | 52.7109 | 1.7351 | 82.5 | 1 | 8 |
| 1 4 4 | 52.7410 | 1.7342 | 270.6 | 5 | 8 |
| 2 6 2 | 52.8123 | 1.7320 | 242.3 | 4 | 8 |
| 5 0 1 | 52.8555 | 1.7307 | 280.3 | 3 | 4 |
| 4 1 3 | 52.8858 | 1.7298 | 121.2 | 1 | 8 |
| 3 1 4 | 53.3946 | 1.7145 | 162.9 | 2 | 8 |
| 3 6 1 | 54.7548 | 1.6751 | 197.8 | 3 | 8 |

Table 6.11: continued.

| Index | $2\theta$ (°) | $d$-spacing (Å) | $\|F\|$ | Relative intensity (%) | Multiplicity |
|---|---|---|---|---|---|
| 3 2 4 | 54.9839 | 1.6686 | 176.9 | 2 | 8 |
| 4 4 2 | 55.1098 | 1.6651 | 333.9 | 8 | 8 |
| 2 7 0 | 55.1235 | 1.6647 | 160.7 | 1 | 4 |
| 4 5 0 | 55.2931 | 1.6600 | 194.8 | 1 | 4 |
| 1 6 3 | 55.3226 | 1.6592 | 258.8 | 4 | 8 |
| 2 4 4 | 55.9564 | 1.6419 | 290.5 | 5 | 8 |
| 1 2 5 | 56.6693 | 1.6229 | 166.9 | 2 | 8 |
| 2 1 5 | 58.2395 | 1.5829 | 131.8 | 1 | 8 |
| 0 8 0 | 58.6587 | 1.5726 | 586.0 | 5 | 2 |
| 4 0 4 | 60.1713 | 1.5366 | 266.9 | 2 | 4 |

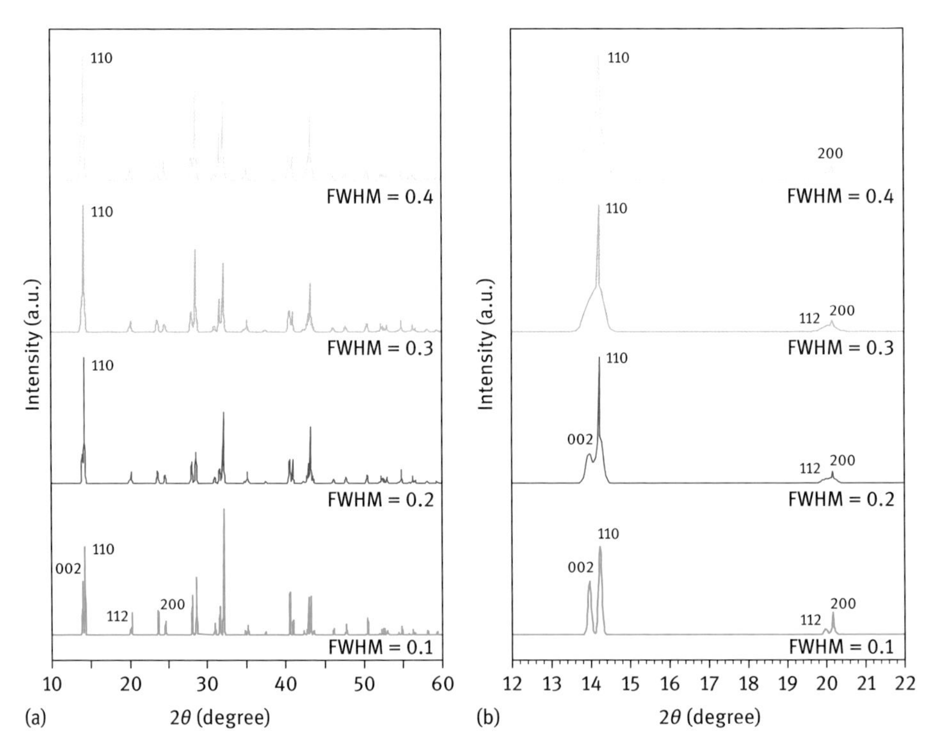


Fig. 6.4: (a) Calculated X-ray diffraction patterns of tetragonal $CH_3NH_3PbI_3$ with various FWHM values. (b) Enlarged diffraction patterns of (a).

increase, and different peak intensities are observed in Fig. 6.4(a). Fig. 6.4(b) is an enlarged calculated X-ray diffraction patterns of $CH_3NH_3PbI_3$. With increasing FWHM values, the diffraction peaks of 002 and 110 of a tetragonal system seem to be combined into 100 of a cubic system, which could mislead the results of XRD structural analysis, and it should be very careful for that.

## 6.4 Electron diffraction of $CH_3NH_3PbI_3$

When the sample amount, sample area or film thickness is small, it is difficult to obtain the necessary diffraction amplitude by XRD. Since TEM observation can handle such small samples, only TEM observation may be applied to obtain structural data. To obtain information on the fundamental atomic arrangements, electron diffraction patterns should be taken along the various directions of the crystal, and the fundamental crystal system and lattice constants may be estimated. Then, high-resolution TEM observation and composition analysis by energy dispersive X-ray spectroscopy are performed, and an approximate model of the atomic structure is constructed. Most materials have similar structures to known materials, and the structures will be estimated if a database on the known structures is available. For example, lots of new structures were found for high-Tc superconducting oxides, which have basic perovskite structures, and approximate atomic structure models can be constructed from high-resolution TEM images, electron diffraction patterns, and composition analysis of the elements [23–25].

If the structure of a TEM specimen is known, the observation direction of the crystal should be selected, and the electron diffraction pattern along the direction should be estimated. Any regions selected by the selected area aperture can be observed in electron diffraction patterns, and the structure can easily be analyzed by comparing TEM images with electron diffraction patterns. When an electron diffraction pattern is observed in the selected area, the diffraction pattern is often inclined from the aimed direction, which can be seen from the asymmetry of the electron diffraction pattern. The sample holder can usually be tilted along two directions, and the specimen should be tilted as the diffraction pattern shows center symmetry. Atomic structure models of cubic $CH_3NH_3PbI_3$ observed along various directions are shown in Fig. 6.5, which were drawn by Vesta [34]. Note that the atomic positions of $CH_3$, $NH_3$ and I are disordered as observed in the structural models. Corresponding electron diffraction patterns of cubic $CH_3NH_3PbI_3$ calculated along the [100], [110], [111] and [210] directions are shown in Fig. 6.6.

Atomic structure models of tetragonal $CH_3NH_3PbI_3$ observed along [001], [100], [021], [221] and [110] are shown in Fig. 6.7, which correspond to [001], [110], [111], [210] and [100] of the cubic phase in Fig. 6.6, respectively. Atomic positions of I are fixed for the tetragonal phase, and only atomic positions of $CH_3$ and $NH_3$ are disordered. For the tetragonal phase, the crystal symmetries are lowered as indicated by arrows

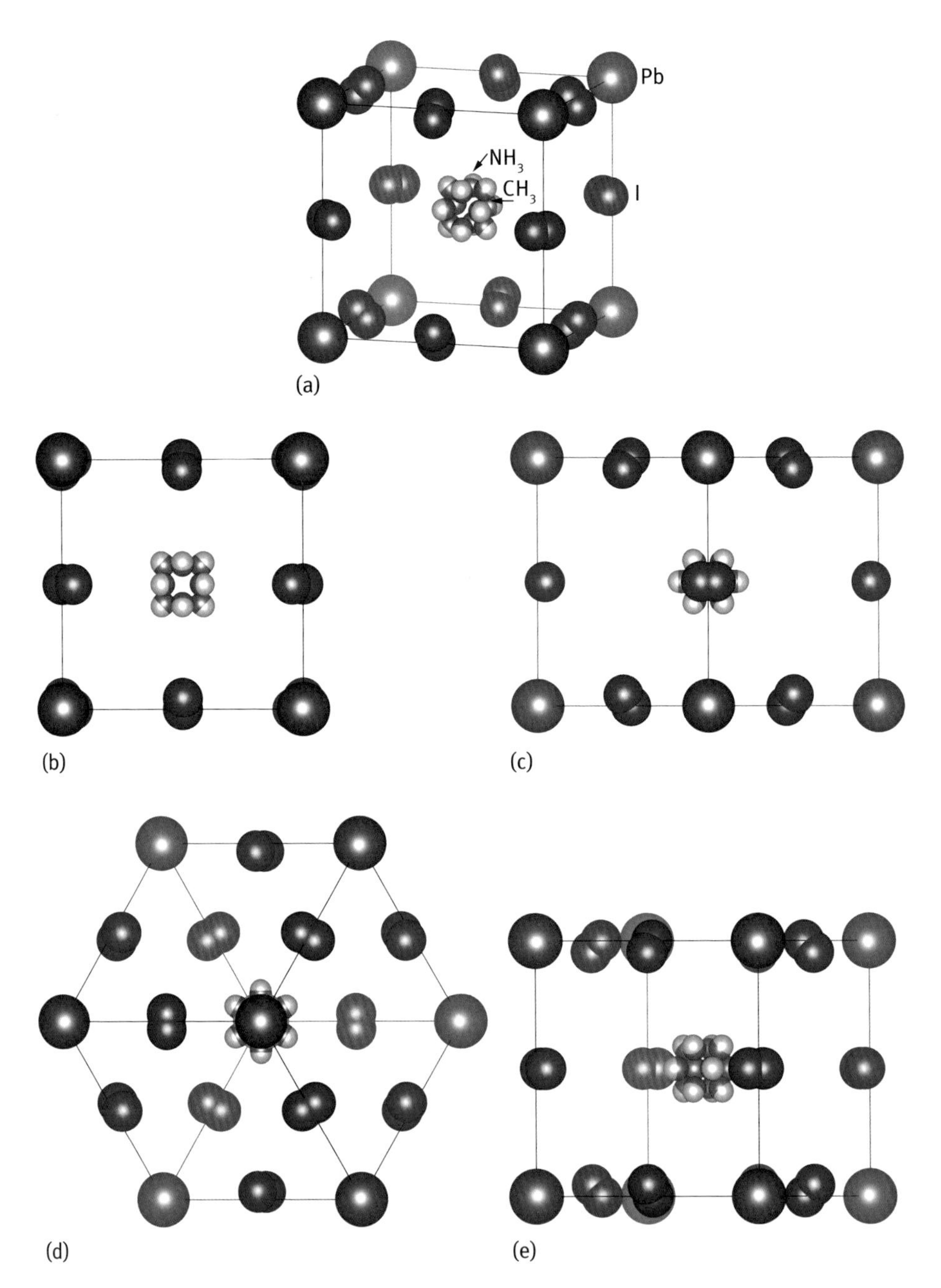


Fig. 6.5: Atomic structure models of cubic $CH_3NH_3PbI_3$ observed along (a) perspective view, (b) [100], (c) [110], (d) [111] and (e) [210].

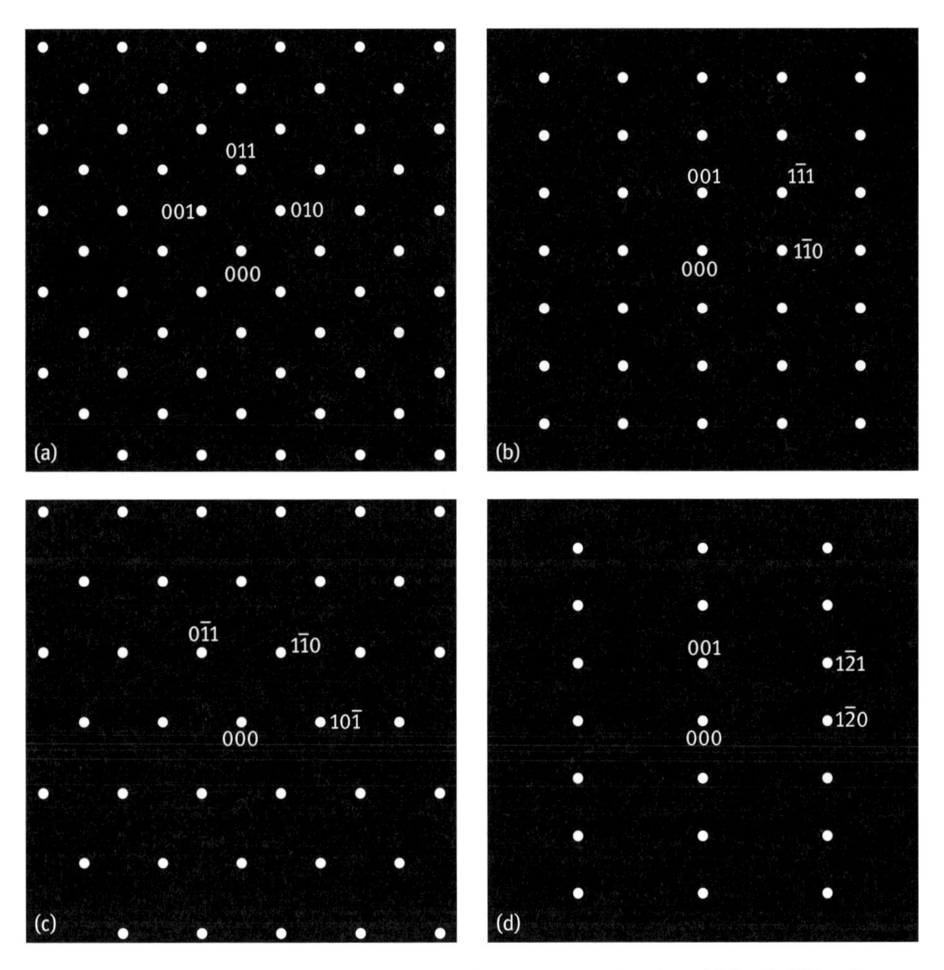


Fig. 6.6: Calculated electron diffraction patterns of cubic $CH_3NH_3PbI_3$ along (a) [100], (b) [110], (c) [111] and (d) [210].

in Fig. 6.7(c) and (e). Several diffraction spots in Fig. 6.7 have different diffraction intensities compared with Fig. 6.6, which is due to the different crystal symmetry of the $CH_3NH_3PbI_3$ compound.

High-resolution TEM observations have been performed for the perovskite materials [24], and the nanostructures have been discussed. Although TEM is a powerful tool for observing the nanostructure of materials, sample damage by electron beam irradiation should be avoided because $CH_3NH_3PbI_3$ is known to be unstable during the process of annealing at elevated temperatures. Several TEM results have been reported for $CH_3NH_3PbI_3$ and $CH_3CH_2NH_3PbI_3$, and the structures were determined by electron diffraction and high-resolution images in these works [26, 35–37].

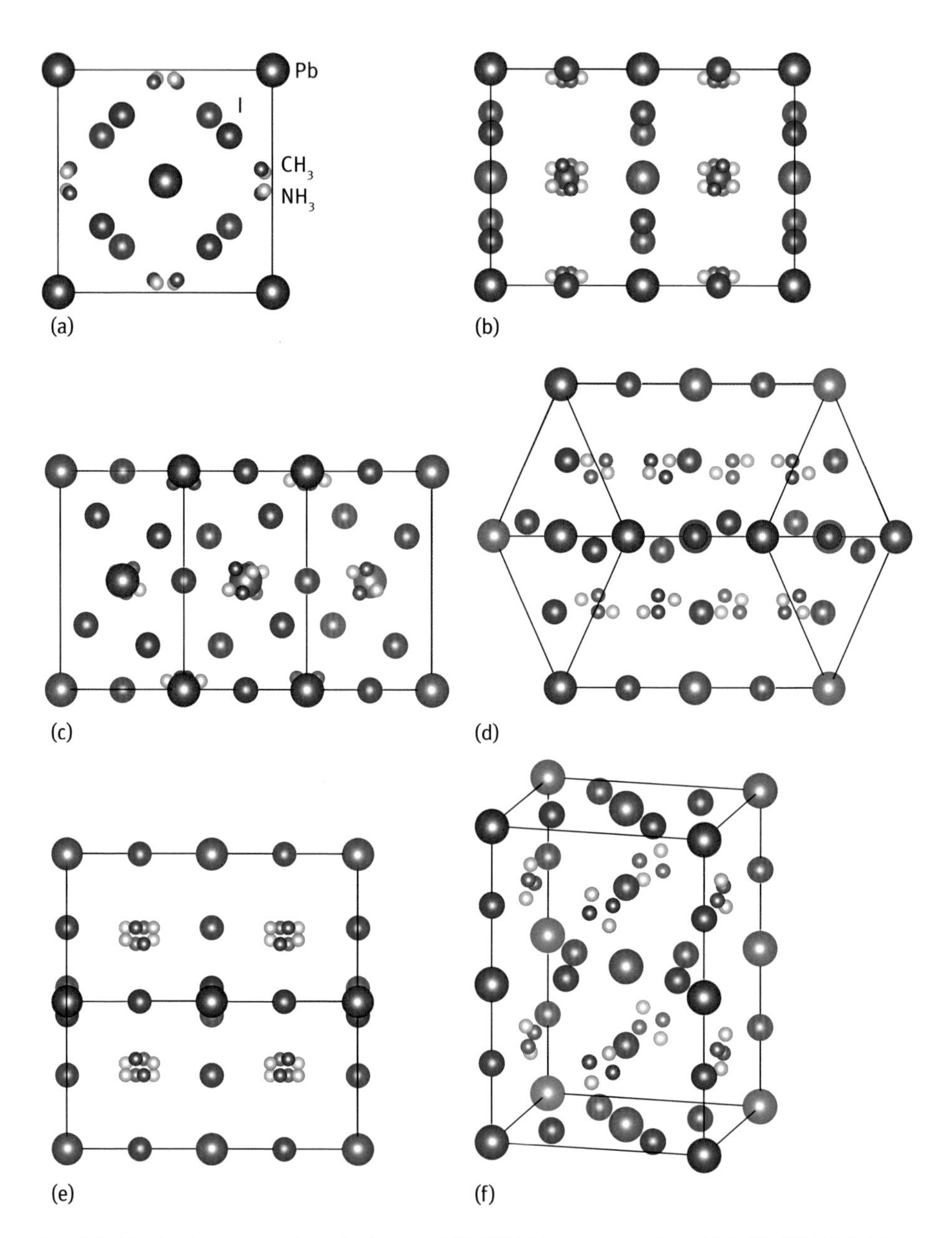


Fig. 6.7: Atomic structure models of tetragonal $CH_3NH_3PbI_3$ observed along (a) [001], (b) [100], (c) [021], (d) [221] and (e) [110] and (f) perspective view.

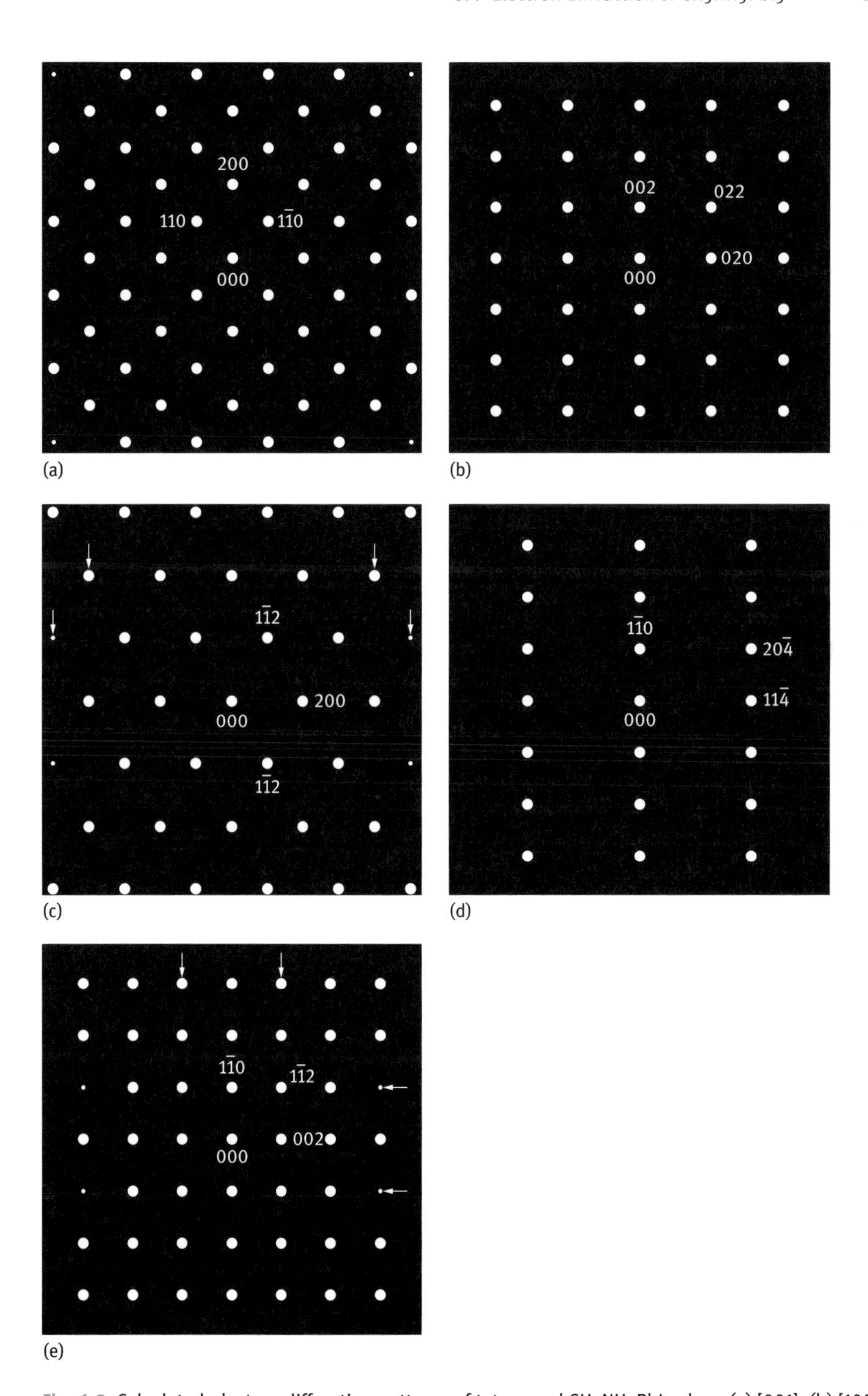


Fig. 6.8: Calculated electron diffraction patterns of tetragonal $CH_3NH_3PbI_3$ along (a) [001], (b) [100], (c) [021], (d) [221] and (e) [110].

## 6.5 Other perovskite structures for solar cells

In addition to $CH_3NH_3PbX_3$ (X = Cl, Br, or I) compounds, various perovskite compounds with perovskite structures for solar cells have been reported on and summarized [21, 26]. The crystal systems and temperatures of $CsSnI_3$ are listed in Table 6.12 and 6.13, which has very similar structures and phase transitions [38, 39] to $CH_3NH_3PbX_3$. The crystal parameters of $CsGeI_3$ for high temperature phase are also listed in Table 6.13 [40], and structural models of $CsSnI_3$ and $CsGeI_3$ are shown in Fig. 6.9. Solar cells with F-doped $CsSnI_{2.95}F_{0.05}$ provided a photo-conversion efficiency of 8.5 % [41].

Table 6.12: Crystal systems and temperatures of $CsSnI_3$.

| **Temperature (K)** | **300** | **350** | **478** |
|---|---|---|---|
| Crystal system | Orthorhombic | Tetragonal | Cubic |
| Space group | *Pnma* | *P4/mbm* | $Pm\bar{3}m$ |
| *Z* | 4 | 2 | 1 |
| Lattice parameters | $a$ = 8.6885 Å<br>$b$ = 12.3775 Å<br>$c$ = 8.3684 Å | $a$ = 8.7182 Å<br>$c$ = 6.1908 Å | $a$ = 6.1057 Å |

Table 6.13: Structural parameters of cubic $CsSnI_3$ and $CsGeI_3$. Space group $Pm\bar{3}m$ ($Z$ = 1). $a$ = 6.219 Å at 446 K for $CsSnI_3$, and $a$ = 6.05 Å at 573 K for $CsGeI_3$, respectively.

| **Atom** | **site** | *x* | *y* | *z* |
|---|---|---|---|---|
| Cs | 1*b* | 0.5 | 0.5 | 0.5 |
| Sn | 1*a* | 0 | 0 | 0 |
| I | 3*d* | 0.5 | 0 | 0 |
| Cs | 1*b* | 0.5 | 0.5 | 0.5 |
| Ge | 1*a* | 0 | 0 | 0 |
| I | 3*d* | 0.5 | 0 | 0 |

Similar structures of $CH_3NH_3GeCl_3$ and $CH_3NH_3SnCl_3$ are shown in Table 6.14 and 6.15, respectively [42, 43]. The ion radii of Ge and Sn ions are listed in Table 6.8, and they can be substituted for the Pb atoms in $CH_3NH_3PbX_3$. Lead-free $CH_3NH_3SnI_3$ solar cells were developed, and provided 5.7 % efficiency [44]. $(CH_3CH_2NH_3)PbI_3$ was found to have a 2H perovskite structure and provided 2.4 % efficiency [35].

The hydrogen positions in the orthorhombic $CH_3NH_3PbI_3$ were determined at 4 K by neutron diffraction [45], as shown in Fig. 6.10(a). In addition to the $MAPbI_3$, $[HC(NH_2)_2]PbI_3$ (formamidinium lead iodide, $FAPbI_3$) provided high conversion

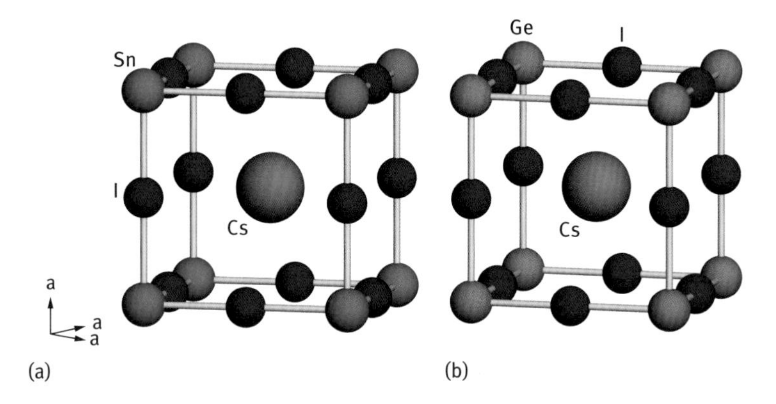


Fig. 6.9: Structural models of (a) $CsSnI_3$ and (b) $CsGeI_3$ with cubic structures.

Table 6.14: Crystal systems and temperatures of $CH_3NH_3GeCl_3$.

| **Temperature (K)** | **2** | **250** | **370** | **475** |
|---|---|---|---|---|
| Crystal system | Monoclinic | Orthorhombic | Trigonal | Cubic |
| Space group | $P2_1/n$ | $Pnma$ | $R3m$ | $Pm\bar{3}m$ |
| $Z$ | 4 | 4 | 1 | 1 |
| Lattice parameters | $a = 10.9973$ Å<br>$b = 7.2043$ Å<br>$c = 8.2911$ Å<br>$\alpha = 90.470°$ | $a = 11.1567$ Å<br>$b = 7.3601$ Å<br>$c = 8.2936$ Å | $a = 5.6784$ Å<br>$\alpha = 90.945°$ | $a = 5.6917$ Å |

Table 6.15: Crystal systems and temperatures of $CH_3NH_3SnCl_3$.

| **Temperature (K)** | **297** | **318** | **350** | **478** |
|---|---|---|---|---|
| Crystal system | Triclinic | Monoclinic | Trigonal | Cubic |
| Space group | $P1$ | $Pc$ | $R3m$ | $Pm\bar{3}m$ |
| $Z$ | 4 | 4 | 1 | 1 |
| Lattice parameters | $a = 5.726$ Å<br>$b = 8.227$ Å<br>$c = 7.910$ Å<br>$\alpha = 90.40°$<br>$\beta = 93.08°$<br>$\gamma = 90.15°$ | $a = 5.718$ Å<br>$b = 8.236$ Å<br>$c = 7.938$ Å<br>$\beta = 93.08°$ | $a = 5.734$ Å<br>$\alpha = 91.90°$ | $a = 5.760$ Å |

efficiencies [18, 19]. Structural parameters including hydrogen positions were also determined at 300 K by neutron diffraction [46], and a structural model is shown in Fig. 6.10(b).

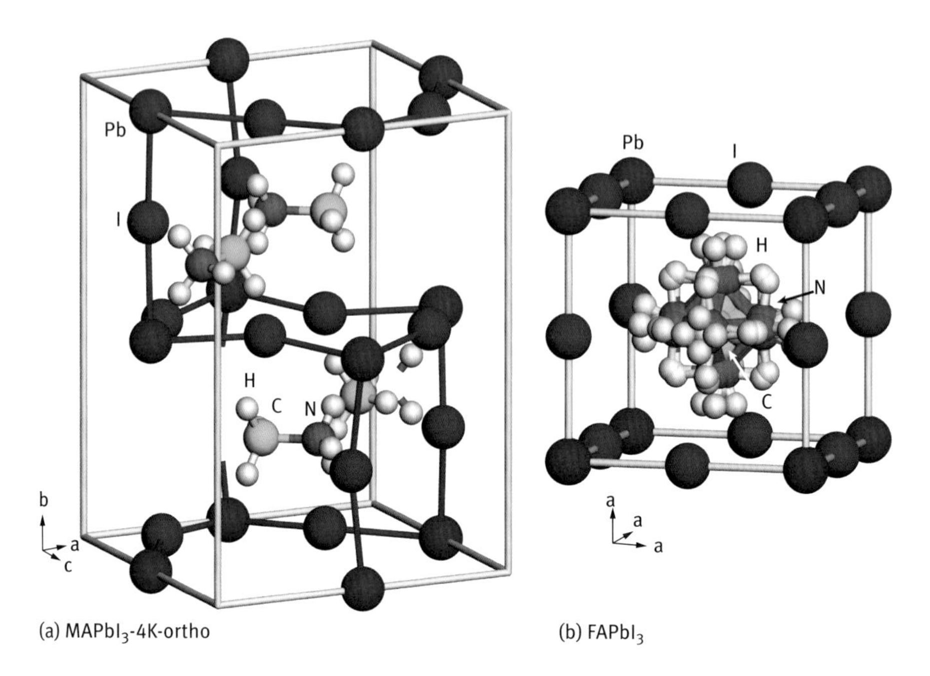


(a) $MAPbI_3$-4K-ortho (b) $FAPbI_3$

Fig. 6.10: Structural models of $CH_3NH_3PbI_3$ and $[HC(NH_2)_2]PbI_3$ with hydrogen positions.

Perovskite oxides such as $[KNbO_3]_{0.9}[BaNi_{0.5}Nb_{0.5}O_{3-x}]_{0.1}$ were found to have an energy gap of ~1.4 eV, implying that they might also be expected to be usable as solar cell materials [47]. With the use of calculations, it has been predicted that various other perovskite compounds would also be useful for solar material design [48].

## 6.6 Basic device structures

A typical fabrication process for $TiO_2/CH_3NH_3PbI_3$ photovoltaic devices is described here [8, 36, 37, 49]. F-Doped tin oxide (FTO) substrates were cleaned using an ultrasonic bath with acetone and methanol and dried under nitrogen gas. A 0.30 M $TiO_x$ precursor solution was prepared from titanium diisopropoxide bis(acetyl acetonate) (0.11 mL) with 1-butanol (1 mL), and the $TiO_x$ precursor solution was spin-coated on the FTO substrate at 3000 rpm for 30 s and annealed at 125 °C for 5 min. This process was performed two times, and the FTO substrate was sintered at 500 °C for 30 min to form the compact $TiO_2$ layer. After that, mesoporous $TiO_2$ paste was coated onto the substrate by a spin-coating method at 5000 rpm for 30 s. For the mesoporous $TiO_2$ layer, the $TiO_2$ paste was prepared with $TiO_2$ powder (Aerosil, P-25) with poly(ethylene glycol) in ultrapure water. The solution was mixed with acetylacetone and triton X-100 for 30 min. The cells were annealed at 120 °C for 5 min and at 500 °C for 30 min. For the preparation of pigment with a perovskite structure, a solution of $CH_3NH_3I$ and $PbI_2$ with a mole ratio of 1 : 1 in γ-butyrolactone (0.5 mL) was mixed at 60 °C. The solution of $CH_3NH_3I$ and $PbI_2$ was then introduced into the $TiO_2$ mesopores using a spin-coating method and annealed at 100 °C for 15 min. Then, the hole transport layer (HTL) was prepared by spin coating. As the HTLs, a solution of spiro-OMeTAD (36.1 mg) in chlorobenzene (0.5 mL) was mixed with a solution of lithium bis(trifluoromethylsulfonyl) imide (Li-TFSI) in acetonitrile (0.5 mL) for 12 h. The former solution with 4-tert-butylpyridine (14.4 μL) was mixed with the Li-TFSI solution (8.8 μL) for 30min at 70 °C. Finally, gold (Au) metal contacts were evaporated as top electrodes. The layered structures of the photovoltaic cells are denoted as FTO/$TiO_2$/$CH_3NH_3PbI_3$/HTL/Au, as shown in Fig. 6.11.

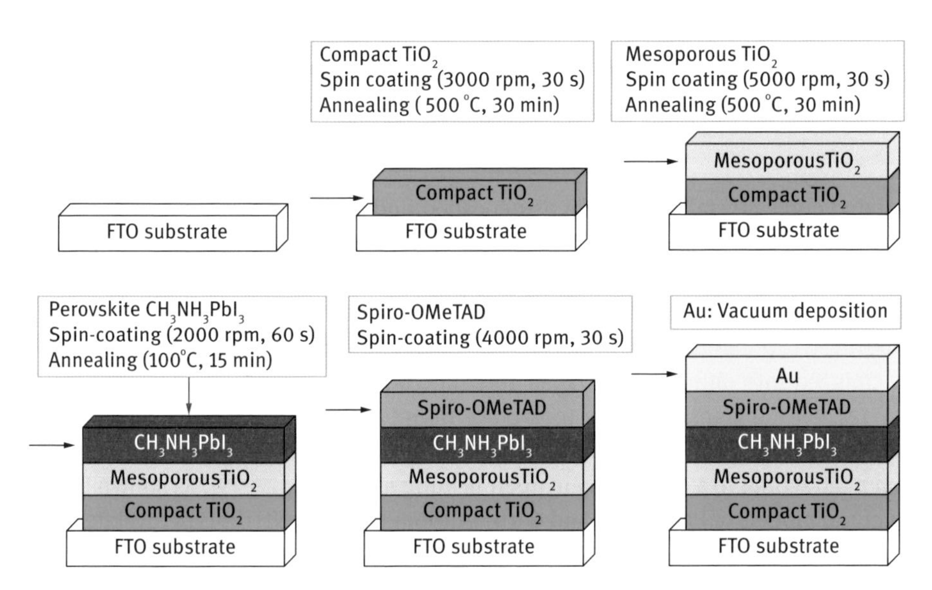


Fig. 6.11: Schematic illustration for the fabrication of $CH_3NH_3PbI_3$ photovoltaic cells.

Typical *J*–*V* characteristics of the $TiO_2$/$CH_3NH_3PbI_3$/spiro-OMeTAD photovoltaic cells under illumination are shown in Fig. 6.12(a), which indicates the annealing effect of the $CH_3NH_3PbI_3$ layer. The as-deposited $CH_3NH_3PbI_3$ cell provided a conversion efficiency of 2.83 %. The $CH_3NH_3PbI_3$ cell annealed at 100 °C for 15 min provided better photovoltaic properties compared with the as-deposited one, as shown in Fig. 6.12(a). The highest efficiency was obtained for the annealed $CH_3NH_3PbI_3$ cell, which provided a power conversion efficiency of 5.16 %, a fill factor of 0.486, a short-circuit current density of 12.9 mA $cm^{-2}$ and an open-circuit voltage of 0.827 V [37]. Figure 6.12(b) shows the results of a multiple spin coating of $CH_3NH_3PbI_3$, which will be described later.

XRD patterns of $CH_3NH_3PbI_3$ thin films on the glass substrate before and after annealing are shown in Fig. 6.15(a). The diffraction peaks can be indexed by tetrag-

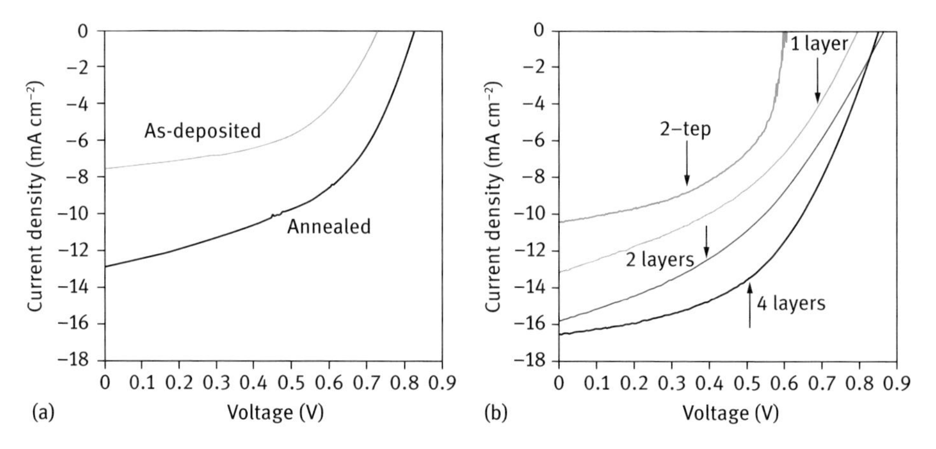


Fig. 6.12: *J–V* characteristic of $TiO_2/CH_3NH_3PbI_3$ photovoltaic cells. (a) As-deposited and annealed samples. (b) $CH_3NH_3PbI_3$ layers prepared by multiple spin-coating.

onal and cubic crystal systems for as-deposited and annealed films, respectively. Although diffraction peaks of the as-deposited film were due to the perovskite phase, broader diffraction peaks due to a $PbI_2$ compound appeared after annealing, as shown in Fig. 6.15(a). Figure 6.15(b) and (c) are enlarged XRD patterns at $2\theta$ of ~ 14 and ~ 28 °, respectively. Split diffraction peaks of 002–110 and 004–220 for the as-deposited sample changed into diffraction peaks of 100 and 200 after annealing, which indicates that a structural transformation took place from the tetragonal to cubic crystal system. The $CH_3NH_3PbI_3$ crystals have perovskite structures and structurally transition from a tetragonal to a cubic system upon heating at ~ 330 K [27–29], as shown in the structural models of Figs. 6.1(a) and 6.1(b). Although the $CH_3NH_3PbI_3$ perovskite crystal has cubic symmetry for the highest temperature phase, the $CH_3NH_3$ ion is polar and has $C_{3v}$ symmetry, which results in a disordered cubic phase. In addition to the disordering of the $CH_3NH_3$ ion, the iodine (I) ions are also disordered in the cubic phase, as shown in Fig. 6.1(a). Site occupancies were set as 1/4 for I and 1/12 for C and N. The $CH_3NH_3$ ion provides 12 equivalent orientations on the $C_2$ axis, and hydrogen atoms have two kinds of configurations on the $C_2$ axis [29]. As the temperature decreased, the cubic phase was transformed into the tetragonal phase, and I ions became ordered, which resulted in a lower symmetry than in the cubic phase, as shown in Fig. 6.1(b). Site occupancies were set as 1/4 for C and N for tetragonal $CH_3NH_3PbI_3$. In the high temperature phase, the unit cell volume of the cubic structure is 261.0 Å$^3$, which is bigger than that of the tetragonal structure (245.6 Å$^3$), as listed in Table 6.16. This could be due to both thermal expansion of the unit cell and atomic disordering of I in the cubic phase. As the temperature lowers, the tetragonal phase is transformed into the orthorhombic systems because of the ordering of $CH_3NH_3$ ions in the unit cell [26].

The XRD results in Fig. 6.13 indicate the phase transformation of the $CH_3NH_3PbI_3$ perovskite structure from a tetragonal to cubic system by partial separation of $PbI_2$ from the $CH_3NH_3PbI_3$ phase during annealing, which corresponds to the decrease of the unit cell volume of the perovskite phase from 248.27 to 246.78 Å$^3$, as listed in Table 6.16. In addition to the I atoms, Pb atoms might also be disordered and deficient, and site occupancies of Pb atoms could be less than 1. It should be noted that the struc-

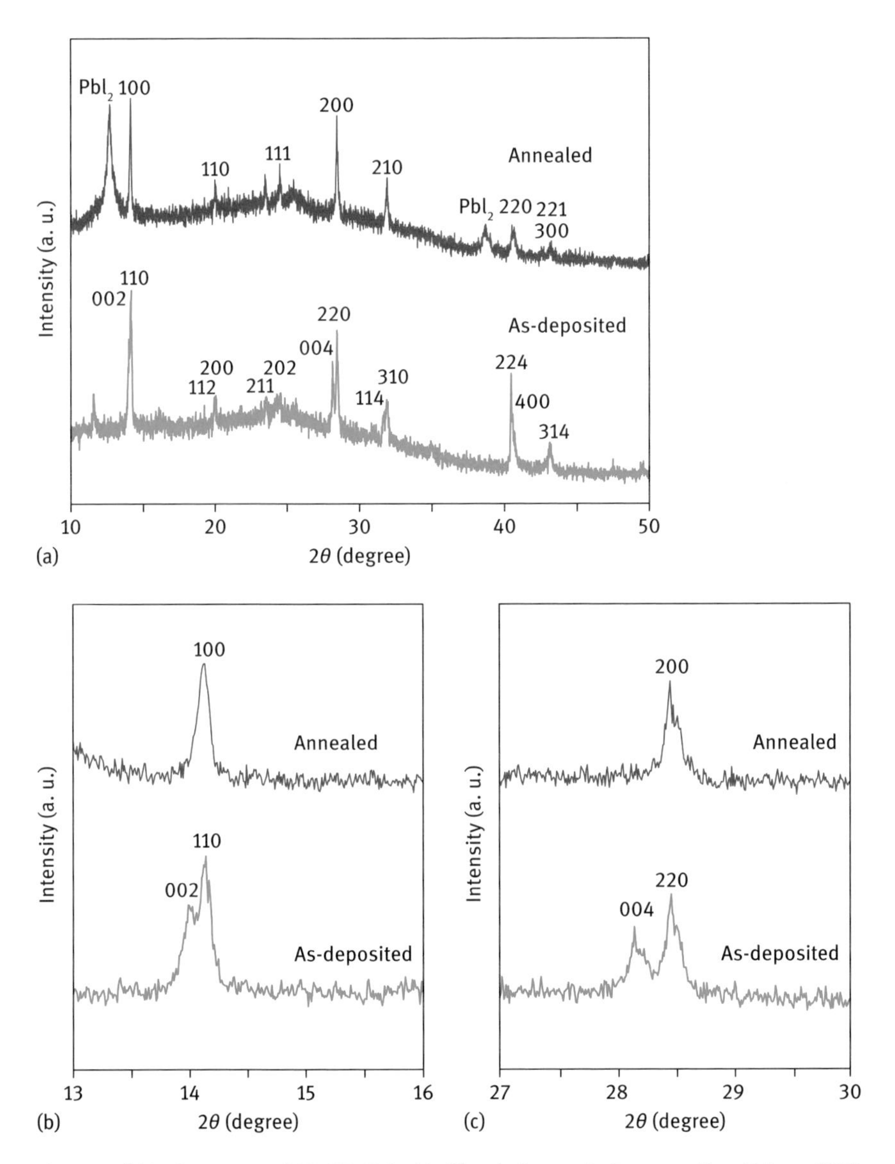


Fig. 6.13: (a) XRD patterns of $CH_3NH_3PbI_3$ thin films before and after annealing. Enlarged XRD patterns at $2\theta$ of (b) ~ 14 ° and (c) ~ 28 °.

Table 6.16: Measured and reported structural parameters of $CH_3NH_3PbI_3$. *V*: unit cell volume. *Z*: number of chemical units in the unit cell.

| Samples | Crystal system | Lattice constants (Å) | *V* (Å$^3$) | *Z* | *V*/*Z* (Å$^3$) |
|---|---|---|---|---|---|
| As-deposited | Tetragonal | $a = 8.8620$<br>$c = 12.6453$ | 993.10 | 4 | 248.27 |
| Annealed | Cubic | $a = 6.2724$ | 246.78 | 1 | 246.78 |
| Ref. 32 (220 K) | Tetragonal | $a = 8.800$<br>$c = 12.685$ | 982.33 | 4 | 245.6 |
| Ref. 31 (330 K) | Cubic | $a = 6.391$ | 261.0 | 1 | 261.0 |

tural transition from the tetragonal to cubic structure presented in this work resulted from the partial separation of $PbI_2$ from the $CH_3NH_3PbI_3$ phase, which is different from the ordinary tetragonal–cubic transition at 330 K [27, 28].

Figure 6.14(a) and (b) are a TEM image and an electron diffraction pattern of $TiO_2/CH_3NH_3PbI_3$, respectively. The TEM image shows $TiO_2$ nanoparticles with sizes of ~ 50 nm, and the polycrystalline $CH_3NH_3PbI_3$ phase shows dark contrast, which is due to the large atomic number of Pb. The electron diffraction pattern of Fig. 6.14(b) shows Debye–Scherrer rings from the nanocrystalline $TiO_2$ particles, which are indexed as the 101, 004, and 200 spacings of the anatase $TiO_2$. The thickness of the mesoporous $TiO_2$ layer was measured to be ~ 300 nm by using atomic force mi-

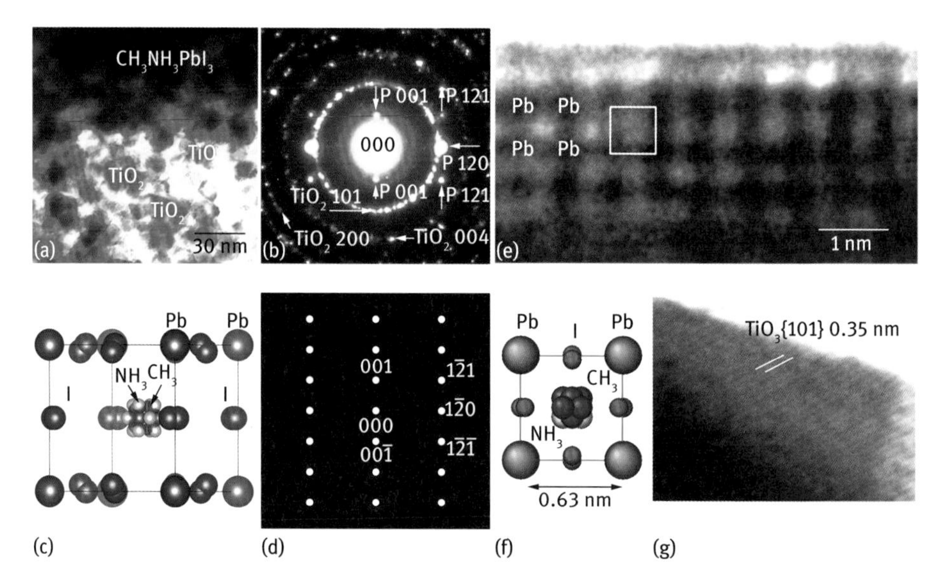


Fig. 6.14: (a) TEM image and (b) electron diffraction pattern of $TiO_2/CH_3NH_3PbI_3$. “P” indicates $CH_3NH_3PbI_3$ perovskite phase. (c) Structural model and (d) its calculated electron diffraction pattern of cubic $CH_3NH_3PbI_3$ projected along the [210] direction. (e) High-resolution TEM image and (f) structural model of $CH_3NH_3PbI_3$. (g) Lattice image of $TiO_2$.

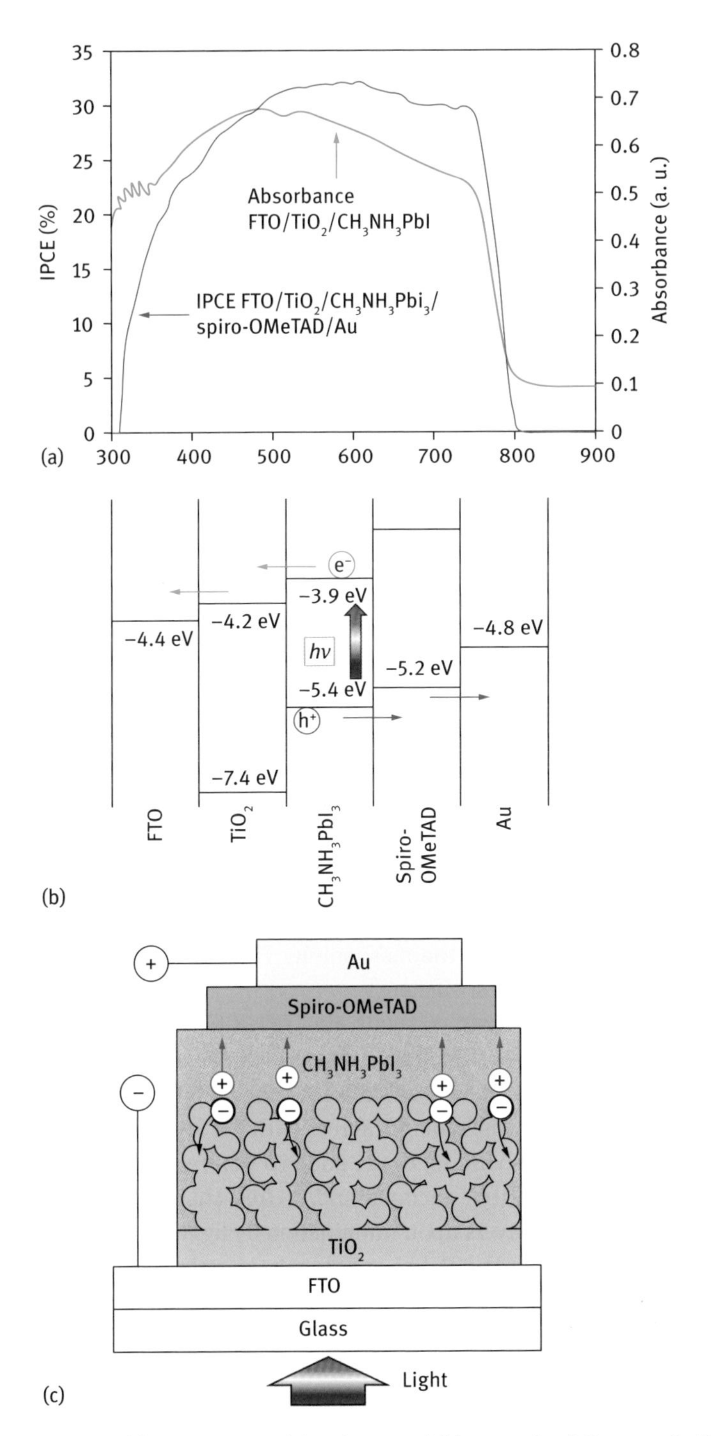


Fig. 6.15: (a) IPCE spectrum/absorbance and (b) energy level diagram of a $TiO_2/CH_3NH_3PbI_3$ cell. (c) Model of interfacial structure.

croscopy. In addition to Debye–Scherrer rings, diffraction spots corresponding to the $CH_3NH_3PbI_3$ crystal with a cubic structure [33] were observed and indexed, as shown in Fig. 6.16(b). Other diffraction spots, except for the Debye–Scherrer rings of $TiO_2$, are also from the $CH_3NH_3PbI_3$ nanoparticles. A structure model and its calculated electron diffraction pattern for a cubic $CH_3NH_3PbI_3$ phase projected along the [210] direction are shown in Figs. 6.14(c) and (d), respectively. The calculated electron diffraction pattern agrees with the observed pattern of Fig. 6.14(b).

Figure 6.14(e) is a high-resolution TEM image of $CH_3NH_3PbI_3$ taken along the $a$-axis. The images of thinner crystals represent the projection of the crystal structure. The darkness or size of dark spots corresponding to Pb positions can be directly identified, and the atomic positions of iodine (I) also indicate weak contrast, as compared with a projected structure model of $CH_3NH_3PbI_3$ along the [100] direction in Fig. 6.14(f). $NH_3$ and $CH_3$ molecules are not represented as dark spots in the image due to the smaller atomic number of N and C. Fig. 6.14(g) is a high-resolution image of the surface of a $TiO_2$ nanoparticle, which indicates {101} lattice fringes.

The $J$–$V$ characteristics of $TiO_2/CH_3NH_3PbI_3$/spiro-OMeTAD photovoltaic cells prepared by multiple spin-coatings of $CH_3NH_3PbI_3$ are shown in Fig. 6.12(b). Figure 6.12(b) indicates the effect of spin-coating times of $CH_3NH_3PbI_3$ on photovoltaic properties. The highest efficiency was obtained for a cell that was spin-coated four times, which provided an $\eta$ of 6.96%, an $FF$ of 0.496, a $J_{SC}$ of 16.5 mA cm$^{-2}$, and a $V_{OC}$ of 0.848 V. Further spin-coating lowered the efficiencies of the cells. Although 2-step deposition (spin-coating $PbI_2$ and dipping in the $CH_3NH_3I$ solution) was also performed in air [8], the efficiency was lower compared with that achieved through multiple spin-coatings, as observed in Fig. 6.12(b). It is believed that the $CH_3NH_3PbI_3$ phase permeated into the mesoporous $TiO_2$ layer during the 1 or 2 spin-coatings. After the pores inside of the mesoporous $TiO_2$ were fully filled with the perovskite phase, the perovskite-only layer formed on the mesoporous $TiO_2$ layer during 3 or 4 spin-coatings, which provided the highest efficiency.

The IPCE spectrum of the photovoltaic cell with the $TiO_2/CH_3NH_3PbI_3$/spiro-OMeTAD structure is shown in Fig. 6.15(a). The measurement region is in the range of 300 to 900 nm, and the perovskite $CH_3NH_3PbI_3$ structure exhibits photo-conversion efficiencies between 300 and 800 nm, which almost agrees with the reported energy gaps of 1.51 [26] and 1.61 eV [50] (corresponding to 821 and 770 nm, respectively) for the $CH_3NH_3PbI_3$ phase. This indicates that excitons and/or free charges could be effectively generated in the perovskite layers upon illumination by light.

An energy level diagram of $TiO_2/CH_3NH_3PbI_3$ photovoltaic cells is shown in Fig. 6.15(b). The electronic charge generation is caused by light irradiation from the FTO substrate side. The $TiO_2$ layer receives the electrons from the $CH_3NH_3PbI_3$ crystal, and the electrons are transported to the FTO. The holes are transported to an Au electrode through spiro-OMeTAD. Experimental evidence for photo-generated free carriers, which would enhance the carrier transport, was reported [51]. In the present work, the samples were prepared in air, which may have resulted in the reduction of their photostability and thermal stability. Higher quality perovskite crystals should be prepared in future works.

From the TEM results, the size distributions of $TiO_2$ nanoparticles were observed, indicating a microcrystalline structure, as shown in Fig. 6.14(b), and there seems to be no special crystallographic relation at the interface [37]. The interface between the $TiO_2$ and $CH_3NH_3PbI_3$ phases would not be perfectly connected over the large area. The cell prepared using 4 spin-coatings provided the highest efficiency, and had the interfacial microstructure shown in Fig. 6.15(c). In the cell prepared using 10 spin-coatings, the layer thickness of the $CH_3NH_3PbI_3$ phase was too large, which resulted in an increase in the inner electronic resistance and a decrease in the efficiency.

In summary, structural analysis of $TiO_2/CH_3NH_3PbI_3$ indicated phase transformation of the perovskite structure from the tetragonal to the cubic system by partial separation of $PbI_2$ from the $CH_3NH_3PbI_3$ phase during annealing, which was confirmed by the reduction of the unit cell volume of the perovskite phase and resulted in improvement in the photovoltaic properties of the devices. The effects of a multiple spin-coating method were also investigated, which increased the efficiency in the case of 4 spin-coatings. This improvement could be due to the perfect coverage and optimum thickness of the perovskite phase on the porous $TiO_2$ layer. Moreover, the lattice constant and crystallite sizes of the $CH_3NH_3PbI_3$ increased and decreased, respectively, which indicates the microstructural differences in the perovskite phase between the inside of and above the porous $TiO_2$.

## 6.7 Enlargement of cell

Enlargement the cell area will be especially necessary in enabling the use of such perovskite devices as commercial solar cell panels [52]. The photovoltaic properties of perovskite-type solar cells with a substrate size of 70 mm × 70 mm were investigated [53]. The photovoltaic devices consisted of a $CH_3NH_3PbI_3$ compound layer, $TiO_2$ electron transport layers and a spiro-OMeTAD hole-transport layer, prepared by a simple spin-coating technique. The effect of the distance from the center of the cell on conversion efficiency was investigated based on light-induced *J-V* curves and IPCE measurements. A photograph of a perovskite solar cell measuring 70 mm × 70 mm and a schematic illustration of the arrangement of Au electrodes on the substrate are shown in Fig. 6.16(a) and (b), respectively.

Measurements were taken of the short-circuit current density, open-circuit voltage, fill factor and photoconversion efficiency of the present $TiO_2/CH_3NH_3PbI_3$ cell as a function of the distance from the center of the cell and are shown in Fig. 6.17(a–d). The highest efficiency was obtained for the electrode at 12.7 mm from the cell center, which provided a photoconversion efficiency of 3.15 %, $J_{SC}$ of 13.0 mA cm$^{-2}$, an open-circuit voltage $V_{OC}$ of 0.653 V and a fill factor *FF* of 0.371. As a result of a long exciton diffusion length [54], the $J_{SC}$ values were almost constant (~ 12 mA cm$^{-2}$) for all elec-

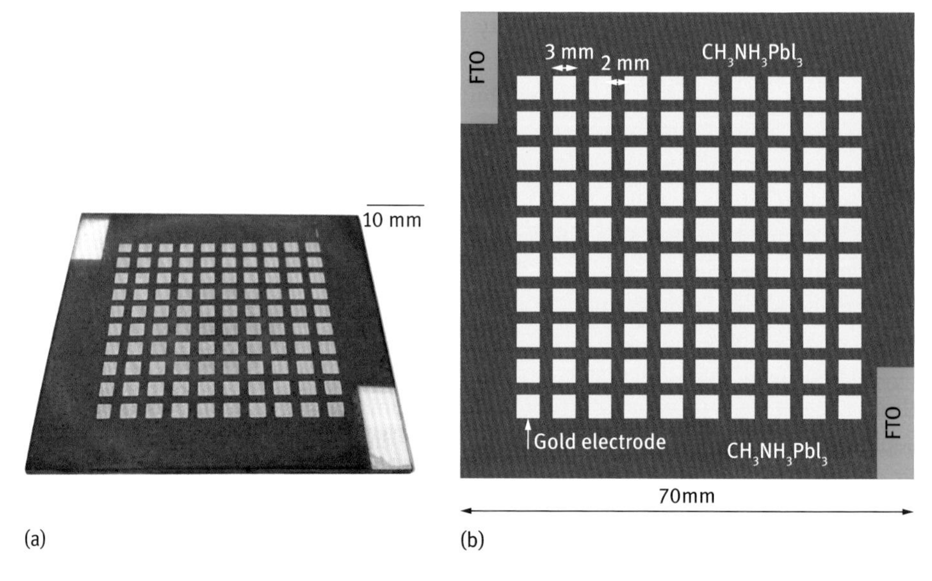


Fig. 6.16: (a) Photograph of perovskite solar cell measuring 70 mm × 70 mm. (b) Schematic illustration of the arrangement of Au electrodes on the substrate.

trodes on the cell, as is shown in Fig. 6.17(a). Although the *FF* value slightly decreased as *d* increased, the deviation was not large, as is shown in Fig. 6.17(c). In contrast, the value of $V_{OC}$ was fairly dependent on *d*, as is shown in Fig. 6.17(b), which resulted in decreased efficiency, as shown in Fig. 6.17(d). The dependency of $V_{OC}$ on *d* values might be related to the thickness of the $CH_3NH_3PbI_3$ prepared on the large substrate by spin-coating. The low *FF* and $V_{OC}$ values could also be related with the coverage ratio of $CH_3NH_3PbI_3$ at the $TiO_2/CH_3NH_3PbI_3$ interface, and the application of further multiple spin-coatings of $CH_3NH_3PbI_3$ layers on the mesoporous $TiO_2$ layer would improve the coverage of $CH_3NH_3PbI_3$ on the $TiO_2$, which would lead to the increase of the efficiency of the cells.

Fig. 6.17(e) shows the IPCE spectra of electrodes at 4.2, 12.7 and 22.8 mm from the cell center. All of the spectra show similar changes with wavelength, which agrees with the $J_{SC}$ results shown in Fig. 6.17(a). The measurement region was in the range of 300–800 nm, and the perovskite $CH_3NH_3PbI_3$ structure showed photoconversion within the whole range, which almost agrees with the reported energy gaps for the $CH_3NH_3PbI_3$ phase. Control of the energy levels of the conduction band and valence band is important for carrier transport in the cell. The efficiencies obtained for the present cells are lower than the previously reported values. It seems to be difficult to control the uniformity of layer thickness and interfacial structure using spin-coating techniques. In the present work, the samples were prepared in air, which may have also resulted in a reduction in the efficiency of the present cells, and perovskite crystals of a higher quality and with a smooth surface should be prepared in future works.

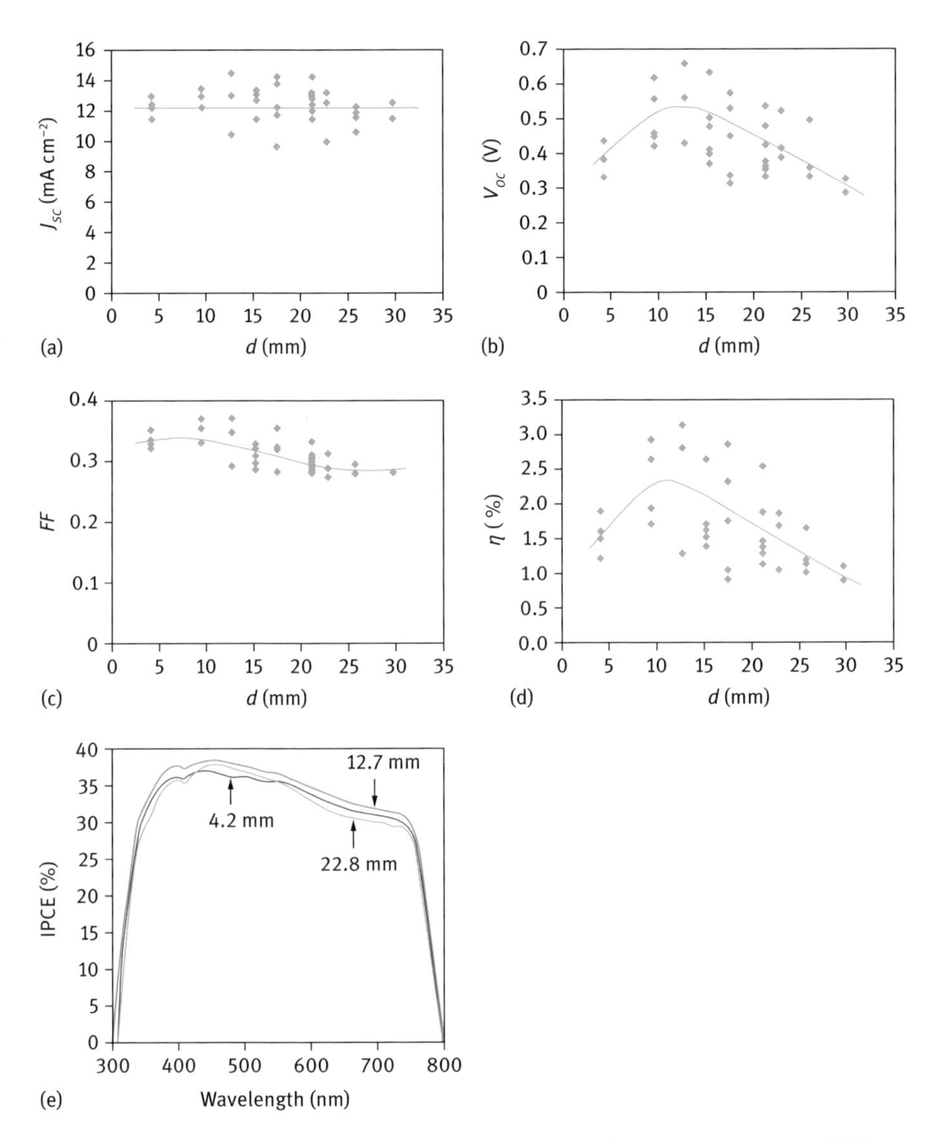


Fig. 6.17: Measurements for the (a) short-circuit current density; (b) open-circuit voltage; (c) fill factor and (d) conversion efficiency of $TiO_2/CH_3NH_3PbI_3$ cell as a function of the distance from the center of the cell. (e) IPCE spectra of the same cell.

In summary, $TiO_2/CH_3NH_3PbI_3$-based photovoltaic devices with a substrate size of 70 mm were fabricated by a spin-coating method using a mixed solution. The photovoltaic properties of the cells and the size effects were investigated by looking at $J$–$V$ characteristics and IPCE measurements, and the dependence of their photoconversion efficiencies on the distance from the center of the cell was investigated. Almost constant values for short-circuit current density were obtained over a large area, owing to the long exciton diffusion length of the $CH_3NH_3PbI_3$ compound. The open-circuit voltage was fairly dependent on the distance from the center of the cell, which resulted in a change in conversion efficiency. Smoothing and optimizing the layer thickness and structure could be important for improving the device's performance.

## 6.8 Electron transport layers

Electron-transport layers (ETLs) such as $TiO_2$ are also important for $CH_3NH_3PbI_3$-based photovoltaic devices. Here, niobium(V) ethoxide was selected as an additive for $TiO_2$ [55]. When niobium (Nb) with 5 valence electrons is introduced to Ti with 4 valence electrons, excess electrons are introduced in the 3*d* band and can play the role of a donor. Since the impurity level of $TiO_2$ is shallow in the energy band gap, transparency can be preserved after Nb doping [56–60]. In addition, the radius of the Nb ion is close to that of the Ti ion, which results in a solid solution of Ti and Nb in the anatase $TiO_2$ structure. Nb-added $TiO_2$ is denoted as $Ti(Nb)O_2$ here.

The XRD patterns of $TiO_2$ and $Ti(Nb)O_2$ thin films on the FTO substrate are shown in Fig. 6.18(a). The diffraction peaks of $TiO_2$ 101 were observed, and it was found that its intensity increased upon Nb-doping. The XRD data indicates that the d-spacing of $Ti(Nb)O_2$ (1.802 Å) is almost the same as that of $TiO_2$ (1.807 Å). The crystallite size seems to increase slightly upon the addition of Nb (28 nm) to $TiO_2$ (24 nm). A scanning electron microscopy (SEM) image of the $Ti(Nb)O_2$ thin film is shown in Fig. 6.18(b), and the image indicates several particles with sizes of ca. 1 µm on the smooth surface. Elemental mapping images of Ti and Nb using SEM with energy-dispersive X-ray spectroscopy (EDX) are shown in Fig. 6.18(c) and 6-18(d), respectively, and they indicate that Ti and Nb are widely distributed in the thin films. The composition ratio of Ti:Nb was measured to be 1.00 : 0.10 from the EDX analysis. The particles observed in Fig. 6.18(b) correspond to the Nb-rich phase, as observed in Fig. 6.18(d), which resulted in an Nb-rich (*ca.* 9 atomic %) composition compared with the preparation composition (*ca.* 5 atomic %). From XRD analysis, no diffraction peak corresponding to Nb and $Nb_2O_5$ was observed.

The sheet resistances of $TiO_2$ and $Ti(Nb)O_2$ thin films were measured to be $1.70 \times 10^6$ and $4.17 \times 10^4$ Ω/□, respectively. The sheet resistance significantly decreased upon the addition of Nb. The $J$–$V$ characteristics of $Ti(Nb)O_2/CH_3NH_3PbI_3$/spiro-OMeTAD photovoltaic cells under illumination are shown in Fig. 6.19(a). The detailed parameters of the best device are listed in Table 6.17. The $Ti(Nb)O_2/CH_3NH_3PbI_3$ photovoltaic cell provided an $\eta$ of 6.63 %, an *FF* of 0.416, a $J_{SC}$ of 20.8 mA cm$^{-2}$, and a $V_{OC}$ of 0.768V. The $J_{SC}$ value was especially improved through the addition of Nb, which resulted in increased conversion efficiency. The averaged efficiency ($\eta_{ave}$) of the three electrodes on the cells was 6.46 %, as listed in Table 6.17.

Differential absorption spectra of $FTO/TiO_2$ and $FTO/Ti(Nb)O_2$ thin films after subtracting the spectrum of the FTO substrate are shown in Fig. 6.19(b). The two absorp-

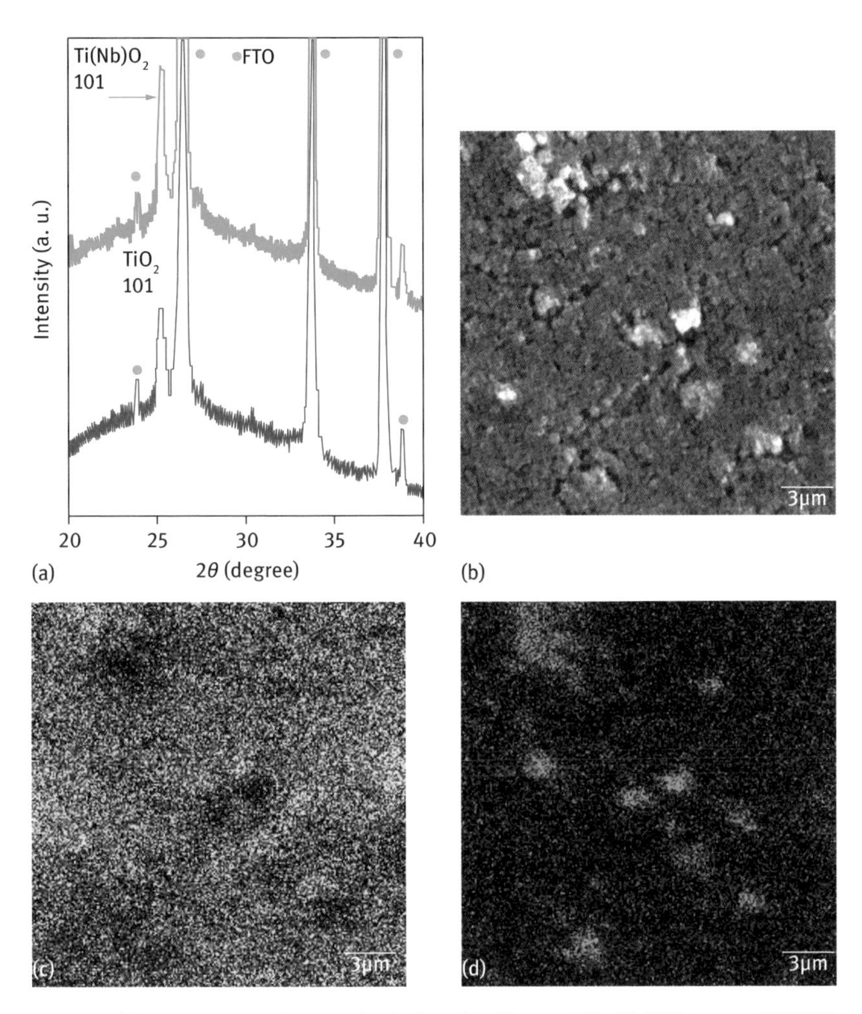


Fig. 6.18: (a) XRD patterns of $TiO_2$ and $Ti(Nb)O_2$ thin films on FTO. (b) SEM image of $Ti(Nb)O_2$ thin film. Elemental mapping of (c) Ti (L$\alpha$) and (d) Nb (L$\alpha$).

Table 6.17: Measured parameters of $Ti(Nb)O_2/CH_3NH_3PbI_3$ cells.

| ETL | $J_{SC}$ (mA cm$^{-2}$) | $V_{OC}$ (V) | $FF$ | $\eta$ (%) | $\eta_{Ave}$ (%) |
|---|---|---|---|---|---|
| $TiO_2$ | 14.6 | 0.796 | 0.478 | 5.56 | 5.03 |
| $Ti(Nb)O_2$ | 20.8 | 0.768 | 0.416 | 6.63 | 6.46 |

tion spectra seem to be nearly identical. Based on the indirect energy band structure [61], the energy gaps for $TiO_2$ and $Ti(Nb)O_2$ were measured to be 3.54 and 3.52 eV from Fig. 6.19(b), respectively, which indicates that the energy gaps are almost same for $TiO_2$ and $Ti(Nb)O_2$.

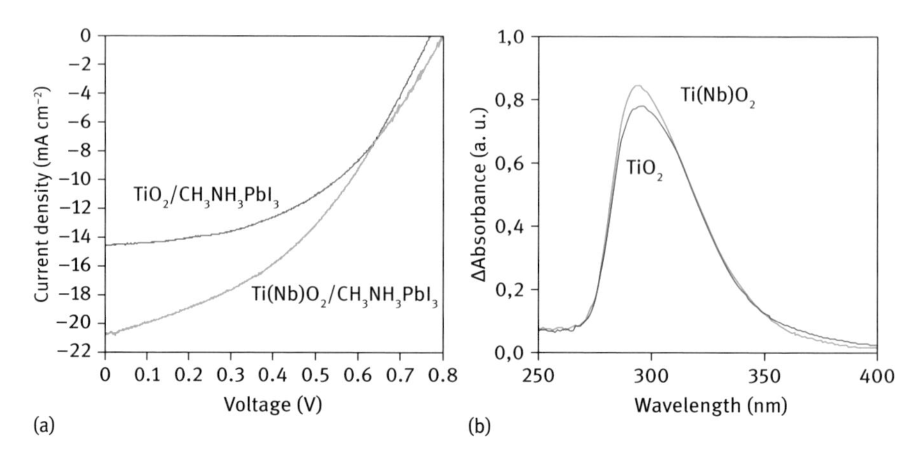


Fig. 6.19: (a) *J–V* characteristic of $Ti(Nb)O_2/CH_3NH_3PbI_3$ photovoltaic cells. (b) Differential absorption spectra of $TiO_2$ and Ti(Nb)O thin films.

The IPCE of the cells was also investigated, and the external quantum efficiency (EQE) and internal quantum efficiency (IQE) were measured using a spectral response system. The EQE spectra of the photovoltaic cells with a $Ti(Nb)O_2/CH_3NH_3PbI_3$/spiro-OMeTAD structure are shown in Fig. 6.20(a). The perovskite $CH_3NH_3PbI_3$ phase shows photoconversion efficiencies between 300 and 800 nm. The addition of Nb into the $TiO_2$ layer resulted in the perovskite $CH_3NH_3PbI_3$ structure showing high EQE values of ca. 60 % at 500 ~ 600 nm and ca. 5 % at 800 nm, while the EQE was 0 % for ordinary $TiO_2$ at 800 nm. The IQE spectra of $Ti(Nb)O_2/CH_3NH_3PbI_3$/spiro-OMeTAD cells were calculated from the EQE and reflectance, as shown in Fig. 6.20(b). The IQE of both cells increased in the 500 ~ 800 nm range, which indicates that the suppression of light reflection in the 500 ~ 800 nm range would increase the conversion efficiencies of the cells. High IQE values of ca. 70 % are observed in the 500~ 600 nm range upon the addition of Nb in the $TiO_2$ layer.

Two mechanisms could be considered to explain the decrease in the sheet resistances of $TiO_2$ upon the addition of Nb. The first is Nb doping at the Ti sites in the $TiO_2$ structure. From the XRD and differential absorption results of Fig. 6.18(a) and (b), the $TiO_2$ phase still preserved the crystal structure, energy gap, and transparency of anatase $TiO_2$. In addition, a small amount of Nb atoms are widely distributed in the $TiO_2$ phase, as can be observed in the SEM-EDX of Fig. 6.18(c) and (d), which could indicate a solid solution of Ti and Nb in the anatase $TiO_2$ structure. Then, the excess electrons of Nb could be introduced into the 3d band of Ti and act as a donor [60]. Another possible mechanism is carrier transport enhancement by the formation of Nb-rich particles in the $TiO_2$ layer, as observed by SEM-EDX. Nanoparticles in the electron-transport and hole-transport layers could enhance the carrier transport [62, 63], and the Nb-rich particles might also contribute to carrier transport. Both

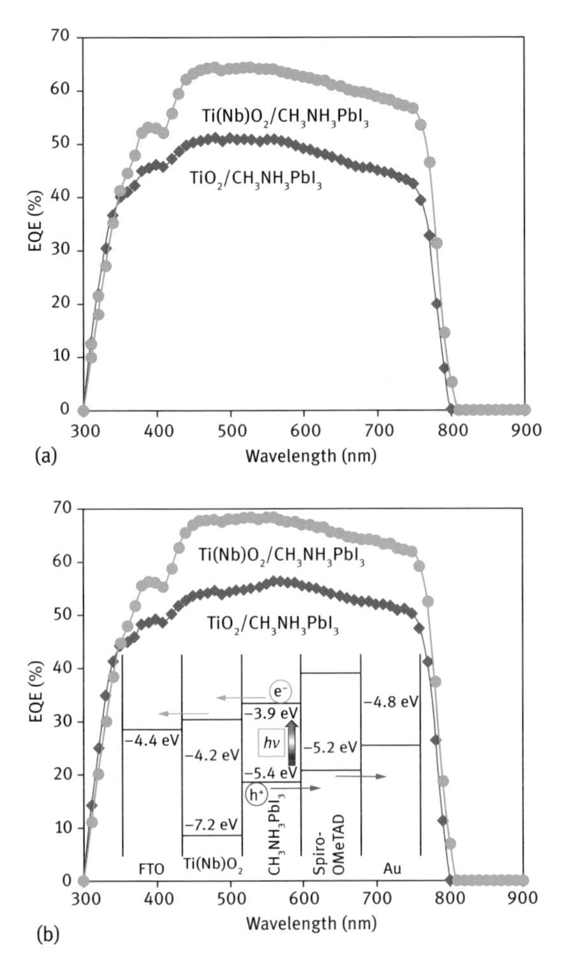


Fig. 6.20: (a) EQE and (b) IQE spectra of $Ti(Nb)O_2/CH_3NH_3PbI_3$ cells.

mechanisms would result in an increase in carrier concentration, and in the improvement of photoconversion efficiency through the increase in $J_{SC}$.

In summary, $Ti(Nb)O_2/CH_3NH_3PbI_3$-based photovoltaic devices were fabricated by a spin-coating method using a mixture solution of niobium(V) ethoxide, and the effects of Nb addition into the $TiO_2$ layer were investigated. By adding a simple solution of niobium(V) ethoxide to the $TiO_2$ precursor solutions, the sheet resistance of the $Ti(Nb)O_2$ thin film decreased, and the $J_{SC}$ value increased, which resulted in an increase in conversion efficiency.

## 6.9 Halogen doping to $CH_3NH_3PbI_3$

The effects of Cl-doping $CH_3NH_3PbI_3$ using a mixture solution of perovskite compounds on microstructures and photovoltaic properties have been investigated in previous work [64]. The $J$–$V$ characteristics of $TiO_2/CH_3NH_3PbI_{3-x}Cl_x$/spiro-OMeTAD photovoltaic cells under illumination are shown in Fig. 6.21(a), which indicates the effect of Cl-doping to the $CH_3NH_3PbI_3$ layer. The photovoltaic parameters measured for $TiO_2/CH_3NH_3PbI_{3-x}Cl_x$ cells are summarized as Table 6.18. The $CH_3NH_3PbI_3$ cell provided a power conversion efficiency ($\eta$) of 6.16 %, and the averaged efficiency ($\eta_{ave}$) of four electrodes on the cells was 5.53 %, as listed in Table 6.18. The highest efficiency was obtained for the $CH_3NH_3PbI_{2.88}Cl_{0.12}$ cell, which provided an $\eta$ of 8.16 %, an $FF$ of 0.504, a $J_{SC}$ of 18.6 mA cm$^{-2}$, and a $V_{OC}$ of 0.869 V. As the Cl composition increased, the $J_{SC}$ and $V_{OC}$ decreased, as shown in Fig. 6.21(b) and Table 6.18. The energy gaps ($E_g$) of $CH_3NH_3PbI_3$, $CH_3NH_3PbI_{2.88}Cl_{0.12}$ and $CH_3NH_3PbI_{1.8}Cl_{1.2}$ were estimated to be 1.578, 1.590 and 1.593, respectively, from the optical absorption, which indicated that the energy gap of $CH_3NH_3PbI_3$ increased by the Cl-doping.

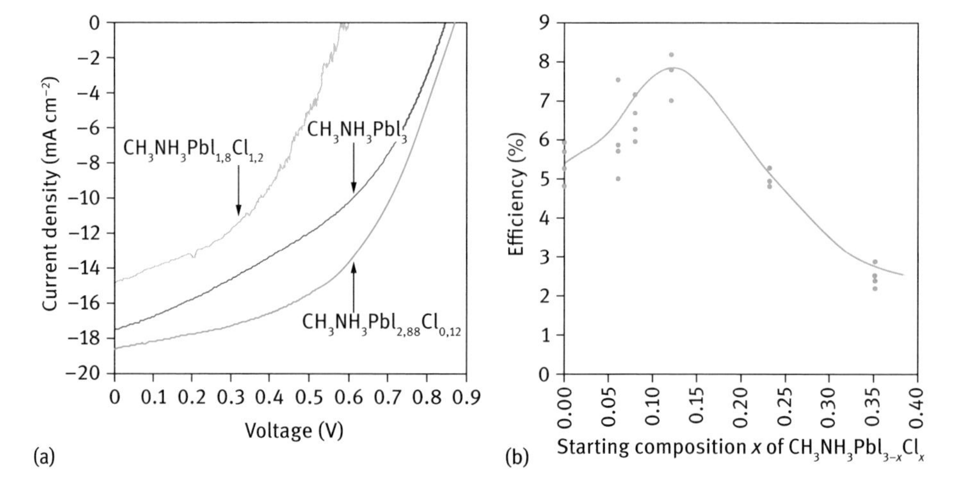


Fig. 6.21: (a) $J$–$V$ characteristic of $TiO_2/CH_3NH_3PbI_{3-x}Cl_x$ photovoltaic cells. (b) The highest and averaged efficiencies of the four electrodes of the cells.

The XRD patterns of $CH_3NH_3PbI_{3-x}Cl_x$ thin films on the FTO/$TiO_2$ are shown in Fig. 6.22(a). The temperature at which the XRD measurements were taken was ~ 292 K. The diffraction peaks can be indexed by cubic and tetragonal crystal systems for $CH_3NH_3PbI_3$ and $CH_3NH_3PbI_{1.8}Cl_{1.2}$ films, respectively. Although the deposited films are a single perovskite phase, broader diffraction peaks due to $PbI_2$ compound appeared in the $CH_3NH_3PbI_3$ film, as shown in Fig. 6.22(a). Figure 6.22(b) shows enlarged XRD patterns at $2\theta$ of ~ 28.5 °. A diffraction peak of 200 for the $CH_3NH_3PbI_3$ split into

Table 6.18: Measured photovoltaic parameters of $TiO_2/CH_3NH_3PbI_{3-x}Cl_x$ cells.

| Preparation composition | $J_{SC}$ (mA cm$^{-2}$) | $V_{OC}$ (V) | *FF* | $\eta$ (%) | $\eta_{ave}$ (%) |
|---|---|---|---|---|---|
| $CH_3NH_3PbI_3$ | 17.5 | 0.844 | 0.416 | 6.16 | 5.53 |
| $CH_3NH_3PbI_{2.94}Cl_{0.06}$ | 17.7 | 0.871 | 0.487 | 7.53 | 6.02 |
| $CH_3NH_3PbI_{2.92}Cl_{0.08}$ | 18.1 | 0.825 | 0.478 | 7.14 | 6.52 |
| $CH_3NH_3PbI_{2.88}Cl_{0.12}$ | 18.6 | 0.869 | 0.504 | 8.16 | 7.77 |
| $CH_3NH_3PbI_{2.77}Cl_{0.23}$ | 13.9 | 0.865 | 0.440 | 5.29 | 4.97 |
| $CH_3NH_3PbI_{2.65}Cl_{0.35}$ | 11.7 | 0.709 | 0.347 | 2.87 | 2.51 |
| $CH_3NH_3PbI_{1.80}Cl_{1.20}$ | 14.8 | 0.598 | 0.436 | 3.87 | 2.00 |

diffraction peaks of 004/220 for the $CH_3NH_3PbI_{1.8}Cl_{1.2}$ by the heavy Cl-doping, which indicates that a structural transformation took place from the cubic to tetragonal crystal system [21]. The heavy Cl-doping suppressed the formation of $PbI_2$, and no $PbCl_2$ was detected for the $CH_3NH_3PbI_{1.8}Cl_{1.2}$. For the $CH_3NH_3PbI_{2.88}Cl_{0.12}$, a small shoulder is observed just to the left of the 200 reflection in Fig. 6.22(b), which could be due to the pseudo-cubic structure between the cubic and tetragonal phases. The structural parameters measured for $CH_3NH_3PbI_{3-x}Cl_x$ are summarized in Table 6.19.

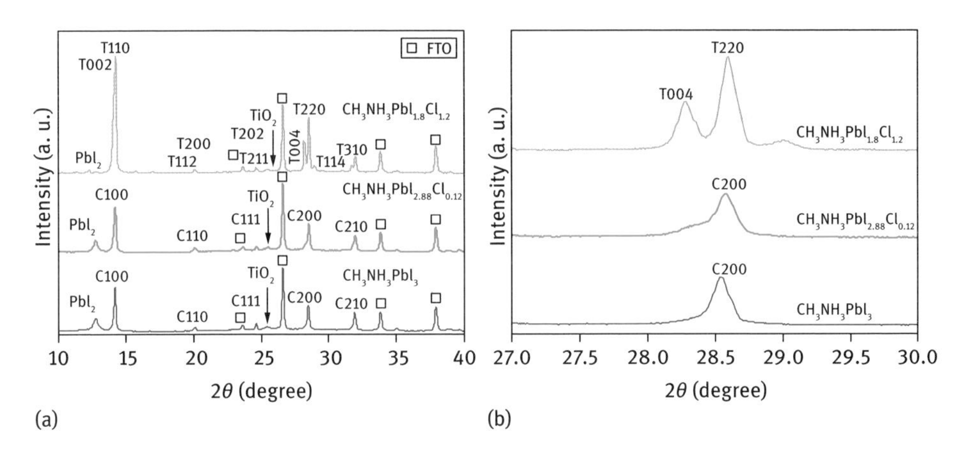


Fig. 6.22: (a) XRD patterns of $CH_3NH_3PbI_{3-x}Cl_x$ thin films. (b) Enlarged XRD patterns at $2\theta$ of ~ 28.5 °

Figure 6.23(a) is a SEM image of $TiO_2/CH_3NH_3PbI_{2.88}Cl_{0.12}$, which indicates particles of ~ 10 μm. Elemental mapping images of Pb, I, and Cl by SEM-EDX are shown in Fig. 6.23(b–d). The elemental mapping images indicate that the particles observed in Fig. 6.23(a) correspond to the $CH_3NH_3PbI_{3-x}Cl_x$ phase, and the composition ratio of Pb:I:Cl was calculated to be 1.00 : 2.70 : 0.11 from the EDX spectrum using the Pb M$\alpha$, I L$\alpha$ and Cl K$\alpha$ lines with background correction by normalizing the spectrum peaks on the atomic concentration of Pb. This result indicates that I might be deficient from the starting composition of $CH_3NH_3PbI_{2.88}Cl_{0.12}$, and the deficient I might increase the

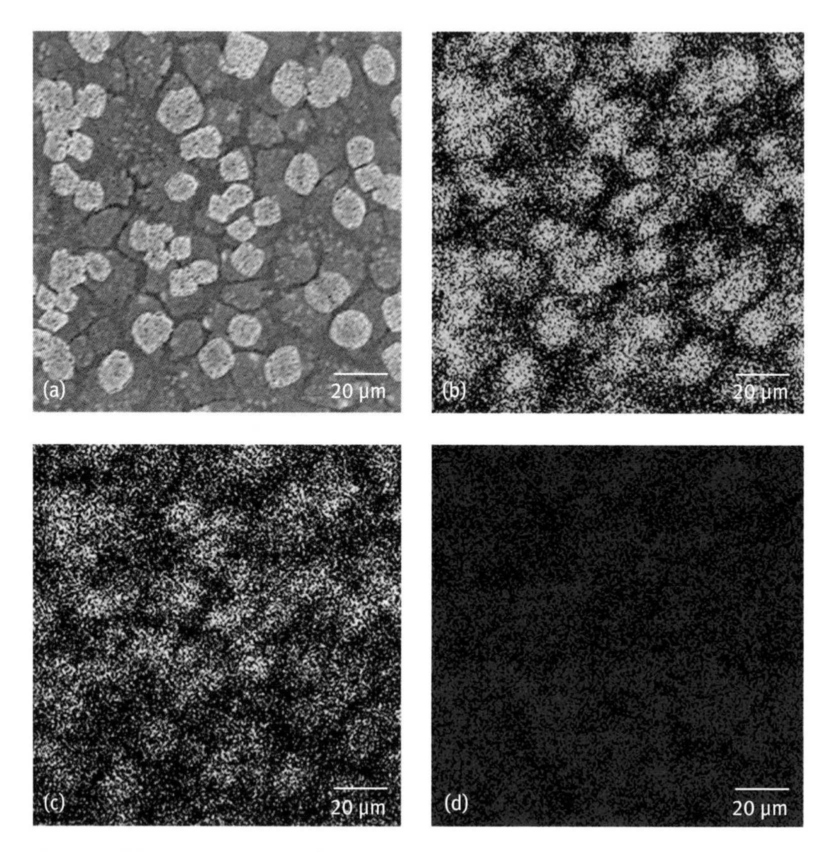


Fig. 6.23: (a) SEM image of $TiO_2/CH_3NH_3PbI_{2.88}Cl_{0.12}$. Elemental mapping images of (b) Pb Mα line, (c) I Lα line and (d) Cl Kα line.

hole concentration. The $CH_3NH_3PbI_3$ crystals have perovskite structures, and structurally transition from the tetragonal to cubic system upon heating at ~ 330 K [27–29].

The XRD results in Fig. 6.22 indicate a phase transformation of the $CH_3NH_3PbI_3$ perovskite structure from tetragonal to cubic system by partial separation of $PbI_2$ from $CH_3NH_3PbI_3$ phase during annealing [37], which corresponds to the decrease of the unit cell volume of the cubic perovskite phase from the ordinary 261.0 to the present 244.42 Å$^3$, as listed in Table 6.19. The SEM-EDX results show that the site occupancies of I atom could be less than 1, which might also reduce the unit cell volume of the crystal. The conversion efficiencies were reported to be improved by the tetragonal to cubic transformation [37].

The XRD pattern indicates that a splitting of diffraction reflections of 200 to 004/220 occurs from heavy Cl-doping. This indicates a reduction in crystal symmetry from the cubic to tetragonal phase, resulting in the reduction of conversion efficiencies. When a small amount of Cl was doped in the $CH_3NH_3PbI_3$ phase, the cubic structure was still preserved as the pseudo-cubic phase. The doped Cl atoms could

Table 6.19: Measured and reported structural parameters of $CH_3NH_3PbI_{3-x}Cl_x$. $V$: unit cell volume. $Z$: number of chemical unit in the unit cell.

| Preparation composition | Crystal system | Lattice constant (Å) | $V$ (Å$^3$) | $Z$ | $V/Z$ (Å$^3$) |
|---|---|---|---|---|---|
| $CH_3NH_3PbI_3$ | Cubic | $a = 6.2524$ | 244.42 | 1 | 244.42 |
| $CH_3NH_3PbI_{2.88}Cl_{0.12}$ | Pseudo-cubic | $a = 6.2446$ | 243.51 | 1 | 243.51 |
| $CH_3NH_3PbI_{1.80}Cl_{1.20}$ | Tetragonal | $a = 8.8255$<br>$c = 12.6180$ | 982.81 | 4 | 245.70 |

lengthen the diffusion length of exitons [7, 54]. which would result in an increase in efficiency.

The EQE spectra of a photovoltaic cell with the $TiO_2/CH_3NH_3PbI_{3-x}Cl_x$/spiro-OMeTAD structure is shown in Fig. 6.24(a). The perovskite $CH_3NH_3PbI_3$ phase shows photo-conversion efficiencies between 300 and 800 nm. In the study presented here, the energy gap of the $CH_3NH_3PbI_3$ phase increased from 1.578 to 1.590 eV through Cl-doping, which could contribute the increase of open-circuit voltage. The IQE spectra of $TiO_2/CH_3NH_3PbI_3$ and $TiO_2/CH_3NH_3PbI_{2.92}Cl_{0.08}$ were calculated from the EQE and transmittance, as shown in Fig. 6.24(b). The IQE of both cells increased to over 40 % in the range of 500 ~ 800 nm, which indicates that an increase of the light absorption in the range of 500 ~ 800 nm would increase the conversion efficiencies of $TiO_2/CH_3NH_3PbI_{3-x}Cl_x$/spiro-OMeTAD cells.

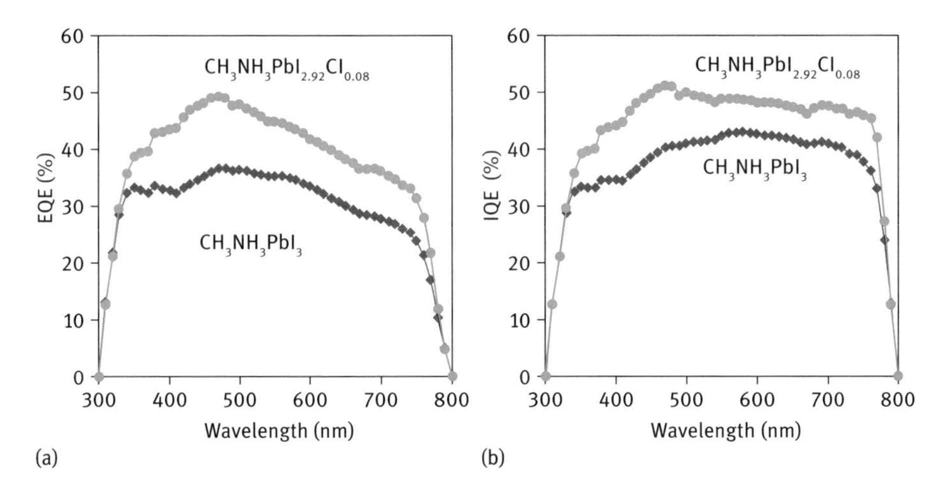


Fig. 6.24: (a) EQE and (b) IQE spectra of $CH_3NH_3PbI_3$ and $CH_3NH_3PbI_{2.92}Cl_{0.08}$ cells.

In summary, $TiO_2/CH_3NH_3PbI_{3-x}Cl_x$-based photovoltaic devices were fabricated by a spin-coating method using a mixture solution, and the effects of the addition of $PbCl_2$ to perovskite $CH_3NH_3PbI_3$ precursor solutions on the material's photovoltaic properties were investigated. The microstructure analysis showed phase transformation of the perovskite structure from cubic to tetragonal system by heavy Cl-doping to the $CH_3NH_3PbI_3$ phase. A small amount of Cl-doping ($CH_3NH_3PbI_{2.9}Cl_{0.1}$) at the I sites improved the efficiencies to ~ 8 %, which could be due to the preservation of the cubic structure and expansion of energy gap and diffusion length of exitons. Both EQE and IQE improved to ~ 40 % in the range of 300 ~ 800 nm by a small amount of Cl-doping, and the IQE results indicate that suppressing reflection in the range of 500 ~ 800 nm would further improve the efficiency.

## 6.10 Metal doping to $CH_3NH_3PbI_3$

The photovoltaic properties of solar cells strongly depend on the crystal structures and the compositions of the perovskite compounds. Halogen and metal atom doping such as chlorine (Cl)/bromine (Br) and tin (Sn) at the iodine (I) and lead (Pb) sites in the perovskite compounds have been studied in previous works [13–15, 65, 66]. In particular, studies on metal atom doping at Pb sites have been carried out for Pb-free devices [66], and detailed research into metal doping at Pb sites is interesting for both Pb-free devices and effects on photovoltaic properties.

Table 6.20: Measured photovoltaic parameters of $TiO_2/CH_3NH_3Pb_{1-x}Sb_xI_3$ cells. Preparation compositions of Sb are indicated by $x$.

| Sb ($x$) | $J_{SC}$ (mA cm$^{-2}$) | $V_{OC}$ (V) | $FF$ | $\eta$ (%) | $\eta_{Ave}$ (%) |
|---|---|---|---|---|---|
| 0.00 | 17.0 | 0.758 | 0.509 | 6.56 | 6.37 |
| 0.01 | 16.0 | 0.789 | 0.534 | 6.74 | 6.41 |
| 0.02 | 16.9 | 0.792 | 0.518 | 6.94 | 6.72 |
| 0.03 | 19.2 | 0.843 | 0.560 | 9.07 | 8.47 |
| 0.05 | 15.7 | 0.755 | 0.575 | 6.82 | 5.61 |
| 0.07 | 14.7 | 0.692 | 0.502 | 5.11 | 4.07 |
| 0.10 | 12.1 | 0.630 | 0.476 | 3.63 | 3.27 |
| 0.15 | 13.1 | 0.570 | 0.402 | 3.00 | 2.85 |

The purpose of the study presented here is to investigate the photovoltaic properties and microstructures of photovoltaic devices with perovskite-type $CH_3NH_3Pb_{1-x}Sb_xI_3$ compounds, which were prepared using a simple spin-coating technique in air. Antimony (Sb) is a group 15 element, and was expected to work as an electronic donor at the sites of the group 14 element lead (Pb). The effects of adding $SbI_3$ using a mixture solution of perovskite compounds on the photovoltaic properties and microstructures of the perovskite crystal were investigated [67, 68].

The $J$–$V$ characteristics of the $TiO_2/CH_3NH_3Pb_{1-x}Sb_xI_3$/spiro-OMeTAD photovoltaic cells under illumination are shown in Fig. 6.25(a), indicating the effects

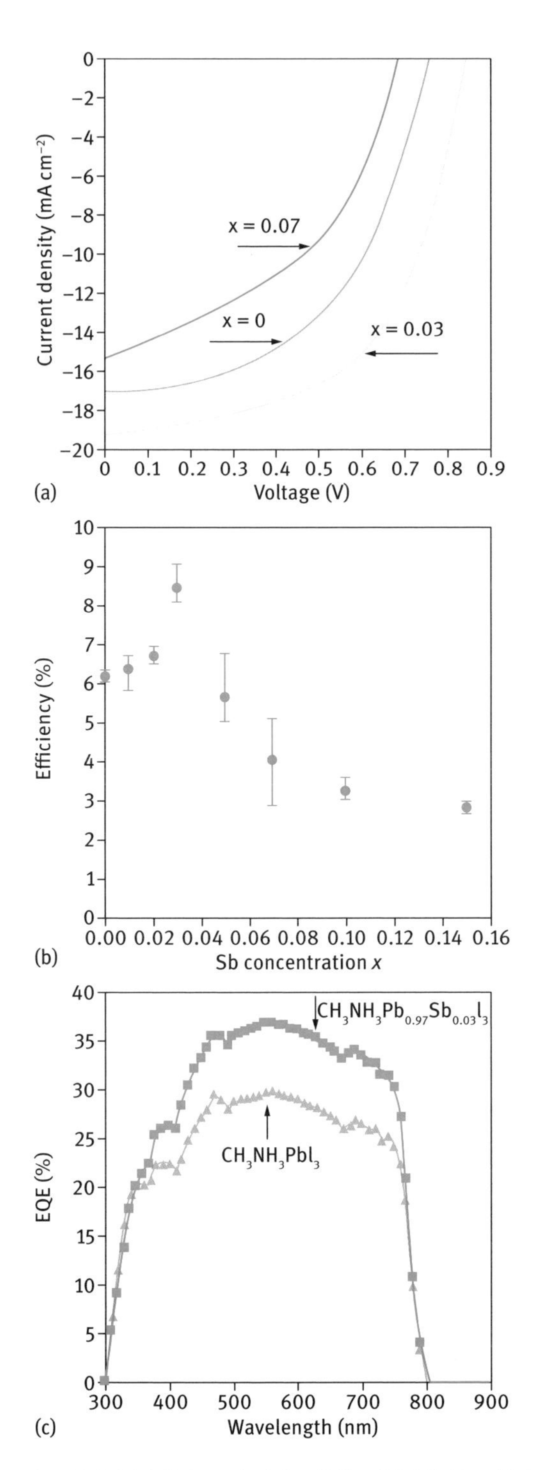


Fig. 6.25: (a) *J–V* characteristic of $TiO_2/CH_3NH_3Pb_{1-x}Sb_xI_3$ photovoltaic cells. (b) Conversion efficiencies of $CH_3NH_3Pb_{1-x}Sb_xI_3$ as a function of Sb concentration. (c) IPCE spectra of $CH_3NH_3PbI_3$ and $CH_3NH_3Pb_{0.97}Sb_{0.03}I_3$ cells.

of the addition of Sb to $CH_3NH_3PbI_3$. The measured photovoltaic parameters of $TiO_2/CH_3NH_3Pb_{1-x}Sb_xI_3$ cells are summarized in Table 6.20. The $CH_3NH_3PbI_3$ cell provided a power conversion efficiency of 6.56 %, and the averaged efficiency of the four electrodes on the cells was 6.37 %, as listed in Table 6.20. The highest efficiency was obtained for the $CH_3NH_3Pb_{0.97}Sb_{0.03}I_3$ cell, which provided an $\eta$ of 9.07 %, an *FF* of 0.560, a $J_{SC}$ of 19.2 mA $cm^{-2}$, and a $V_{OC}$ of 0.843V. As the $x$ value (preparation composition of Sb) increased, the efficiencies decreased, as shown in Fig. 6.25(b) and Table 6.20. IPCE spectra of the $CH_3NH_3PbI_3$ and $CH_3NH_3Pb_{0.97}Sb_{0.03}I_3$ cells are shown in Fig. 6.25(c). The perovskite $CH_3NH_3Pb_{1-x}Sb_xI_3$ demonstrated a photoconversion efficiency between 300 and 800 nm. The IPCE improved over ~ 30 % in the range of 400~ 750 nm by adding a small amount of Sb.

XRD patterns of $CH_3NH_3Pb_{1-x}Sb_xI_3$ cells on the $FTO/TiO_2$ are shown in Fig. 6.26(a). The diffraction peaks can be indexed by a cubic crystal system ($Pm\bar{3}m$) for the $CH_3NH_3Pb_{1-x}Sb_xI_3$ thin films. Although the deposited films are a single perovskite structure, broader diffraction peaks due to the $PbI_2$ compound appeared in the $CH_3NH_3PbI_3$ film, as shown in Fig. 6.26(a). Most $PbI_2$ was not detected for $CH_3NH_3Pb_{1-x}Sb_xI_3$ ($x > 0.05$) cells, which indicated the addition of Sb suppressed the formation of $PbI_2$ with. Fig. 6.26(b) shows the lattice constants $a$ of $CH_3NH_3Pb_{1-x}Sb_xI_3$ as a function of Sb concentration. A small increase in the lattice constants $a$ are observed for $x = 0.03$ and 0.05, and the further addition of Sb decreased the lattice constants. The XRD results for $CH_3NH_3PbI_3$ in Fig. 6.26(a) showed the existence of $PbI_2$ after annealing at 100 °C for 15 min. This could indicate a partial separation of $PbI_2$ from $CH_3NH_3PbI_3$ after annealing, which might also correspond to the smaller lattice constant $a$ (6.266 Å) of the cubic perovskite structure, in comparison to that (6.391 Å) of $CH_3NH_3PbI_3$ single crystal reported in [31].

Two mechanisms could be considered for increasing photoconversion efficiencies. The first is the doping effect of Sb at the Pb sites. Since the ionic valence of Sb is 3, which is higher than that of $Pb^{2+}$, the excess charge of $Sb^{3+}$ would produce carriers in the perovskite structure, which would increase $J_{SC}$. In the second mechanism, $I^-$ ions would be attracted to the I sites by $Sb^{3+}$ with more ionic valence compared with that of $Pb^{2+}$, which would result in the suppression of $PbI_2$ elimination from $CH_3NH_3PbI_3$ and in the increase of lattice constant $a$ in $CH_3NH_3PbI_3$. The suppression of $PbI_2$ would improve the $TiO_2/CH_3NH_3PbI_3$ interfacial structure, which also would improve $V_{OC}$. As the amount of the Sb added increases, the lattice constants would be decreased as an effect of Sb whose ionic size is smaller in comparison to Pb. Since direct evidence of Sb doping at the Pb site or suppression of $PbI_2$ separation was not obtained in the present work, further studies will be needed to determine the precise structure of the perovskite structure. Other elemental dopings at the Pb sites were also reported such as Ge, Tl and In [69, 70].

In summary, $TiO_2/CH_3NH_3Pb_{1-x}Sb_xI_3$-based photovoltaic devices were fabricated, and the effects of the addition of $SbI_3$ to perovskite $CH_3NH_3PbI_3$ precursor solutions on photovoltaic properties were investigated. The microstructures of the devices in-

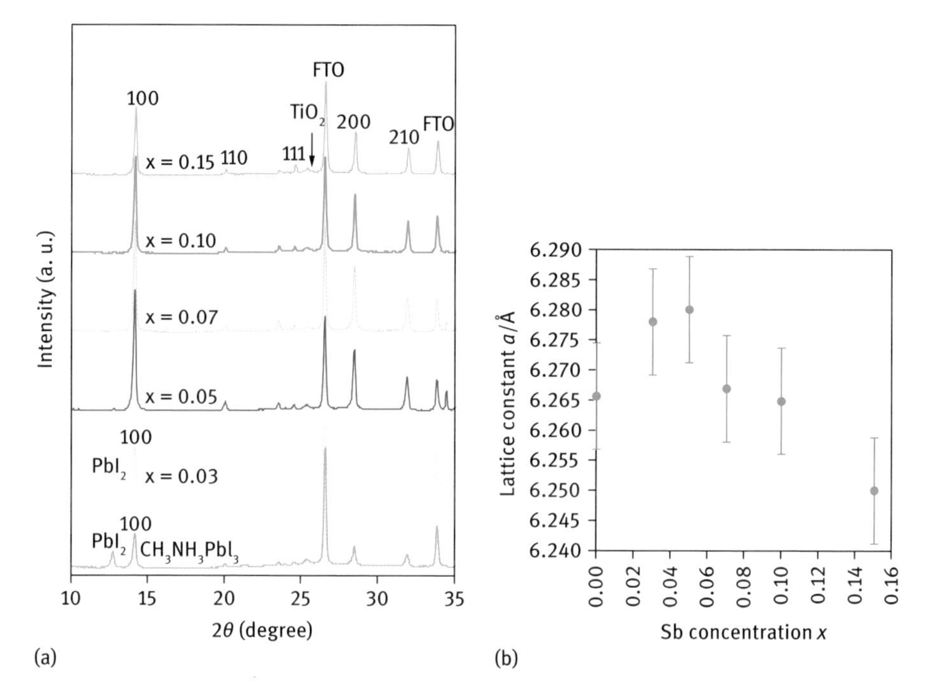


Fig. 6.26: (a) XRD patterns of $CH_3NH_3Pb_{1-x}Sb_xI_3$ solar cells. (b) Lattice constants $a$ of $CH_3NH_3Pb_{1-x}Sb_xI_3$ as a function of Sb concentration.

dicated that the lattice constant of $CH_3NH_3Pb_{1-x}Sb_xI_3$ slightly increased and that the formation of $PbI_2$ was suppressed by the addition of a small amount of Sb, which resulted in an increase in the conversion efficiencies to *ca.* 9 %. IPCE was also improved in the range of 350~770 nm by a small amount of Sb.

## 6.11 Bibliography

[1] Kojima A, Teshima K, Shirai Y, Miyasaka T. Organometal halide perovskites as visible-light sensitizers for photovoltaic cells. *J Am Chem Soc*. 2009; 131: 6050–6051.

[2] Im JH, Lee CR, Lee JW, Park SW, Park NG. 6.5 % efficient perovskite quantum-dot-sensitized solar cell. *Nanoscale*. 2011; 3: 4088–4093.

[3] Kim HS, Lee CR, Im JH, Lee KB, Moehl T, Marchioro A, Moon SJ, Yum JH, Humphry-Baker R, Moser JE, Grätzel M, Park NG. Lead iodide perovskite sensitized all-solid-state submicron thin film mesoscopic solar cell with efficiency exceeding 9 %. *Sci Rep*. 2012; 2: 591–1–7.

[4] Kojima A, Ikegami M, Teshima K, Miyasaka T. Highly luminescent lead bromide perovskite nanoparticles synthesized with porous alumina media. *Chem Lett*. 2012; 41: 397–399.

[5] Lee MM, Teuscher J, Miyasaka T, Murakami TN, Snaith HJ. Efficient hybrid solar cells based on meso-superstructured organometal halide perovskites. *Science*. 2012; 338: 643–647.

[6] Chung I, Lee B, He JQ, Chang RPH, Kanatzidis MG. All-solid-state dye-sensitized solar cells with high efficiency. *Nature*. 2012; 485: 486–489.

[7] Stranks SD, Eperon GE, Grancini G, Menelaou C, Alcocer MJP, Leijtens T, Herz LM, Petrozza A, Snaith HJ. Electron-hole diffusion lengths exceeding 1 micrometer in an organometal trihalide perovskite absorber. *Science*. 2013; 342: 341–344.
[8] Burschka J, Pellet N, Moon SJ, Humphry-Baker R, Gao P, Nazeeruddin MK, Grätzel M. Sequential deposition as a route to high-performance perovskite-sensitized solar cells. *Nature*. 2013; 499: 316–320.
[9] Liu M, Johnston MB, Snaith HJ. Efficient planar heterojunction perovskite solar cells by vapour deposition. *Nature*. 2013; 501: 395–398.
[10] Liu D, Kelly TL. Perovskite solar cells with a planar heterojunction structure prepared using room-temperature solution processing techniques. *Nat Photonics*. 2014; 8: 133–138.
[11] Wang, JTW, Ball JM, Barea EM, Abate A, Alexander-Webber JA, Huang J, Saliba M, Mora-Sero I, Bisquert J, Snaith HJ et al. Low-temperature processed electron collection layers of graphene/$TiO_2$ nanocomposites in thin film perovskite solar cells. *Nano Lett*. 2014; 14: 724–730.
[12] Wojciechowski K, Saliba M, Leijtens T, Abate A, Snaith HJ. Sub-150 °C processed meso-superstructured perovskite solar cells with enhanced efficiency. *Energy Environ Sci*. 2014; 7: 1142–1147.
[13] Zhou H, Chen Q, Li G, Luo S, Song TB, Duan HS, Hong Z, You J, Liu Y, Yang Y. Interface engineering of highly efficient perovskite solar cells. *Science*. 2014; 345: 542–546.
[14] Jeon NJ, Noh JH, Yang WS, Kim YC, Ryu S, Seo J, Seok SI. Compositional engineering of perovskite materials for high-performance solar cells. *Nature*. 2015; 517: 476–480.
[15] Nie W, Tsai H, Asadpour R, Blancon JC, Neukirch AJ, Gupta G, Crochet JJ, Chhowalla M, Tretiak S, Alam MA, Wang HL, Mohite AD. High-efficiency solution-processed perovskite solar cells with millimeter-scale grains. *Science*. 2015; 347: 522–525.
[16] Yang WS, Noh JH, Jeon NJ, Kim YC, Ryu S, Seo J, Seok SI. High-performance photovoltaic perovskite layers fabricated through intramolecular exchange. *Science*. 2015; 348: 1234–1237.
[17] Saliba M, Orlandi S, Matsui T, Aghazada S, Cavazzini M, Correa-Baena JP, Gao P, Scopelliti R, Mosconi E, Dahmen KH, De Angelis F, Abate A, Hagfeldt A, Pozzi G, Graetzel M, Nazeeruddin MK. A molecularly engineered hole-transporting material for efficient perovskite solar cells. *Nat Energy*. 2016; 1: 15017–1–7.
[18] Bi D, Tress W, Dar MI, Gao P, Luo J, Renevier C, Schenk K, Abate A, Giordano F, Baena JPC, Decoppet JD, Zakeeruddin SM, Nazeeruddin MK, Grätzel M, Hagfeldt A. Efficient luminescent solar cells based on tailored mixed-cation perovskites. *Sci Adv*. 2016; 2: e1501170-1-7.
[19] Saliba M, Matsui T, Seo JY, Domanski K, Correa-Baena JP, Nazeeruddin MK, Zakeeruddin SM, Tress W, Abate A, Hagfeldtd A, Grätzela M. Cesium-containing triple cation perovskite solar cells: improved stability, reproducibility and high efficiency. *Energy Environ Sci*. 2016; 9: 1989–1997.
[20] Miyasaka T. Organo-lead halide perovskite: rare functions in photovoltaics and optoelectronics. *Chem Lett*. 2015; 44: 720–729.
[21] Oku T. Crystal structures of $CH_3NH_3PbI_3$ and related perovskite compounds used for solar cells, in Solar Cells – New Approaches and Reviews, Edit. Kosyachenko LA. *InTech*. 2015; 77–102.
[22] Oku T, Takeda A, Nagata A, Kidowaki H, Kumada K, Fujimoto K, Suzuki A, Akiyama T, Yamasaki Y, Ōsawa E. Microstructures and photovoltaic properties of $C_{60}$-based solar cells with copper oxides, $CuInS_2$, phthalocyanines, porphyrin, PVK, nanodiamond, germanium and exciton-diffusion blocking layers. *Mater.Technol*. 2013; 28: 21–39.
[23] Oku T. Direct structure analysis of advanced nanomaterials by high-resolution electron microscopy. *Nanotechnol Rev*. 2012; 1: 389–425.
[24] Oku T. High-resolution electron microscopy and electron diffraction of perovskite-type superconducting copper oxides. *Nanotechnol Rev*. 2014; 3: 413–444.

[25] Oku T. Structure analysis of advanced nanomaterials: nanoworld by high-resolution electron microscopy. Walter De Gruyter Inc, Germany. 2014.

[26] Baikie T, Fang Y, Kadro JM, Schreyer M, Wei F, Mhaisalkar SG, Gräetzel M, White TJ. Synthesis and crystal chemistry of the hybrid perovskite $(CH_3NH_3)PbI_3$ for solid-state sensitised solar cell applications. *J Mater Chem A*. 2013; 1: 5628–5641.

[27] Poglitsch A, Weber D. Dynamic disorder in methylammonium trihalogenoplumbates (II) observed by millimeter–wave spectroscopy. *J Chem Phys*. 1987; 87: 6373–6378.

[28] Weber D. $CH_3NH_3PbX_3$, ein Pb(II)–system mit kubischer perowskitstruktur. *Z Naturforsch B*. 1978; 33: 1443–1445.

[29] Onoda–Yamamuro N, Matsuo T, Suga H. Calorimetric and IR spectroscopic studies of phase transitions in methylammonium trihalogenoplumbates (II). *J Phys Chem Solids*. 1990; 51: 1383–1395.

[30] Hahn T. International tables for crystallography, Volume A. Kluwer Academic Publishers, The Netherlands.1995.

[31] Mashiyama H, Kurihara Y, Azetsu T. Disordered cubic perovskite structure of $CH_3NH_3PbX_3$ (X = Cl, Br, I). *J Kor Phys Soc*. 1998; 32: S156–S158.

[32] Kawamura Y, Mashiyama H, Hasebe K, Structural study on cubic–tetragonal transition of $CH_3NH_3PbI_3$. *J Phys Soc Jpn*. 2002; 71: 1694–1697.

[33] Mashiyama H, KawamuraY, Magome E, Kubota Y. Displacive character of the cubic-tetragonal transition in $CH_3NH_3PbX_3$. *J Kor. Phys Soc*. 2003, 42, S1026–S1029.

[34] Momma K, Izumi F. VESTA 3 for three-dimensional visualization of crystal, volumetric and morphology data. *J Appl Crystallogr*. 2011; 44: 1272–1276.

[35] Im JH, Chung J, Kim SJ, Park NG. Synthesis, structure, and photovoltaic property of a nanocrystalline 2H perovskite–type novel sensitizer $(CH_3CH_2NH_3)PbI_3$. *Nanoscale Res Lett*. 2012; 7: 353–1–7.

[36] Zushi M, Suzuki A, Akiyama T, Oku T. Fabrication and characterization of $TiO_2/CH_3NH_3PbI_3$-based photovoltaic devices. *Chem Lett*. 2014; 43: 916–918.

[37] Oku T, Zushi M, Imanishi Y, Suzuki A, Suzuki K. Microstructures and photovoltaic properties of perovskite-type $CH_3NH_3PbI_3$ compounds. *Appl Phys Express*. 2014; 7: 121601–1–4.

[38] Yamada K, Funabiki S, Horimoto H, Matsui T, Okuda T, Ichiba S. Structural phase transitions of the polymorphs of $CsSnI_3$ by means of Rietveld analysis of the X-ray diffraction. *Chem Lett*. 1991; 20: 801–804.

[39] Chung I, Song JH, Im J, Androulakis J, Malliakas CD, Li H, Freeman AJ, Kenney JT, Kanatzidis MG. $CsSnI_3$: semiconductor or metal? High electrical conductivity and strong near–infrared photoluminescence from a single material. High hole mobility and phase–transitions. *J Am Chem Soc*. 2012; 134: 8579–8587.

[40] Thiele G, Rotter HW, Schmidt KD. Kristallstrukturen und Phasentransformationen von Caesiumtrihalogenogermanaten(II) $CsGeX_3$ (X = Cl, Br, I). *Z Anorg Allg Chem*. 1987; 545: 148–156.

[41] Chung I, Lee B, He J, Chang RPH, Kanatzidis MG. All-solid-state dye-sensitized solar cells with high efficiency. *Nature*. 2012; 485: 486–489.

[42] Yamada K, Mikawa K, OkudaT, Knight KS. Static and dynamic structures of $CD_3ND_3GeCl_3$ studied by TOF high resolution neutron powder diffraction and solid state NMR. *J Chem Soc Dalton Trans*. 2002: 2112–2118.

[43] Yamada K, Kuranaga Y, Ueda K, Goto S, Okuda T, Furukawa Y. Phase transition and electric conductivity of $ASnCl_3$ (A = Cs and $CH_3NH_3$). *Bull Chem Soc Jpn*. 1998; 71: 127–134.

[44] Hao F, Stoumpos CC, Cao DH, Chang RP, Kanatzidis MG. Lead-free solid-state organic-inorganic halide perovskite solar cells. *Nat Photonics*. 2014; 8: 489–494.

[45] Chen T, Benjamin JF, Ipek B, Tyagi M, Copley JRD, Brown CM, Choi JJ, Lee S.H. Rotational dynamics of organic cations in the $CH_3NH_3PbI_3$ perovskite. *Phys Chem Chem Phys*. 2015; 17: 31278–31286.
[46] Weller MT, Weber OJ, Frost JM, Walsh A. Cubic perovskite structure of black formamidinium lead iodide, $\alpha$-[HC($NH_2$)$_2$]$PbI_3$, at 298 K. *J Phys Chem Lett*. 2015; 6: 3209–3212.
[47] Grinberg I, West DV, Torres M, Gou G, Stein DM, Wu L, Chen G,. Gallo EM, Akbashev AR, Davies PK, Spanier JE, Rappe AM. Perovskite oxides for visible-light-absorbing ferroelectric and photovoltaic materials. *Nature*. 2013; 503: 509–512.
[48] Körbel S, Marques MAL, Botti S. Stability and electronic properties of new inorganic perovskites from high-throughput ab initio calculations. *J Mater Chem C*. 2016; 4: 3157–3167.
[49] Oku T, Kakuta N, Kobayashi K, Suzuki A, Kikuchi K. Fabrication and characterization of $TiO_2$-based dye-sensitized solar cells. *Prog Nat Sci*. 2011; 21: 122–126.
[50] Yamada Y, Nakamura T, Endo M, Wakamiya A, Kanemitsu Y. Near-band-edge optical responses of solution-processed organic-inorganic hybrid perovskite $CH_3NH_3PbI_3$ on mesoporous $TiO_2$ electrodes. *Appl Phys Express*. 2014; 7: 032302–1–4.
[51] Yamada Y, Nakamura T, Endo M, Wakamiya A, Kanemitsu Y. Photocarrier recombination dynamics in perovskite $CH_3NH_3PbI_3$ for solar cell applications. *J Am Chem Soc*. 2014; 136: 11610–11613.
[52] Li X, Bi D, Yi C, Décoppet JD, Luo J, Zakeeruddin SM, Hagfeldt A, Grätzel M. A vacuum flash–assisted solution process for high-efficiency large-area perovskite solar cells. *Science*. 2016; 353: 58–62.
[53] Oku T, Matsumoto T, Suzuki A, Suzuki K, Fabrication and characterization of a perovskite-type solar cell with a substrate size of 70 mm. *Coatings* 2015; 5: 646–655.
[54] Dong Q, Fang Y, Shao Y, Mulligan P, Qiu J, Cao L, Huang J. Electron-hole diffusion lengths >175 μm in solution-grown $CH_3NH_3PbI_3$ single crystals. *Science*. 2015; 347: 967–970.
[55] Oku, T, Iwata T, Suzuki A. Effects of niobium addition into $TiO_2$ layers on $CH_3NH_3PbI_3$-based photovoltaic devices. *Chem Lett*. 2015; 44: 1033–1035.
[56] Miyagi T, Kamei M, Sakaguchi I, Mitsuhashi T, Yamazaki A, Photocatalytic property and deep levels of Nb-doped anatase $TiO_2$ film grown by metalorganic chemical vapor depostion. *Jpn J Appl Phys*. 2004; 43: 775–776.
[57] Emeline AV, Furubayashi Y, Zhang X, Jin M, Murakami T, Fujishima A. Photoelectrochemical behavior of Nb-doped $TiO_2$ electrodes. *J Phys Chem B*. 2005; 109: 24441–24444.
[58] Hirano M, Matsushima K. Photoactive and adsorptive niobium-doped anatase ($TiO_2$) nanoparticles: influence of hydrothermal conditions on their morphology, structure, and properties. *J Am Ceram Soc*. 2006; 89: 110–117.
[59] Zhang SX, Kundaliya DC, Yu W, Dhar S, Young SY, Salamanca-Riba LG, Ogale SB, Vispute RD, Venkatesan T. Niobium doped $TiO_2$: Intrinsic transparent metallic anatase versus highly resistive rutile phase. *J Appl Phys*. 2007; 102: 013701.
[60] Maghanga CM, Jensen J, Niklasson GA, Granqvist CG, Mwamburi M. Transparent and conducting $TiO_2$:Nb films made by sputter deposition: Application to spectrally selective solar reflectors. *Sol Energy Mater Sol Cells*. 2010; 94: 75–79.
[61] Emori M, Sugita M, Ozawa K, Sakama H. Electronic structure of epitaxial anatase $TiO_2$ films: angle-resolved photoelectron spectroscopy study. *Phys Rev B*. 2012; 85: 035129.
[62] Tian Y, Tatsuma T, Mechanisms and applications of plasmon-induced charge separation at $TiO_2$ films loaded with gold nanoparticles. *J Am Chem Soc*. 2005; 127: 7632–7637.
[63] Matsumoto T, Oku T, Akiyama T, Incorporation effect of silver nanoparticles on inverted type bulk-heterojunction organic solar cells. *Jpn J Appl Phys*. 2013; 52: 04CR13–1–5.
[64] Oku T, Suzuki K, Suzuki A. Effects of chlorine addition to perovskite-type $CH_3NH_3PbI_3$ photovoltaic devices. *J Ceram Soc Jpn*. 2016; 124: 234–238.

[65] Shi D, Adinolfi V, Comin R, Yuan M, Alarousu E, Buin A, Chen Y, Hoogland S, Rothenberger A, Katsiev K, Losovyj Y, Zhang X, Dowben PA, Mohammed OF, Sargent EH, Bakr OM. Low trap-state density and long carrier diffusion in organolead trihalide perovskite single crystals. *Science*. 2015; 347: 519–522.

[66] Hao F, Stoumpos CC, Cao DH, Chang RPH, Kanatzidis MG, Lead-free solid-state organic–inorganic halide perovskite solar cells. *Nat Photonics*. 2014; 8: 489–494.

[67] Oku T, Ohishi Y, Suzuki A. Effects of antimony addition to perovskite-type $CH_3NH_3PbI_3$ photovoltaic devices. *Chem Lett*. 2016; 45: 134–136.

[68] Oku T, Ohishi Y, Suzuki A, Miyazawa Y. Effects of Cl addition to Sb-doped perovskite-type $CH_3NH_3PbI_3$ photovoltaic devices. *Metals*. 2016; 6: 147–1–13.

[69] Krishnamoorthy T, Ding H, Yan C, Leong WL, Baikie T, Zhang Z, Sherburne M, Li S, Asta M, Mathews N, Mhaisalkarac SG. Lead-free germanium iodide perovskite materials for photovoltaic applications. *J Mater Chem A*. 2015; 3: 23829–23832.

[70] Ohishi Y, Oku T, Suzuki A. Fabrication and characterization of perovskite-based $CH_3NH_3Pb_{1-x}Ge_xI_3$, $CH_3NH_3Pb_{1-x}Tl_xI_3$ and $CH_3NH_3Pb_{1-x}In_xI_3$ photovoltaic devices. *AIP Conf Proc*. 2016; 1709: 020020–1–8.

# 7 Future solar cells

## 7.1 Next generation solar cells

Conceptual figures for a new type of solar cells whose target is conversion efficiencies of ~ 40 % are shown in Fig. 7.1. The absorption wavelength range of one kind of semiconductor is constant. For multiple junctions, sunlight with wider wavelengths can be absorbed by multiplying the absorption wavelength. Solar cells with multiple bands have energy levels (mini-band), and are excited even by low energy light to cover a wide wavelength [1, 2].

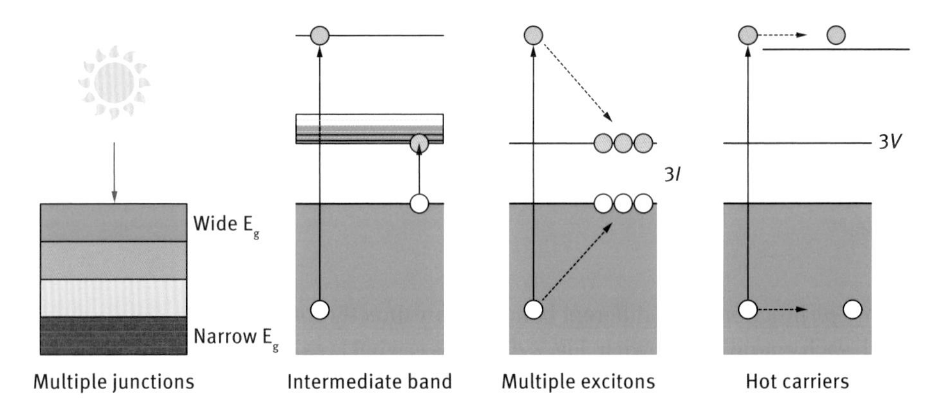


Fig. 7.1: Conceptual figures of next generation solar cells.

To realize efficiencies higher than 40 %, several device structures such as tandem-type, interband-type, multi-exciton-type and hot carrier-type have been proposed. The energy relaxation time in quantum dots is long, which is due to the interaction of phonons. To utilize this characteristic, a multi-exciton type that generates several electrons from one photon has been proposed, and a hot carrier-type that directly uses high energy has also been proposed.

## 7.2 Multi-junction

Since the photo-conversion efficiencies of *pn* junctions depend on the bandgap energy of the semiconductors. An efficiency of ~ 30 % is the upper limit for a single-junction cell. To improve the efficiency, multi-junction solar cells are being developed [3, 4]. The spectrum of sunlight contains a wide variety of wavelengths, from ultraviolet to infrared light, and only one energy can be used for the single-junction cell, as shown in Fig. 7.2(a).

DOI 10.1515/9783110298505-010

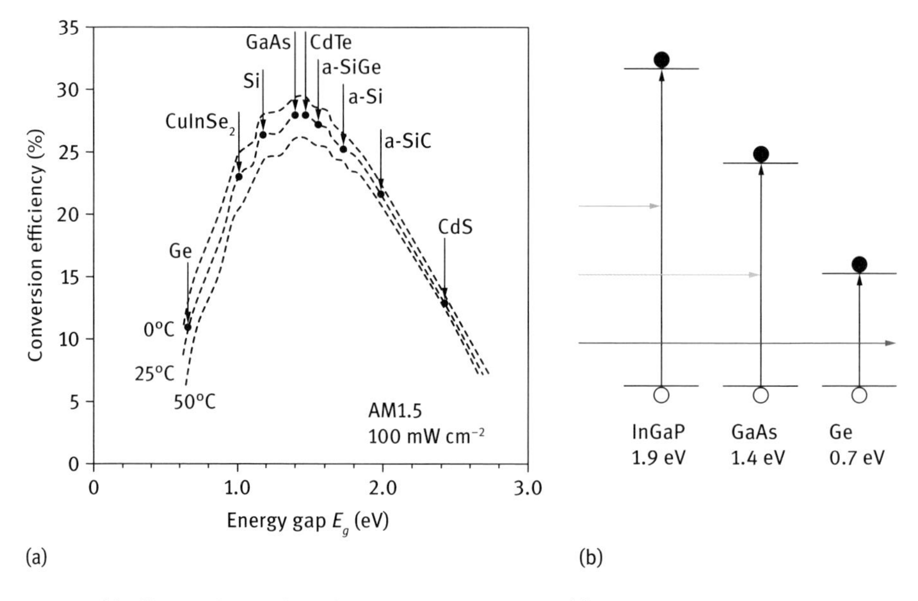


Fig. 7.2: (a) Efficiency limits of single-junction solar cells and (b) enhancement of efficiency by multi-junction.

If plural *pn* junctions with different bandgaps are directly connected, the cell is able to absorb each energy, as shown in Fig. 7.2(b). This is what is called a multi-junction solar cell. The cell generally absorbs light from short wavelengths to long wavelengths. The multi-junction solar cells currently being fabricated are GaInP/InGaAs/Ge cells.

As shown in Fig. 7.3(a), lattice-matched III-V type multi-junction solar cells consist of single-crystal $In_{0.49}Ga_{0.51}P/In_{0.01}Ga_{0.99}As/Ge$ thin films epitaxially grown on the Ge substrate. Lattice distances of InGaP top cell ($E_g$ = 1.88 eV) and InGaAs middle cell ($E_g$ = 1.40 eV) are almost equal to that of the Ge bottom cell ($E_g$ = 0.67 eV) of the substrate. The photo-conversion efficiency is ~ 29 % under the AM0 spectrum, which is higher than the 17 % efficiency of Si solar cells for space. Since the total cost of solar cells can be reduced by high efficiency, 97 % of solar cells used in space are such triple-junction compound semiconductor solar cells, as shown in Fig. 7.3(b). Furthermore, the efficiency of ~ 30 % is improved to 40 % when light concentration is increased to 1000 times the intensity of the original light.

If a semiconductor with a bandgap lower than 0.65 eV is added to triple-junction solar cells to fabricate quadruple-junction cells, high efficiencies of 40 and 50 % are expected under direct light and concentrated light, respectively. Multi-junction solar cells have been used in artificial satellites and spacecrafts used on Mars. The characteristics required for the space solar cells are high efficiency, high radiation resistance, temperature stability, high reliability and lightness.

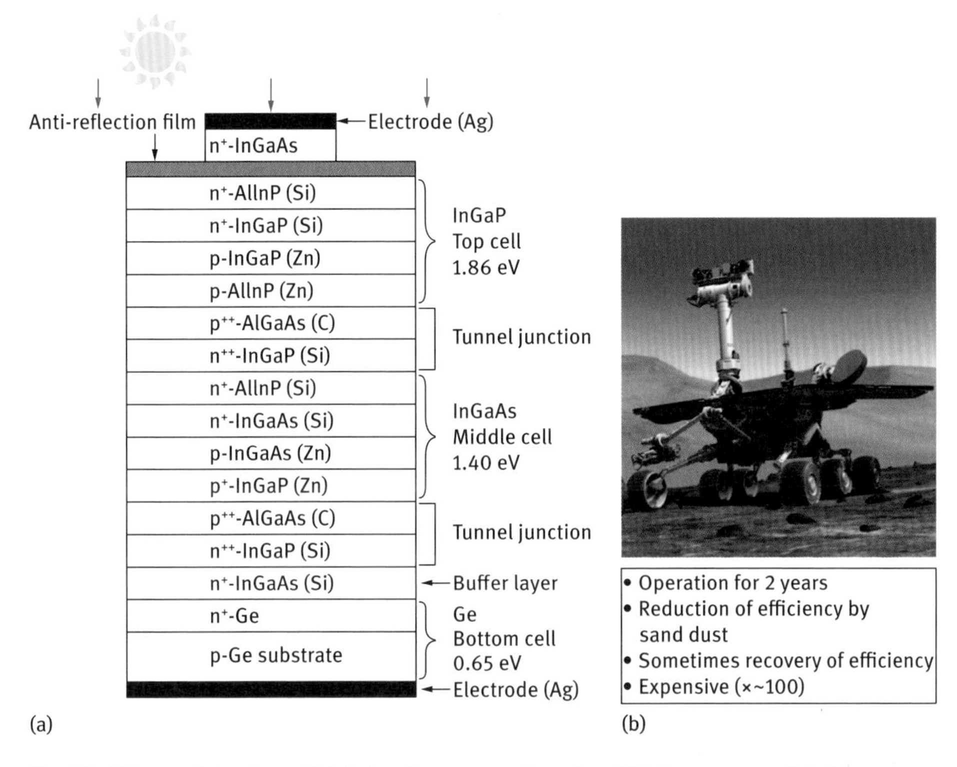


Fig. 7.3: (a) Layered structure of triple-junction space solar cell and (b) Mars spacecraft Spirit (NASA).

## 7.3 Quantum size effect

Electrons carry a negative charge, and are both particles and waves in quantum dynamics. The wavelength of the electrons in materials generally measures tens of nanometers. If the sizes of nanoparticles or nanostructures are below this wavelength, a quantum wave effect appears and the energy gap increases, which is called a quantum size effect. Although energy gaps generally depend on the materials, the energy gaps can be controlled by changing the sizes of the nanoparticles or nanostructures.

If the surfaces of the nanoparticles are covered by materials with wide gap energies such as $SiO_2$ and oxides, the electrons can be confined in the nanoparticles, which is called a quantum confinement effect. The quantum efficiency of electron generation may be improved by the quantum confinement effect. Although Si has a band structure with an indirect transition from photons to electrons, the formation of a band structure with direct transition has been reported for Si nanoparticles.

Such quantum size effects and quantum confinement effects are possible in quantum structures, as shown in Fig. 7.4. The quantum structures with 2-, 1- and 0-dimensionalities are quantum well, quantum wire and quantum dot, respectively.

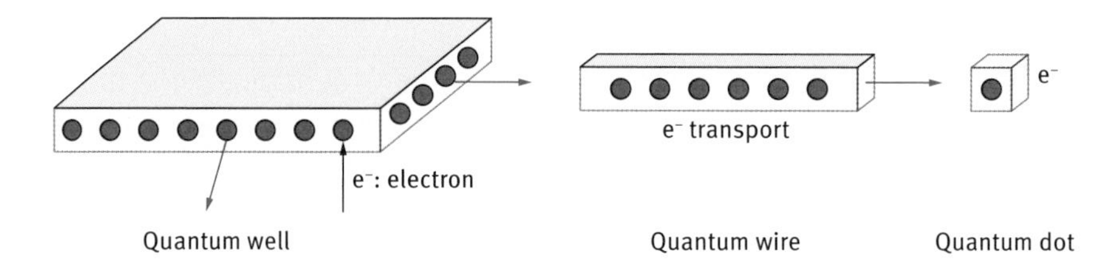


Fig. 7.4: Schematic illustration of a quantum well, quantum wire and quantum dot.

The quantum wells have already been utilized for semiconductor laser diodes and light emitting diodes. Since only one photon can be obtained from the quantum dots, new applications such as in quantum information can be expected. By these effects, the energy gap of a semiconductor can be controlled by only one element such as Si.

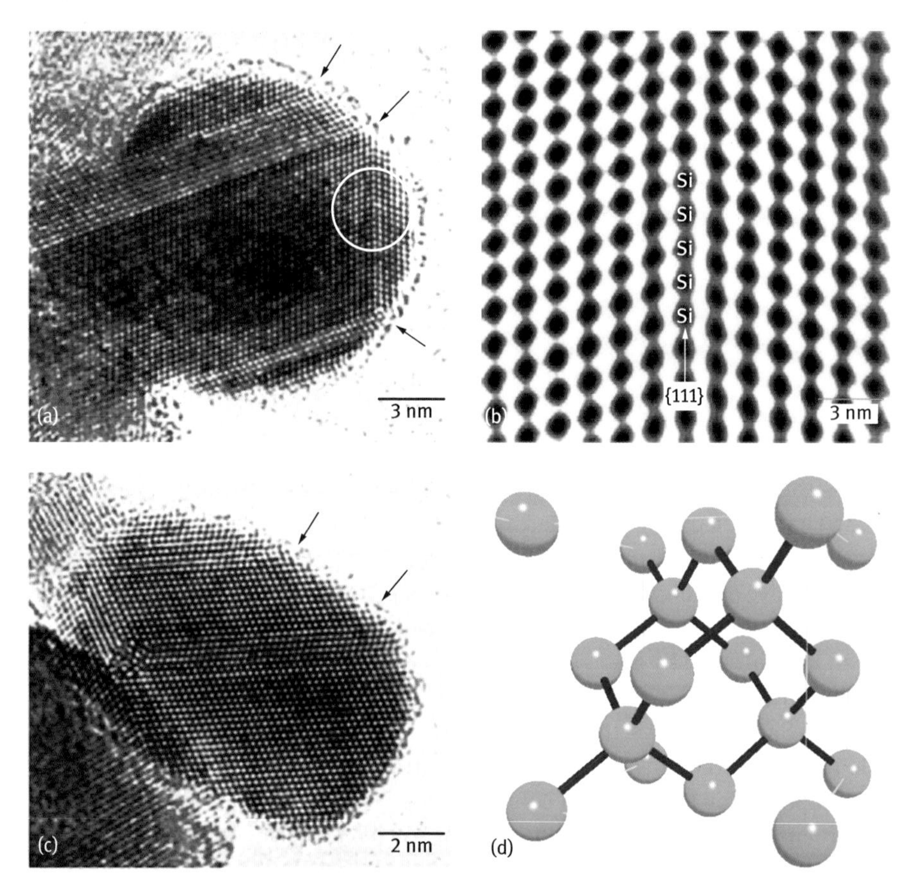


Fig. 7.5: (a) High-resolution electron microscopy image of Si nanoparticle (b) enlarged image of a part of (a). (c) High-resolution electron microscopy image of Ge nanoparticle. (d) Atomic structure model of Si and Ge.

Figure 7.5(a) is a high-resolution electron microscopy image of an Si nanoparticle [5]. At the surface of the Si nanoparticle, a disordering of the atomic arrangements is observed, as indicated by the arrows, and the atomic disordering is a thin oxide layer formed by reacting oxygen in air. Figure 7.5(b) is an enlarged image of (a) after Fourier noise filtering, and the dark dots in the image correspond to an atomic pair of Si. A region indicated by an arrow is a {111} twin boundary. A high-resolution image of Ge nanoparticle is shown in Fig. 7.5(c), and a similar native oxide layer on Si is observed at the nanoparticle's surface. Both Si and Ge have a diamond structure as shown in Fig. 7.5(d), and these nanoparticles had wider bandgaps compared to ordinary bulk semiconductors. If such nanoparticles are applied to solar cells and laser diodes, photo-electron conversion efficiencies could be improved.

## 7.4 Quantum dots

Quantum dots are nanostructures that confine electrons three-dimensionally. The confined electrons occupy discrete energy levels, which is similar to the way electrons behave in an atom. By confining electrons in quantum dots, wave characteristics with quantum dynamics can be utilized.

The wavelengths of light absorbed in the quantum dots depend on the size of the dots, which is called a quantum size effect. Smaller quantum dots absorb light with short wavelengths (blue), and larger dots absorb light with longer wavelengths (red), and quantum dot solar cells that make use of these characteristics have been proposed.

If quantum dots are used for *i*-layer in the *pin*-type solar cell as shown in Fig. 7.6(a), light excitation is possible with lower energy, which is due to energy levels of quantum dots, as shown in Fig. 7.6(b). Excitons generated by quantum dots transport by thermal separation.

Quantum dot solar cells are expected to provide high efficiency for a single semiconductor junction cell [6]. Further improvement of the efficiency would be possible through a tandem structure [7]. Problems that exist for the creation of quantum dot cells are dot materials, control of dot size and their arrangements. In many cases, quantum dot solar cells are multi-band solar cells with quantum dots, which is because new mini-bands are formed through the interaction of many, arranged quantum dots.

## 7.5 Intermediate band type

Intermediate band solar cells are a kind of multi-junction solar cell [8, 9]. If the intermediate band structure is realized by using quantum dots or superlattice structures, the entire range of sunlight can be utilized effectively, as shown in Fig. 7.7(a). Even

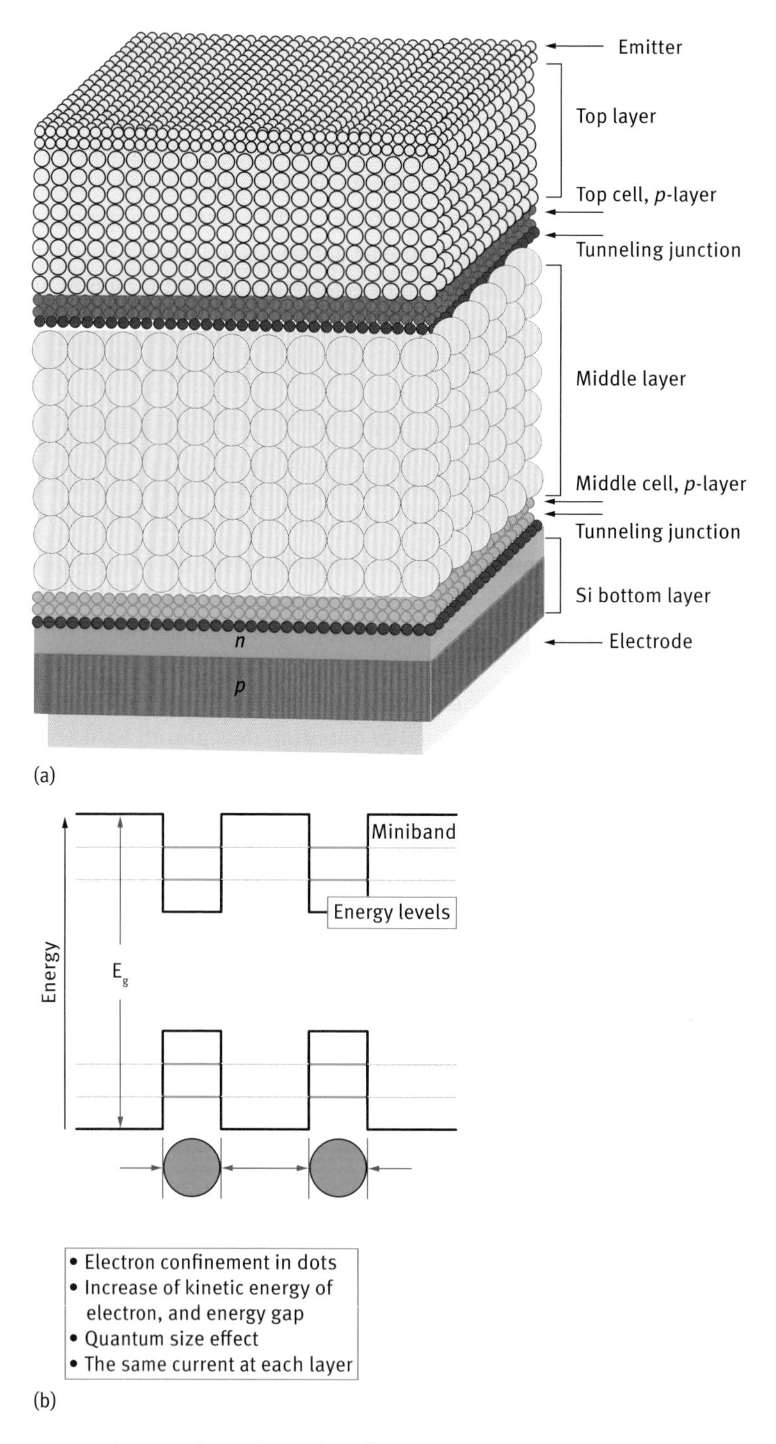


Fig. 7.6: Quantum dot tandem solar cells.

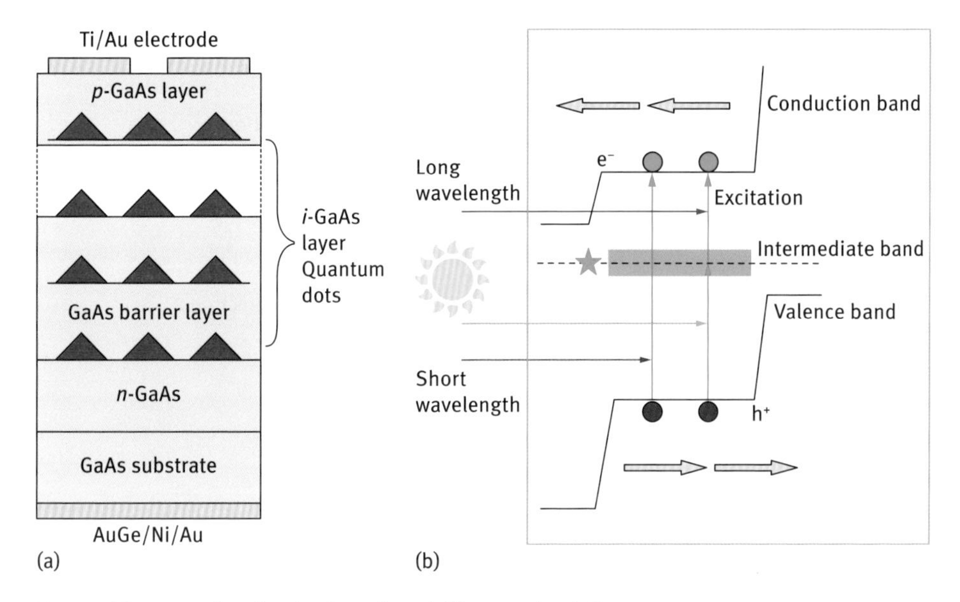


Fig. 7.7: (a) Intermediate band solar cells and (b) energy level diagram.

when the light energy is low, electrons are excited by dividing the energy, as shown in Fig. 7.7(b).

If a lot of semiconductor quantum dots are arranged with great regularity 3-dimensionally, an intermediate band structure can be formed at different energy levels from the ordinary levels due to the binding effect of the quantum dots. Since the intermediate band structure should absorb a wide wavelength-range of light by utilizing optical transitions between plural bands for consistency with sunlight spectra, energy loss due to penetration of sunlight can be reduced for single-junction solar cells. To increase the escape velocity of electrons and holes from the quantum dots compared with the recombination velocity, the thickness of the intermediate layer should be below 10 nm, and structural control is important.

## 7.6 Multi-exciton

Light with high energy is utilized in multi-exciton and hot carrier solar cells in order to reduce thermal energy loss. Energy relaxation of electrons and holes excited at higher energy levels in a bulk crystal occurs through carrier diffusion and phonon emission over a short period of time (~ ps), as shown in Fig. 7.8. Since carriers are confined in the quantum dots, the energy relaxation time of phonon interaction (thermal energy loss) is slow. Therefore, improvement in the conversion efficiency of solar cells could be attempted by multi-exciton generation [10–12]. Since an electron excited by the light excites other electrons, the internal quantum efficiency becomes 200 % or

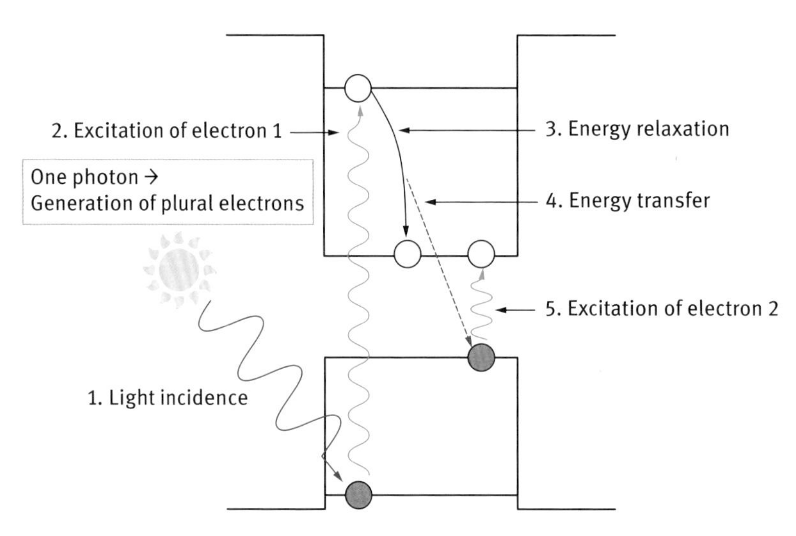


Fig. 7.8: Multi-exciton generation.

300 %. At that time, the light should have two or three times higher energy compared to the bandgap energy. Multi-excitons have been confirmed for PbS, PbSe, and Si. The problem to be solved is how to optimize the device structure by extracting the generated carriers in the quantum dots.

## 7.7 Hot carrier type

The basic structure of hot carrier solar cells is shown in Fig. 7.9 [13]. The light is absorbed in the absorption layer and excites electrons from the valence band to the conduction band. Hot carrier cells should be designed so that excited carriers are not relaxed to the band edge. Since relaxation to the band edge from the excited state occurs in ordinary solar cells, the objective of hot carrier solar cells is to prevent relaxation. In addition, hot carriers should be extracted before the relaxation. Contact is possible only in a very narrow energy range, and the hot carriers cool down quickly. The hot carriers may be extracted by an increase of the carrier relaxation time by quantum dots and tunneling contacts.

Hot carriers lose their excited energy by colliding with atoms in the light absorption layer, which causes phonon vibration. Control of the phonon property is important for hot carrier solar cells [14, 15]. One of the methods for controlling phonons is to arrange the quantum dots regularly. The quantum dots superlattice controls phonon vibration and improves optical properties. The hot carriers could be extracted for s quantum dots with smaller sizes by tunneling contact. Quantum dots are several nanometers in size, and a distance of ~ 15 nm from the electrode is needed for a resonance tunneling junction.

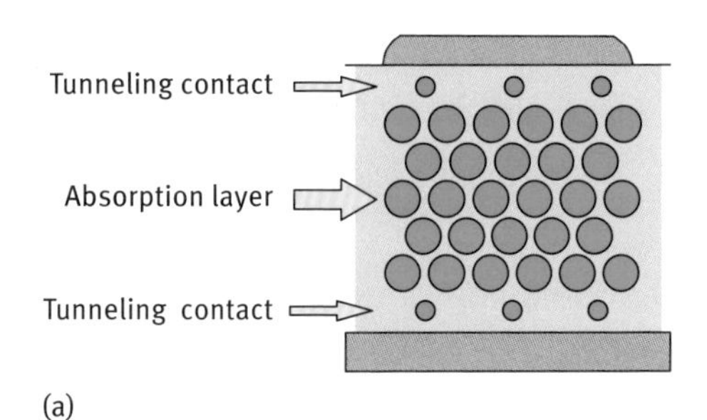


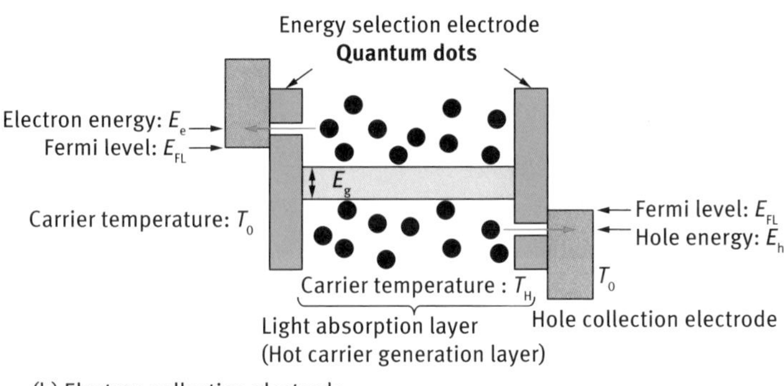


Fig. 7.9: (a) Si hot carrier cell and (b) the energy band diagram.

## 7.8 Space solar power system

A space solar power system (SSPS) is a power generation system that generates solar power in space and then transfers the generated electric power to Earth by a microwave or laser [16]. The electric power is converted to microwave or laser light, which is sent from power transmission antennas to power receiving antennas, and the energy is converted to electric power again, as illustrated in Fig. 7.10.

A significant amount of sunlight is lost to absorption by elements in the air before arriving on Earth, and it is also effected by changes in the weather. If electric power is generated outer space, converted into electromagnetic waves with high air permeability, and transfered to the earth, the efficiency is better because of the low energy loss. If a suitable orbit is selected, the influence of night can be reduced, and the efficiency of an SSPS is ~10 times better compared to the efficiencies of ordinary solar cells on the ground.

The SSPS system consists of a power-generating satellite in space and a power-receiving facility on the ground. Electric power is generated by solar cells set on the satellite orbits, and the power is converted into microwave or laser light, which is sent to a power receiving facility in the desert or at sea, and the energy is then converted to electric power again on the ground. If both power generation satellites and the power transfer satellites are utilized, power can be supplied regularly to the earth even at

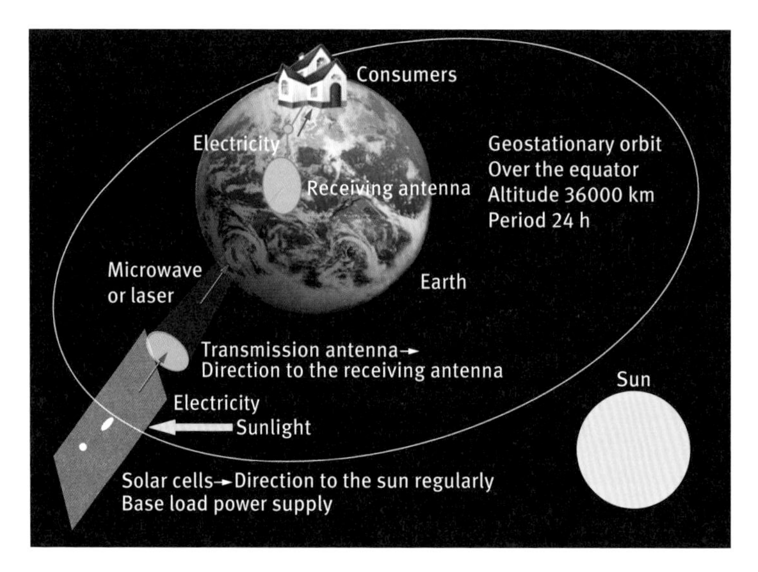


Fig. 7.10: Schematic illustration of space solar power system.

night, and unlimited electric power can be used 24/7. In addition to making use of satellites, a power generation system could also be put on the moon.

SSPS is possible in principle. However, there are several technical problems such as the estimated launch cost, material deterioration and maintenance. Japan Aerospace Exploration Agency (JAXA)[17] set a goal of commercialization in 2020 ~ 2030. An SSPS would have the following characteristics: it would be an unlimited energy source, 1 GW of energy could be received at a power receiving facility of 10 $km^2$, it would provide an energy density of 10 W $m^{-2}$. Disadvantages of the SSPS are the difficulty of repair, attenuation of transferred energy by the atmosphere, possible influences from other satellites, wireless communication, influences of microwave or laser light on living things such as birds, the need for a regularly supply of fuel for orbit control, reduced power generation in the Earth's shadow and radiation damage.

## 7.9 Bibliography

[1] Yamaguchi M, Nishimura KI, Sasaki T, Suzuki H, Arafune K, Kojima N, Ohsita Y, Okada Y, Yamamoto A, Takamoto T, Araki K. Novel materials for high-efficiency III–V multi-junction solar cells. *Sol Energy*. 2008; 82: 173–180.

[2] Green MA, Emery K, Hishikawa Y, Warta W, Dunlop ED. Solar cell efficiency tables (Version 48). *Prog Photovolt*. 2016; 24: 905–913.

[3] Yamaguchi M, Takamoto T, Araki K. Super high-efficiency multi-junction and concentrator solar cells. *Sol Energy Mater Sol Cells*. 2006; 90: 3068–3077.

[4] Dimroth F, Grave M, Beutel P, Fiedeler U, Karcher C, Tibbits TN, Oliva E, Siefer G, Schachtner M, Wekkeli A, Bett AW, Krause R, Piccin M, Blanc N, Drazek C, Guiot E, Ghyselen B, Salvetat T,

Tauzin A, Signamarcheix T, Dobrich A, Hannappel T, Schwarzburg K. Wafer bonded four-junction GaInP/GaAs//GaInAsP/GaInAs concentrator solar cells with 44.7 % efficiency. *Prog Photovoltaics: Res Appl*. 2014; 22: 277–282.
[5] Oku T, Nakayama T, Kuno M, Nozue Y, Wallenberg LR, Niihara K, Suganuma K. Formation and photoluminescence of Ge and Si nanoparticles with oxide layers. *Mater Sci Eng B*. 2000; 74: 242–247.
[6] Cho EC, Park S, Hao X, Song D, Conibeer G, Park SC, Green MA. Silicon quantum dot/crystalline silicon solar cells. *Nanotechnol*. 2008; 19: 245201.
[7] Kamat PV. Quantum dot solar cells. The next big thing in photovoltaics. *J Phys Chem Lett*. 2013; 4: 908–918.
[8] Shoji Y, Akimoto K, Okada Y. Self-organized InGaAs/GaAs quantum dot arrays for use in high-efficiency intermediate-band solar cells. *J Phys D: Appl Phys*. 2013; 46: 024002.
[9] Okada Y, Morioka T, Yoshida K, Oshima R, Shoji Y, Inoue T, Kita T. Increase in photocurrent by optical transitions via intermediate quantum states in direct-doped InAs/GaNAs strain-compensated quantum dot solar cell. *J Appl Phys*. 2011; 109: 024301.
[10] Semonin O, Luther1 JM, Choi S, Chen HY, Gao J, Nozik AJ, Beard MC. Peak external photocurrent quantum efficiency exceeding 100 % via MEG in a quantum dot solar cell. *Science*. 2011; 334: 1530–1533.
[11] Ahsan N, Miyashita N, Islam MM, Yu KM, Walukiewicz W, Okada Y. Two-photon excitation in an intermediate band solar cell structure. *Appl Phys Lett*. 2012; 100: 172111.
[12] Huang X, Han S, Huang W, Liu X. Enhancing solar cell efficiency: the search for luminescent materials as spectral converters. *Chem Soc Rev*. 2013; 42: 173–201.
[13] Green MA. Third generation photovoltaics. Springer. 2003.
[14] Nozik AJ. Quantum dot solar cells. *Physica E*. 2002; 14: 115–120.
[15] Farrell DJ, Sodabanlu H, Wang Y, Sugiyama M, Okada Y. A hot-electron thermophotonic solar cell demonstrated by thermal up-conversion of sub-bandgap photons. *Nat Commun*. 2015; 6: 8685.
[16] McSpadden JO, Mankins JC. Space solar power programs and microwave wireless power transmission technology. *IEEE microwave magazine*. 2002; 12: 46–57.
[17] Mori M, Kagawa H, Saito Y. Summary of studies on space solar power systems of Japan Aerospace Exploration Agency (JAXA). *Acta Astronautica*. 2006; 59: 132–138.

# 8 Nuclear fusion materials

## 8.1 Nuclear fusion in the sun

The sun, which provides us with light and energy that is essential for our lives, is one of $2\times 10^{11}$ fixed stars in the galaxy. The sun is a gigantic hydrogen sphere and has a radius ($R_{Sun} = 70\times 10^4$ km) one hundred times larger than earth's. Energy emission ($L$) from the sun is $3.85\times 10^{26}$ W, and its mass $M$ is $2\times 10^{30}$ kg. When the sun was formed 4.5 billion years ago, gravitational energy $E$ was released by its mass $M$ becoming concentrated in one position.

$$E = \frac{GM^2}{R_{Sun}} = 3.8 \times 10^{41} \text{ (J)} \tag{8.1}$$

$G$ is a constant of gravitation which corresponds to the energy of one ion or an electron of ~ 1 keV (~ 7.7 million K). The current temperature of the core of the sun is estimated to be 16 million K, which is produced by gravitational action. Atoms are divided into electrons and nuclei, which means that the sun is a huge, high-density plasma sphere. The density of the core of the sun is ~ 160 g $cm^{-3}$, and half of the mass of the sun is concentrated within one fourth of the radius of the sun.

The nucleus has a positive charge, and nuclei repulse when close to each other. If a kinetic energy higher than the repulsive force is added to the nucleus, two nuclei can collide and become one nucleus through what is called a nuclear fusion reaction. The difference between the binding energy before and after the reaction is emitted as what is called nuclear fusion energy.

In the sun and other fixed stars, a part of their mass is transformed into energy according to the equation of $E = mc^2$. When two protons (p) are fused, $^2He$ might be formed. However, $^2He$ is unstable and emits a positron ($e^+$) to form a neutron (n), which binds with another proton to form deuteron (D) and to emit energy in the amount of 0.42 MeV and an electron neutrino ($\nu_e$), indicated as follows:

$$p + p \rightarrow D + e^+ + \nu_e + 0.42 \text{ MeV} \tag{8.2}$$

In the first fusion reaction, which took place at the beginning of the universe, deuteron was formed from two proton through the same process that is shown in Fig. 8.1. Since the temperature of the sun is low, this pp fusion reaction is an extremely slow reaction with a time constant of ~ 10 billion years. A positron binds with an electron immediately to emit an energy of 1.02 MeV. The produced deuteron fuses with proton to form $^3He$ and to emit gamma rays ($\gamma$), as follows:

$$p + D \rightarrow {}^3_2He + \gamma + 5.49 \text{ MeV} \tag{8.3}$$

Then, one ${}_2{}^4He$ is formed from two${}_2{}^3He$ nuclei, as follows:

$${}^3_2He + {}^3_2He \rightarrow {}^4_2He + 2p + 12.86 \text{ MeV} \tag{8.4}$$

DOI 10.1515/9783110298505-011

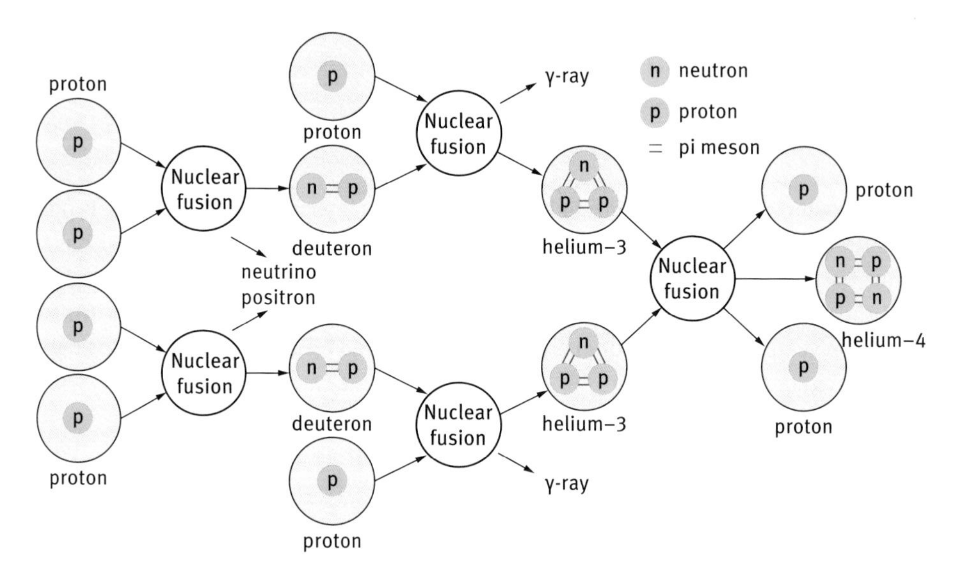


Fig. 8.1: Nuclear fusion at the center of the sun.

In summary, ${}^4_2$He is formed from four protons.

$$4p + 2e^- \rightarrow {}^4_2\mathrm{He} + 26.72\,\mathrm{MeV} \tag{8.5}$$

The energy generated from a proton is 6.55 MeV (26.46/4 MeV) excluding the energy of 0.26 MeV taken away by a neutrino. The enormous number of hydrogen nuclei produced at the beginning of the universe fuse at the center of fixed stars. The combustion efficiency of a proton at the center of the Sun is $L/6.55$ MeV = $3.7\times 10^{38}$ s$^{-1}$, and protons of $6.2\times 10^{11}$ kg s$^{-1}$ are transformed into $^4$He. The mass transformed into energy is $4.4\times 10^9$ kg s$^{-1}$, and 6 % of protons in the Sun have been consumed over the course of 4.5 billion years. The energy generated at the center arrives at the surface of the sun through radiation and convection, and the temperature of the surface is ~ 5800 K. An energy of $6.4\times 10^{11}$ J is obtained from 1 g of protons, which corresponds to 15 t of oil. High energy (~ 5.5 keV) photons generated at the center of the sun produce electrons and positrons near protons and helium nuclei, which also emit photons with lower energy. Thus, photons arrive at the surface of the sun after ~ $10^5$ years of absorption and scattering, and are emitted as radiation energy. Therefore, activity conditions of the core of the Sun cannot be observed by light but by neutrinos.

Since 99 % of the power density of the sun is generated within 24 % of the sun's radius, the power density of the core is ~ 20 W m$^{-3}$, which is much smaller than the ~ 1000 W m$^{-3}$ of human beings (~ 100 W / person). Coincidentally, the power of the human heart and human brain is ~ 1 W and ~ 30 W, which corresponds to power densities of ~ 3000 W m$^{-3}$ and ~ 20000 W m$^{-3}$, respectively.

## 8.2 DT and DD fusion

Three conditions should be fulfilled to achieve nuclear fusion on the Earth: very high temperature on the order of $\sim 10^8$ K; sufficient plasma particle density to increase the collision probability; and sufficient confinement time to hold the plasma within a defined volume. The form of nuclear fusion currently possible in a laboratory is the reaction between two hydrogen isotopes, deuterium (D) and tritium (T). A reaction between deuterium nucleuses would also be possible, called a DD reaction. These reactions are shown in Fig. 8.2, and are represented as follows:

$$^{2}\mathrm{D} + {}^{3}\mathrm{T} \rightarrow {}^{4}\mathrm{He}\,(3.52\,\mathrm{MeV}) + {}^{1}\mathrm{n}\,(14.07\,\mathrm{MeV}) \tag{8.6}$$

$$^{2}\mathrm{D} + {}^{2}\mathrm{D} \rightarrow {}^{3}\mathrm{T}\,(1.01\,\mathrm{MeV}) + {}^{1}\mathrm{p}\,(3.02\,\mathrm{MeV}) \tag{8.7}$$

$$^{2}\mathrm{D} + {}^{2}\mathrm{D} \rightarrow {}^{3}\mathrm{He}\,(0.82\,\mathrm{MeV}) + {}^{1}\mathrm{n}\,(2.45\,\mathrm{MeV}) \tag{8.8}$$

Here, $^{2}$D: deuteron, $^{3}$T: triton, $^{1}$p: proton, and $^{1}$n: neutron. The possibilities of the two DD reactions are almost the same, and the average energy generated by DD fusion is 3.65 MeV.

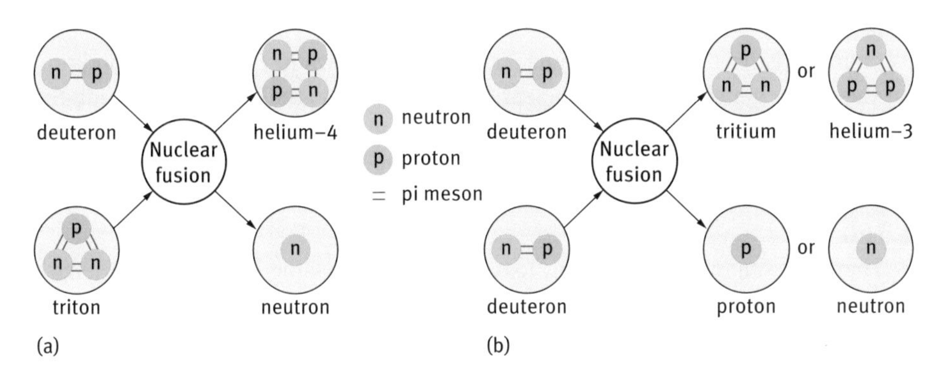


Fig. 8.2: Schematic illustration of (a) DT and (b) DD fusion reactions.

The DT fusion reaction, which is expected to be used in the first practical applications of fusion reactors, produces the highest energy gain at the lowest temperatures. A small amount of tritium is generated only in the upper atmosphere, and tritium is a radioactive isotope with a short life-time, which makes it difficult to extract tritium using contemporary methods. Since fast neutrons are also generated, the design and selection of materials for the reactor is necessary. One of the methods for obtaining tritium would be to utilize the nuclear reaction of lithium with fast neutrons around the reactor.

## 8.3 Fusion conditions

To achieve nuclear fusion, huge energy should be added from outside the reactor. One of the methods to do this involves the collision of atomic nuclei using thermal kinetic energy in what is called a thermonuclear fusion reactor. When a mixture of deuterium and tritium gas is heated above $10^5$ K, they become plasma, which means that they completely separate into ions and electrons – this is often referred to as the fourth state of matter. When the temperature of plasma is increased to $10^8$ K, ions and electrons in the plasma move very fast, which increases the possibility of nuclear fusion caused by the collision of ions. Since the plasma is expanded and scattered at the elevated temperatures, the plasma must be confined during the time required for fusion to take place.

In nuclear fusion, plasma temperature $T$, plasma density (number of nucleus per volume) $n$, and plasma confinement time $\tau$ should be above a certain level. The conditions for nuclear fusion are as follows:

$$\text{DT fusion: } T > 3 \times 10^7\ \text{K}, \quad n\tau > 10^{20}\ \text{s m}^{-3} \tag{8.9}$$

$$\text{DD fusion: } T > 2 \times 10^8\ \text{K}, \quad n\tau > 10^{22}\ \text{s m}^{-3} \tag{8.10}$$

Characteristics of plasma for nuclear fusion are shown in Fig. 8.3. The condition that the power generated by DT nuclear fusion is equal to the power for heating the plasma is a critical condition for plasma confinement, which is indicated in the figure. High ion temperatures and (confinement time) × (plasma density) are mandatory in order for the plasma confinement to bring deuteron close to triton at a distance below $10^{-15}$ m.

There are some sources of energy loss in real fusion reactors, and the condition for the self-continued DT reaction in actual fusion reactors is called Lawson's criterion. Lawson's criterion indicates that the electric energy obtained from the fusion reactor must be larger than the energy needed for heating the plasma.

The tokamak is an experimental apparatus designed to utilize the energy of nuclear fusion, and was first developed by Soviet researchers in the late 1960s. The term "tokamak" comes from a Russian acronym that stands for "toroidal chamber with magnetic coils." The energy produced by nuclear fusion is absorbed as heat in the vessel walls. The fusion power plant uses this heat to produce steam, turning turbines and generators for the generation of electricity, which is a similar process to that used in conventional power plants. The tokamak has a doughnut-shaped vacuum chamber, in which hydrogen plasma is introduced to fuse and yield energy. Charged particles of the plasma can be confined by the huge magnetic coils placed around the vessel to confine the hot plasma away from the vessel walls.

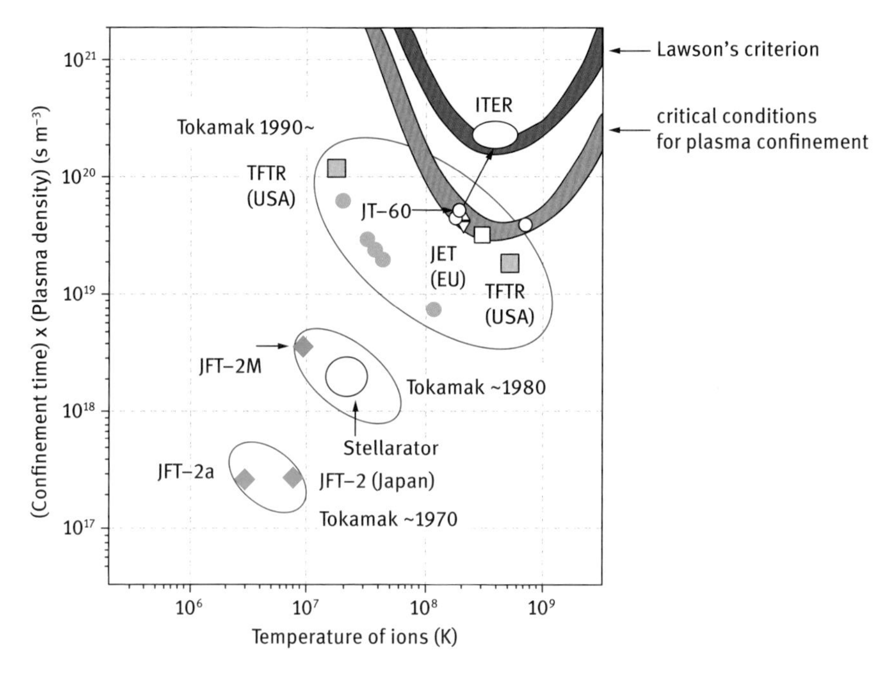


Fig. 8.3: Nuclear fusion and characteristics of plasma.

## 8.4 ITER

ITER, which is shown in Fig. 8.4(a), is one of the most advanced international energy projects in the world [1]. ITER means "The Way" in Latin, and is an acronym for the International Thermonuclear Experimental Reactor. 35 nations including Japan, Korea, China, India, Russia, the United States and the countries of the European Union are collaborating to build the world's largest tokamak-type fusion device in southern France. Thousands of scientists and engineers have contributed to the design of the ITER since the international joint experiment was first launched in 1985. ITER will be the first fusion device to maintain nuclear fusion over long periods of time, and will test the integrated technologies and materials necessary for the commercial production of fusion-based electricity.

The tokamak utilizes magnetic plasma confinement to achieve nuclear fusion, and there are two main problems that still need to be overcome before this method can be implemented commercially. The first is that both deuterium and tritium are isotopes of hydrogen. Tritium is a radioactive isotope that emits beta-rays with a half-life of 12.3 years and forms tritium-heavy water (tritium oxide), causing radioactive pollution. Tritium of about 2 kg is stored in the device. The second is the high energy neutrons generated by DT fusion above $10^8$ K, which leads to the radioactivation of reactor walls and buildings and low level radioactive wastes.

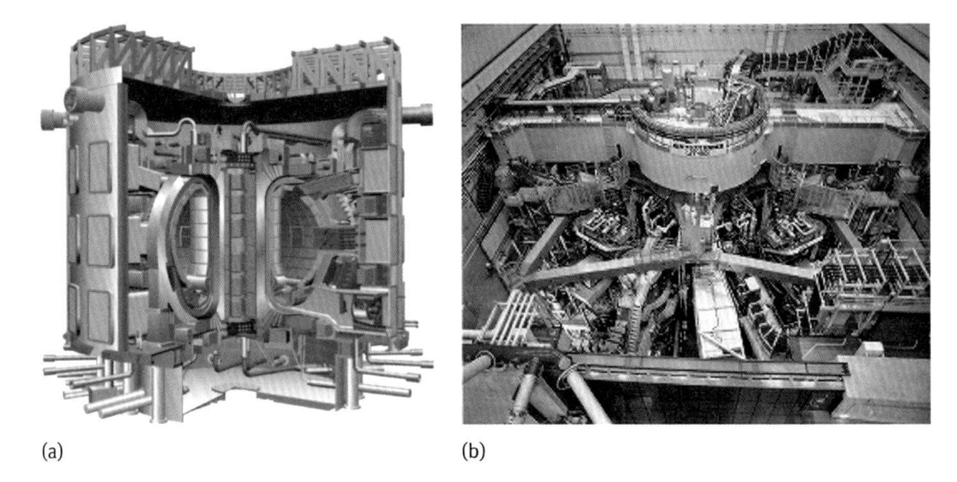

(a) (b)

Fig. 8.4: Images of (a) ITER and (b) JT-60 (ITER and JAEA).

The experimental device ITER is a totally integrated device with various apparatuses including a superconducting magnet for plasma confinement, blanket, diverter, plasma-facing wall, heating, diagnostics, cryogenics and remote maintenance. The ITER is expected to produce 500 MW, making it the first fusion-energy device for practical use. The ITER will bridge the gap between today's smaller-scale experimental fusion devices and the fusion power plants of the future [1]. Scientists will be able to study plasma under extreme conditions similar to those expected in a future power plant. The plasma conditions of ITER and the sun are summarized as listed in Table 8.1.

Table 8.1: Comparison of ITER and Sun.

| | ITER plasma | Sun |
|---|---|---|
| Outer diameter (m) | 16.4 | $1.4 \times 10^9$ km |
| Temperature of center (K) | $2 \times 10^8$ | $1.6 \times 10^7$ |
| Density of center ($m^{-3}$) | $10^{20}$ | $10^{32}$ |
| Pressure of center (atm) | ~ 5 | $2.7 \times 10^{11}$ |
| Energy density ($W\,m^{-3}$) | $6 \times 10^5$ | 0.3 |
| Fusion reaction | DT | pp |
| Mass (g) | 0.35 | $2 \times 10^{33}$ |
| Burning time constant (s) | 500 | $3 \times 10^{17}$ (~ $10^{10}$ years) |

The European tokamak JET produced the world record of 16 MW for fusion power from a total input power of 24 MW, which indicates a fusion energy gain factor ($Q$) of 0.67. The ITER is designed to produce 500 MW of fusion power from 50 MW of input power, which indicates a ten-fold return on energy ($Q = 10$). Although it is necessary to maintain nuclear fusion for long periods of time in real fusion power plants, plasma with enough pressure for practical use could not be maintained, and the JT-60 in Japan achieved 28.6 seconds [2], as shown in Fig. 8.4(b). One of the important goals of the ITER is to sustain a stable fusion state for 500 seconds for $Q = 10$, and to sustain regular, continued fusion for $Q = 5$.

Although it will be necessary to develop materials that are hardly radioactivated by high neutron irradiation, the quantity of neutrons in the ITER is not sufficient for the material development, and therefore, it is necessary to study the materials for nuclear fusion in other facilities in addition to the ITER. The International Fusion Materials Irradiation Facility (IFMIF) was just opened in Japan in 2016, and is an accelerator-based neutron flux source with a peak at 14 MeV equivalent to the conditions of the deuterium-tritium reactions in a fusion power plant.

## 8.5 Muon catalyzed fusion

Muon-catalyzed fusion (μCF) is one of methods used in nuclear fusion to allow fusion to take place even at temperatures significantly lower than the temperatures required for thermonuclear fusion, even at room temperature or lower. When an electron is replaced by a muon around the DD or DT molecule to reduce the Coulomb barrier for fusion, the distance between the nuclei is reduced to 1/200 ($\sim 10^{-10}$ m $\rightarrow 5\times 10^{-13}$ m). Then, the probability of nuclear fusion is significantly increased even at room temperature [3, 4].

Muons are similar to electrons, but they are unstable subatomic particles with 207 times more mass compared with electrons. To create muons, protons or heavy ions with huge energies are irradiated onto metals such as beryllium or copper, which results in the emission of negative or positive charged muons from the metals, as shown in Fig. 8.5. $\pi$-mesons, which have a lifetime of 0.026 μs before transforming into positrons or electrons, are also generated simultaneously. Since the muon has a short lifetime and many muons cannot be obtained from nature, they are produced by a particle accelerator, which limits the usability of muon-catalyzed nuclear fusion as a practical power source.

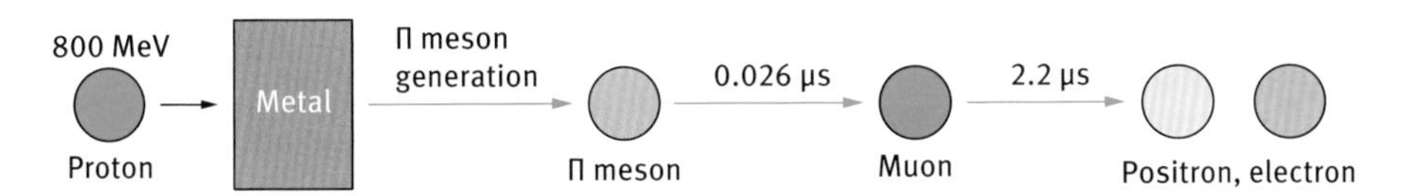


Fig. 8.5: Principle of muon-catalyzed fusion.

Muons become free again after muon-catalyzed fusion, and can be used for the fusion cycle. In addition, ~1% of muons are trapped by $^4$He ($\alpha$-particles). Therefore, the information that is most crucial in controlling the cycle is the number of chain reactions of the μCF caused by the first μCF and the probability of muon trapping by $\alpha$-particles. The fusion probability of one muon is calculated to be ~ 1000 times larger compared to thermonuclear fusion in a tokamak reactor.

Moreover, each muon has about a 1 % chance of "sticking" to the alpha particle produced by the nuclear fusion of deuteron and triton, removing the "stuck" muon from the catalytic cycle, meaning that each muon can only catalyze at most a few hundred deuterium tritium nuclear fusion reactions. So, these two factors, that muons are prohibitively expensive to make and that they too easily stick to alpha particles, limit muon-catalyzed fusion to remaining a laboratory curiosity. To produce practical muon-catalyzed fusion reactors, a more efficient source of muons and an effective muon-catalysis method are needed.

## 8.6 Pyroelectric fusion

In an experiment using a pyroelectric power source using lithium tantalite ($LiTaO_3$) crystal [5], it was reported that nuclear fusion had been achieved in a desktop-like device [6], as shown in Fig. 8.6(a) and (b), respectively. Deuteron beams are generated and accelerated by the pyroelectric effect of $LiTaO_3$ under temperature differences. Pyroelectric fusion is the technique of using pyroelectric crystals to generate high intensity electrostatic fields to accelerate deuterium ions into a metal hydride target containing deuterium with sufficient accelerating energy to cause these ions to produce nuclear fusion. In the experiment, the pyroelectric crystal was heated from −34 to 7 °C, combined with a tungsten needle to produce an electric field of ~ 25 GV

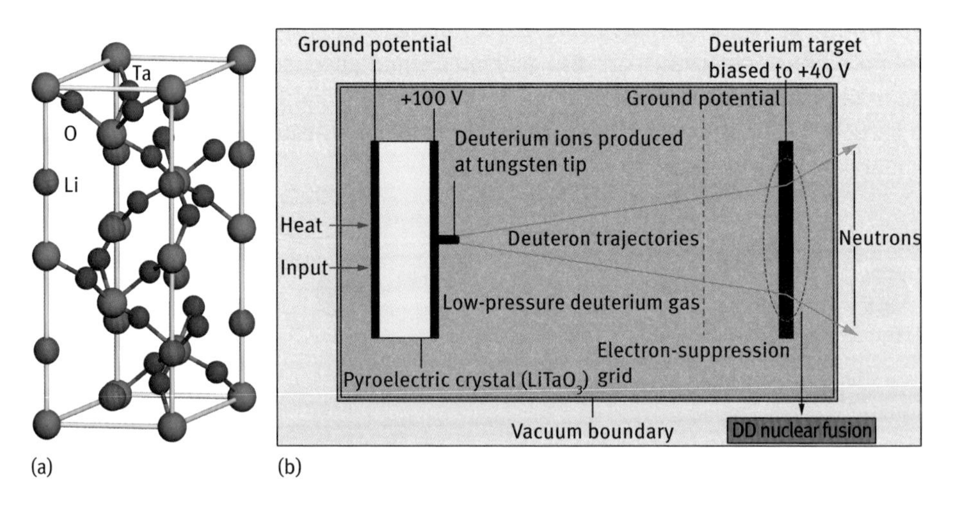


Fig. 8.6: DD nuclear fusion utilizing $LiTaO_3$ pyroelectric crystal.

per meter to ionize and accelerate deuterium nuclei into an erbium deuteride target. Then, the two deuterium nuclei fused together to produce a $^3$He nucleus, a 2.45 MeV neutron and radiation. This technique realized DD fusion without high temperatures and high pressure. Tritium could also be used for the nuclear fusion. Since the apparatus requires far more energy than it produces, it is not suitable for power generation. However, the apparatus is useful as a desktop neutron generator.

## 8.7 Condensation fusion in solids

Nuclear energy is the highest energy density power source on the Earth. In particular, nuclear fusion has no $CO_2$ production and strong radiation, and the nuclear fusion energy is one of the leading candidates for creating an ultimate energy source. Although a plan for ITER has been developed and pushed forward, the ITER presents many engineering problems, and is extremely expensive. On the other hand, the nuclear fusion phenomena produced in solid and liquid phases with deuterium have attracted a great deal of attention. In general, nuclear fusion of hydrogen atoms occurs at high temperatures under high pressure. However, condensation fusion at close to room temperatures was reported, and various fundamental studies are still ongoing. Studies on nuclear fusion for hydrogen storage metallic alloys are currently in progress [7–14]. Nuclear fusion in condensed matter is expected to have potential as a clean energy source because DD fusion produces only helium-4 and low radioactivity.

There are several characteristics of the condensation fusion:

- Smaller amount of neutrons detected in comparison to ordinary nuclear fusion.
- Few $\gamma$-rays detected.
- Deuterium is used for fusion.
- Occurrence of fusion in materials with fcc and hcp structures, and no fusion reaction for bcc structure.
- $^4$He is the main element generated.
- Reproducibility of nucleus change.
- Excess heat of $0.1 \sim 1$ W cm$^{-2}$.

The nuclear reaction in hydrogen storage metallic alloys and a cross section of DD nuclear fusion were summarized here. Metallic alloys for hydrogen storage such as palladium were selected for analysis. Using the results mentioned above, the possibility of nuclear fusion in the solid phase was discussed.

The atomic structure model of hydrogen (deuterium) storage Pd is shown in Fig. 8.7(a) [15]. The deuterium injection into the Pd crystal causes a temporal formation of D-cluster restricted by Platonic regular polyhedrons and tetrahedral symmetric condensate (TSC) [7–9], as shown in Fig. 8.7(b). Four deuterons and four electrons condensate at the T-site by background neutrons in nature, and these background neutrons play a role in promoting the condensation.

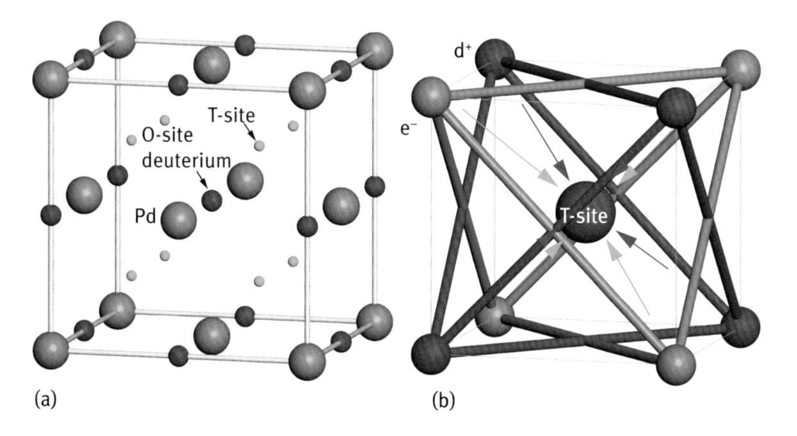


Fig. 8.7: (a) Atomic structure model of PdD. Deuterium atoms are situated at octahedral sites (O-site). Tetrahedral sites (T-site) are also indicated. a = 0.389 nm. (b) Schematic illustration of tetrahedral symmetry condensate of four deuterium ions and four electrons around the T-site.

The time-dependent TSC-cluster trapping potential was calculated as Eq. (8.11) from the Langevin equation for 4D/TSC based on the heavy mass electronic quasi-particle expansion theory calculation for barrier factors and the fusion rate [7–9], and the $R_{dd}$ is expected value of D-D distance.

$$V_{\mathrm{TSC}}(R' : R_{\mathrm{dd}}(t)) = -\frac{11.85}{R_{\mathrm{dd}}(t)} + 6V_s(R_{\mathrm{dd}}(t); m, Z) + 2.2\frac{|R' - R_{\mathrm{dd}}(t)|^3}{[R_{\mathrm{dd}}(t)]^4} \tag{8.11}$$

The fusion rate was calculated by Fermi's golden rule, and the 4D fusion yield per TSC generation was also calculated [9]. The fusion yield and fusion products of deuterium condensate in PdD are summarized in Table 8.2 [7–9]. Although neutrons and tritium are produced by 2D and 3D condensation, respectively, the fusion yield is not higher than that of 4D condensation of TSC. Instead of TSC, an octahedral symmetric condensate (OSC) model has also been proposed. The calculation indicates that OSC is stable at $R_{dd} = 40$ pm as the ground states, and the potential was calculated to be *c.a.* −800 eV.

Table 8.2: Fusion reaction of deuterium condensate in PdD.

| Multibody deuterium | Fusion yield ($fs^{-1}$ $cm^{-3}$) | Fusion product ($ns^{-1}$ $cm^{-3}$) |
|---|---|---|
| 2D | 1.9 | Neutron: 10 |
| 3D | $1.6 \times 10^9$ | Tritium: $8 \times 10^8$ |
| 4D | $3.1 \times 10^{11}$ | Helium: $3 \times 10^{11}$ |

On the other hand, the Langevan calculation indicates a lowest potential for TSC of −3500 eV for $R_{dd} = 3$ fm, as shown in Fig. 8.8. The condensation time was calculated to be 1.4 fs [9]. For $R_{dd}$, the deuterons and electrons condensate into an intermediate $^{8}$Be. After that, the $^{8}$Be collapses into two $^{4}$He. This indicates that the TSC model is more suitable for D-cluster condensation. For actual Pd crystals, the deuteron supply is required to preserve the continuous fusion reaction, and the diffusion mechanism of deuterium in the Pd crystal should be investigated.

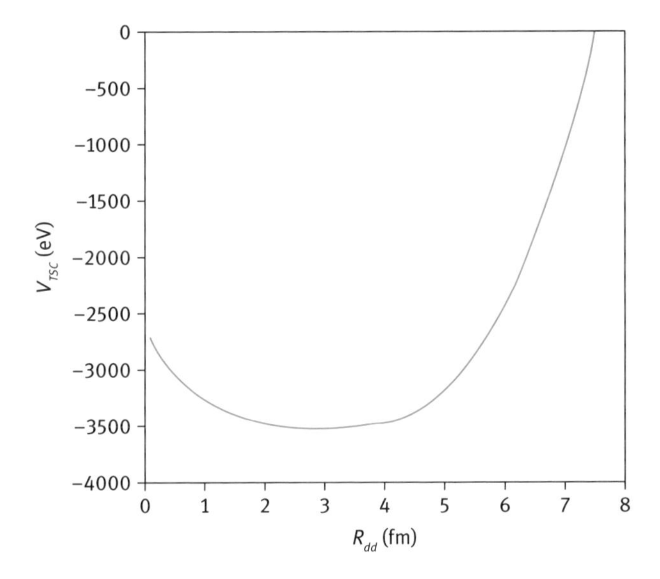


Fig. 8.8: TSC trapping potential at the final stage (TSC-min) of condensation calculated by Langevin equation.

The diffusion coefficient $D$ of deuterium in the Pd crystal was investigated using the data reported in Table 8.3 and the equation (8.12), where $D$: diffusion coefficient, $D_0$: constant, $E_D$: activation energy, $k$: Boltzmann constant and $T$: temperature.

$$D = D_0 \exp\left(\frac{-E_D}{kT}\right) \tag{8.12}$$

Although the reported data for the Pd-D system was obtained at around room temperature, deuteron diffusion at higher temperatures were estimated by the Flynn-Stoneham equation (8.13) based on the thermally activated tunneling mechanism [15–17]. The hopping probability $W$ between the nearest deuterium atomic sites is calculated as follows, where the tunneling matrix element $J$ value was estimated from the experimental data.

$$W = \left(\frac{J^2}{h}\right) \pi \sqrt{\left(\frac{\pi}{E_a kT}\right)} \exp\left(-\frac{E_a}{kT}\right) \tag{8.13}$$

Then, the diffusion coefficients at higher temperatures are calculated from the following relation:

$$D = \frac{d^2 W}{6} \tag{8.14}$$

The nearest D-D distance ($d$) is equal to the distance of the nearest octahedral-sites, which is 0.2751 nm. The diffusion coefficients of deuterium atoms in Pd as a function of temperature are shown in Fig. 8.9(a). The $D$ values increase as the temperature increases. Fig. 8.9(b) shows the diffusion time of deuterium atoms between the nearest O-sites in Pd, and the transport time of deuterium between the nearest neighboring D atoms in Pd crystal lattice is shown in Fig. 8.10 as a function of temperature [18]. The diffusion time decreases exponentially as the temperature increases, and then slowly decreases at higher temperatures near the melting point of Pd (1828 K). Therefore, the decrease in the diffusion time of deuteron caused by elevating the temperature is limited.

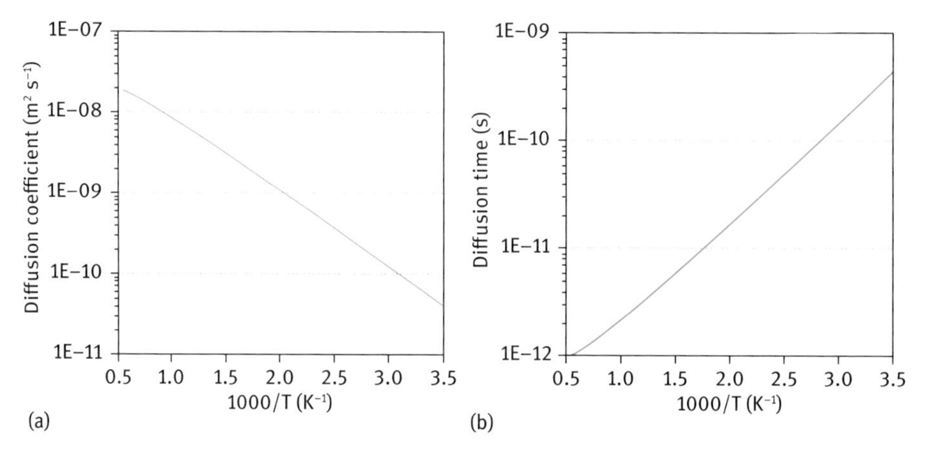


Fig. 8.9: (a) Diffusion coefficient of deuterium atoms in Pd as a function of temperature. (b) Diffusion time of deuterium atoms between the nearest O-sites in Pd.

Table 8.3: Diffusion of deuterium atoms in fcc metals.

| Metal-D | $E_a$ (eV) | $D_0$ ($10^{-7}$ m$^2$ s$^{-1}$) | Temperature (K) |
|---|---|---|---|
| Pd-D | 0.21 | 1.7 | 218–333 |
| Ni-D | 0.40 | 4.2 | 220–1273 |
| Cu-D | 0.38 | 7.3 | 723–1073 |

To reduce the diffusion time, several approaches can be considered: reducing the activation energy, particle mass $m$ and distance $d$. If the other elements are intro-

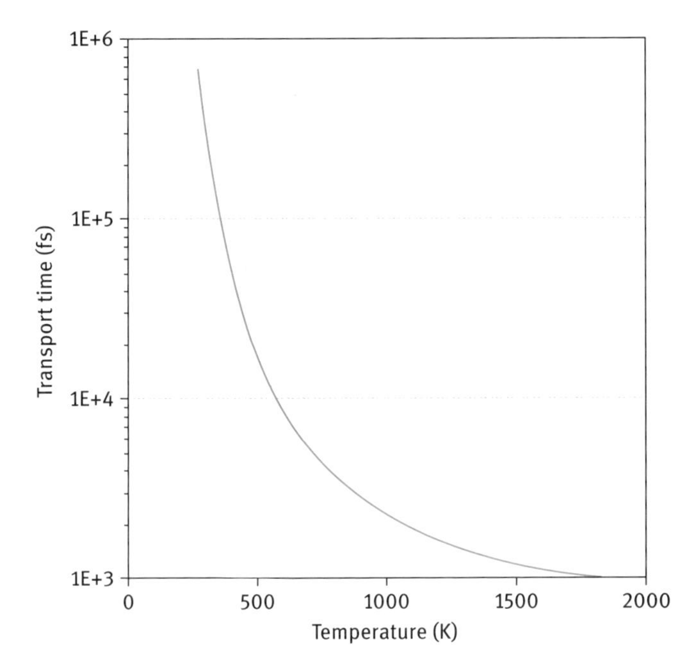


Fig. 8.10: Transport time of deuterium between the nearest neighboring D atoms in Pd crystal lattice as a function of temperature.

duced into the Pd crystal, a reduction of the barrier energy of deuteron diffusion or an increase in the tunneling probability may be expected. Here, silver (Ag) was introduced into the Pd crystal, and the diffusion coefficients and diffusion time were calculated based on the proposed data [19], which indicated almost the same values as for the non-Ag doped Pd. Continuous fusion reaction of deuterium was reported for Pd-$ZrO_2$ nanocomposite materials [20], and further studies on the other elements or other methods might be examined [21–24].

A schematic illustration of tetrahedral symmetric condensate and deuteron diffusion is summarized in Fig. 8.11 [8, 9]. At the beginning of the condensation process, tetrahedral clusters of $4d^+$ and $4e^-$ are formed. After 1.4 fs, an intermediate nucleus $^8$Be forms from the smallest TSC. Then, the $^8$Be collapses into two $^4$He. In order to continue the TSC, the deuterium should be supplied at the deuterium site in the Pd lattice. The diffusion time of deuterium is fairly long (~ 1000 fs) in comparison to the fusion time for the 4D/TSC (1.4 fs), and a continuous fusion reaction may be difficult. Another method should be introduced to promote deuterium diffusion in the Pd crystal for the fusion reaction. Although the Pd element was selected for an analysis of continuous tetrahedral symmetric condensation of deuterium clusters, palladium has the material problem of being scarce and its cost is consequently high. Other hydrogen storage compounds should be investigated in further considering hydrogen diffusion in hydrogen storage materials [25].

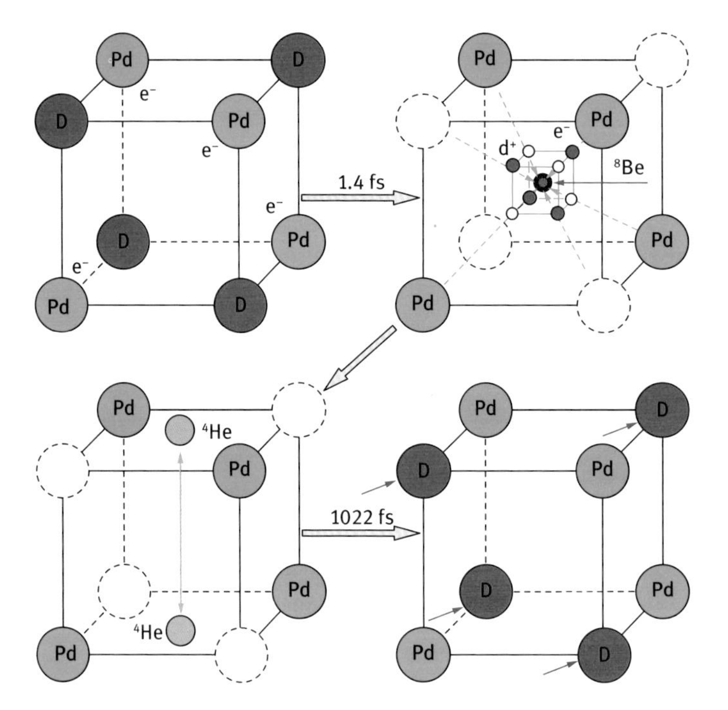


Fig. 8.11: TSC and diffusion model for deuterons in Pd.

The above 4D-TSC model predicted the transitory Bose-Einstein condensate (BEC) and $^4$He production in microscopic space in hydrogen crystals with high symmetry. When the deuterons are introduced into the Pd lattice, the crystal lattice could expand and lose its tetrahedral symmetry, which might cause a decrease in the deuterium condensates in Pd. Although other theoretical predictions on the BEC condensate and deuteron fusion have been reported [26–28], further quantitative calculation and analysis are necessary to describe deuteron fusion.

The possibility of hydrogen storage in Pd was studied using a diffusion calculation and a potential calculation on 4D fusion. The nuclear fusion model for the 4D/TSC and diffusion of deuterium in Pd alloys was investigated. The diffusion time of deuterium is fairly long in comparison to the fusion time for 4D/TSC, and creating a continuous fusion reaction may be difficult. The diffusion time of deuterium at the Pd-Ag alloy surface was almost the same as that of Pd. Enhancing deuterium diffusion in Pd alloys will be key in developing potential future methods for continuous nuclear fusion or for hydrogen gas storage.

## 8.8 Fusion reactor materials

Both excellent erosion resistance to high heat loading from plasma and high resistance to radiation damage by high-energy neutrons are required for the plasma facing components used in fusion reactors such as ITER. Although carbon materials have been extensively used for plasma facing components because of carbon's low atomic number and comparatively high thermal conductivity, its thermal conductivity is quite low at high temperatures such as those that occur during the operation of fusion reactors [29]. Carbon/copper (C/Cu) composite materials have been produced because of the high thermal conductivity of Cu. The thermal conductivity of C/Cu composites was measured and examined at up to 1400 K for fusion applications [30, 31], and these measurements showed some improvement in the composites' thermal conductivity in comparison to C. Although C/Cu composite materials are promising from the standpoint that they have higher thermal conductivity at high temperatures, there are some cases where their thermal conductivity at room temperature becomes smaller than that of the original C/C composite materials. The reason is that the cohesion of the C/Cu interface is poor.

The goals of this study were twofold. The first was to develop carbon-based materials with high thermal conductivity and good stability at high temperatures. For application in fusion devices, Cu was selected as a metallic element because of its high thermal conductivity. The crystal structures of carbon [32] and copper [33] are shown in Fig. 8.12(a) and (b), respectively. In addition, titanium (Ti) was selected in a small amount as a third element because Ti has low enthalpy ($\Delta H$) in alloy formation with C ($\Delta H$ = –156 kJ mol$^{-1}$) and Cu ($\Delta H$ = –40 kJ mol$^{-1}$) [34], as shown in Fig. 8.12(c). In Fig. 9.8, elements with $\Delta H$ values higher than 0 should be removed from the candidates. Although zirconium and hafnium also have low $\Delta H$ with C and Cu, Ti has the best cost performance. The addition of Ti to C/Cu composite materials is expected to result in high thermal conductivity due to strong cohesion at the C/Cu interface. My The second goal of the study was to understand the heat conduction mechanism of carbon materials by analyzing their microstructure using optical microscopy, transmission electron microscopy and high-resolution electron microscopy. In particular, HREM is a powerful method for investigating the interfacial structures of advanced materials.

Nuclear grade fine-grained isotropic graphite (IG-430U, IG) and C/C composite (CX-2002U, CX) were used for the base materials, which are used for the plasma facing components of some fusion devices such as JT-60. Molten Cu and Cu(Ti) were impregnated in these carbon materials at a pressure of 15 MPa. These samples were denoted as IG–Cu, IG–Cu(Ti), CX–Cu and CX–Cu(Ti). The sample compositions were: IG–Cu (33.1 wt.% Cu), IG–Cu(Ti) (36.0 wt.% Cu and 0.8 wt.% Ti), CX–Cu (44.3 wt.% Cu) and CX–Cu(Ti) (48.1 wt.% Cu and 1.1 wt.% Ti) [31]. CX is a felt type C/C composite, and thermal conduction takes place along the fiber plane of the felt composites. The specimens were 10 mm in diameter and 2–4 mm in thickness. Thermal conductivity

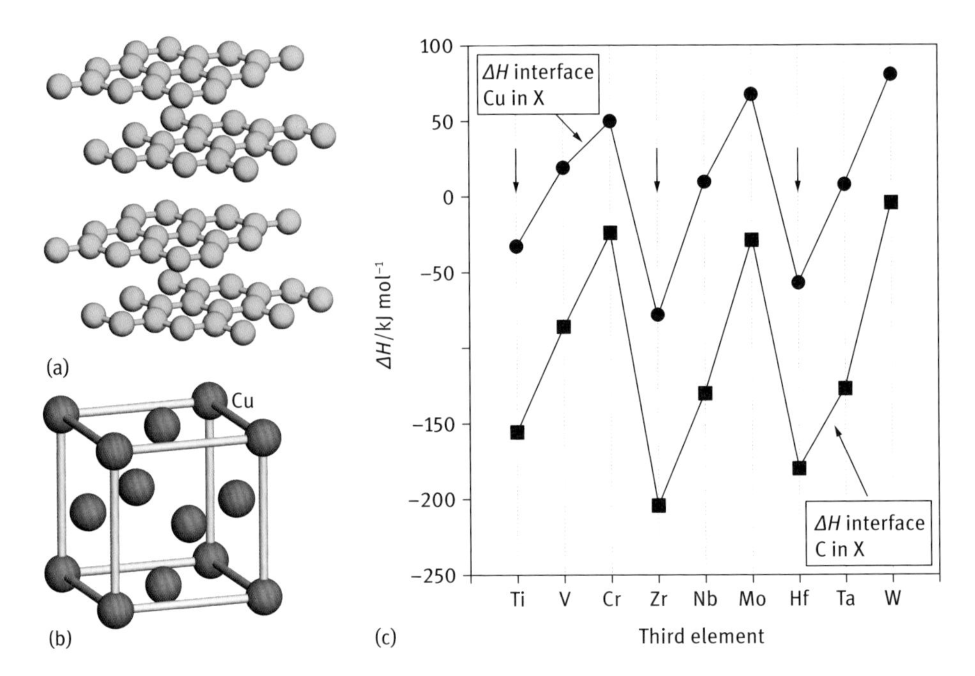


Fig. 8.12: Crystal structures of (a) carbon and (b) copper. (c) Calculated formation enthalpy of refractory elements with C and Cu at the interface.

measurements were performed in the range of 293–1200 K using a laser flush device for thermal diffusivity and specific heat. The measurement conditions were as follows: laser pulse energy of 0.6 J, pulse width of 0.5 μs, temperature increase of 1.8–2.8 K at room temperature and 0.6–0.8 K at 1000 K.

The temperature dependences for thermal conductivity for the two base carbon materials are shown in Fig. 8.13. Although the thermal conductivity of IG–Cu is larger than that of the original sample at higher temperatures, as shown in Fig. 8.13(a), it is smaller than the original one at room temperature, which could be due to poor cohesion at the C/Cu interface. In order to improve the cohesion between C and Cu, a small amount of Ti was added to the IG–Cu and CX–Cu. A clear increase in thermal conductivity over the measured temperature range and a decrease in temperature dependence are observed in Fig. 8.13(a) and (b). CX is a felt-type C/C composite, and its thermal conductivity measurements were performed in the high thermal conductivity plane, i.e. within the felt layer. In this case, the effect of the addition of titanium to the copper-impregnated material on the thermal conductivity was clear. The reason that the thermal conductivity of CX–Cu is smaller than that of the original material is that insufficient cohesion exists between carbon and copper.

To observe the morphology of IG–Cu(Ti), which has high thermal conductivity, an optical micrograph of IG–Cu(Ti) was taken as shown in Fig. 8.14(a). Many small Cu grains (20 μm) with white contrast are distributed in the IG–Cu(Ti) sample homo-

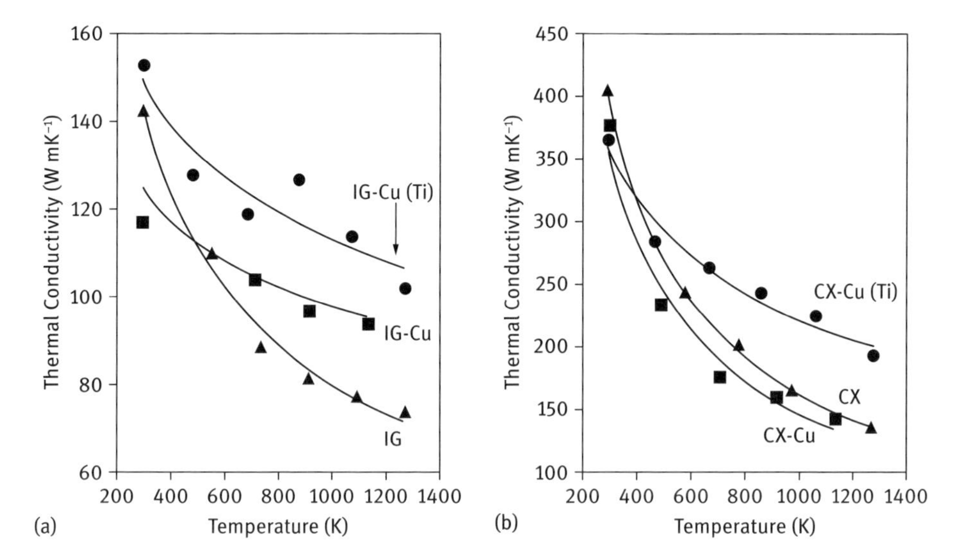


Fig. 8.13: Temperature dependence for thermal conductivity for (a) IG-based composites and (b) CX-based composites.

geneously. No change in the optical micrographs and X-ray diffraction patterns was observed after the addition of Ti to C/Cu composite materials. TEM observations were also carried out, as shown in Fig. 8.14(b). Figure 8.14(b) is a low magnification image of the C/Cu interface in the IG–Cu(Ti) sample. A reaction layer with many grains is observed at the C/Cu interface. The Cu/TiC and TiC/C interfaces are shown in Fig. 8.14(c) and (d), respectively [35]. The lattice fringes of Cu, $CuTi_2$ and TiC, which were determined by the d-spacings of the lattice fringes, are observed. The grain sizes of the TiC and $TiCu_2$ compounds are 5 nm. The electron diffraction patterns of the C/Cu interface also showed Debye–Scherrer rings corresponding to TiC and $CuTi_2$ in addition to the Cu reflections.

Figure 8.15 shows the calculated formation enthalpy of Ti–C and Ti–Cu compounds [34, 35]. The melting points ($T_m$) of the compounds are also indicated in Fig. 8.15(a) [36]. The TiC compound has Ti composition in the range of 32–49 at.%, which correspond to the compounds of $TiC_2$, $Ti_3C_5$, $Ti_2C_3$ and TiC.

In these compositions, the TiC has the lowest $\Delta H$ value as indicated in Fig. 8.15(a), which could result in the formation of TiC. On the other hand, in Ti–Cu compounds, the differences in $\Delta H$ values are not so large. The melting points of the Ti–Cu compounds are indicated in Fig. 8.15(a), and the $Ti_2Cu$ compound has the highest melting point of 1278 K, which could result in the formation of $Ti_2Cu$ at the interface.

The above results indicate that a small amount of titanium reacted with C and Cu to form TiC and $TiCu_2$ compounds, resulting in an increase in thermal conductivity. Fig. 8.15(b) shows a model of thermal conduction for the C/Cu(Ti) composite materials presented in this text. It is believed that thermal energy would pass through

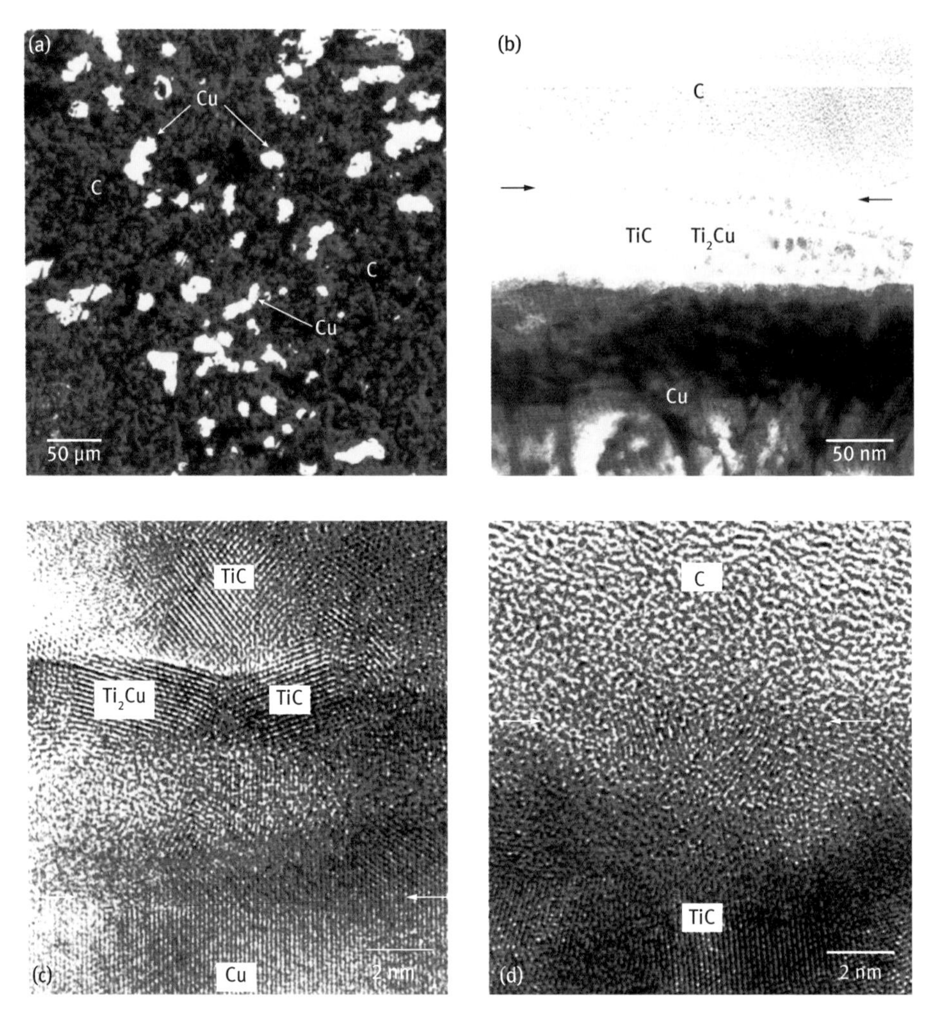


Fig. 8.14: (a) Optical micrograph of IG–Cu(Ti). (b) TEM image, (c), (d) HREM images of the Cu/C interface.

both C and C/Cu/C, and that the heat flux is higher around the Cu particles, as shown in Fig. 8.15(b). In particular, thermal conductivity is increased by the addition of a small amount of titanium, which forms TiC and Ti–Cu compounds at the interface. This compound formation would increase cohesion at the C/Cu interface, resulting in high thermal conductivity at high temperatures. The present study indicates that addition of a small amount of a third element with low formation enthalpy with C and Cu will increase the thermal conductivity at high temperatures. The $\Delta H$ values can be one of the guidelines for material selection [37–41]. Since the element zirconium has the lowest $\Delta H$ value with C and Cu, a further increase in thermal conductivity may be possible through the addition of a small amount of zirconium to C/Cu materials.

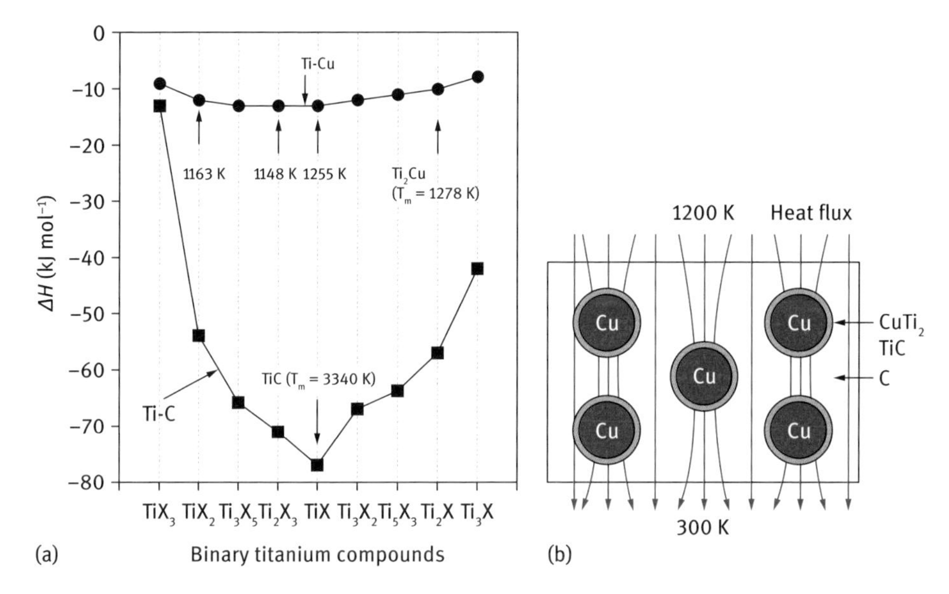


Fig. 8.15: (a) Calculated formation enthalpy of Ti–C and Ti–Cu compounds. The melting points ($T_m$) of compounds are indicated. (b) Schematic illustration of thermal conduction for C/Cu(Ti) composite materials.

In summary, C/Cu-based materials with high thermal conductivity at high temperatures were developed by adding a small amount of titanium. The isotropic fine-grained nuclear grade graphite and felt type C/C composite, which were impregnated with Cu (36–48 wt.%) and Ti (0.8–1.1 wt.%), provided 1.3 times higher thermal conductivity of 110 and 200 W mK$^{-1}$ at 1200 K than the original carbon materials. Microstructural analysis showed that the increase of thermal conductivity was due to the formation of titanium compounds of TiC and $CuTi_2$ at the C/Cu interface, and that the thermal energy could pass through both the C and Cu. The present study indicates that addition of a small amount of a third element with a low enthalpy of alloy formation with C and Cu will further increase the thermal conductivity and the stability of C/Cu-based materials. These carbon-based materials could be a candidate for materials used in the plasma facing components of fusion devices.

## 8.9 Bibliography


[1] ITER. http://www.iter.org/proj/inafewlines

[2] JAEA. http://www.jaea.go.jp/english/index.html

[3] Imao H, Ishida K, Kawamura N, Matsuzaki T, Matsuda Y, Toyoda A, Strasser P, Iwasaki M, Nagamine K. Density effect in d-d muon-catalyzed fusion with ortho- and para-enriched $D_2$. *Phys Lett B*. 2008; 658: 120–124.

[4] Knowles PE, Adamczak A, Bailey JM, Beer GA, Beveridge JL, Fujiwara MC, Huber TM, Jacot-Guillarmod R, Kammel P, Kim SK, Kunselman AR, Marshall GM, Martoff CJ, Mason GR, Mulhauser F, Olin A, Petitjean C.; Porcelli TA, Zmeskal J. Muon catalyzed fusion in 3 K solid deuterium. *Phys Rev A*. 1997; 56: 1970–1982.
[5] Kasatani H, Ootaka H, Aoyagi S, Kimura A, Kuroiwa Y. Charge Density Study on the Ferroelectric Phase in $LiTaO_3$ by Synchrotron Radiation Powder Diffraction. *Ferroelectrics*. 2004; 304: 163–166.
[6] Naranjo B, Gimzewski JK, Putterman S. Observation of nuclear fusion driven by a pyroelectric crystal. *Nature*. 2005; 434: 1115–1117.
[7] Takahashi A, Yabuuchi N. Fusion rates of bosonized condensates. *J Cond Matt Nucl Sci*. 2007; 1: 106–128.
[8] Takahashi A. A theoretical summary of condensed matter nuclear effects. *J Cond Matt Nucl Sci*. 2007, 1, 129–141.
[9] Takahashi A. Dynamic mechanism of TSC condensation motion. *J Cond Matt Nucl Sci*. 2009; 2: 33–44.
[10] Heidenreich A, Jortner J, Last I. Cluster dynamics transcending chemical dynamics toward nuclear fusion. *Proc Natl Acad Sci*. 2006; 103: 10589–10593.
[11] Kayano H, Teshigawara M, Konashi K, Yamamoto T. Derivation of energy generated by nuclear fission-fusion reaction. *Sci Rep Res Ins Tohoku Univ Ser A*. 1994; 40: 13–15.
[12] Konashi K, Kayano H, Teshigawara M. Analysis of heavy-ion-induced deuteron-deuteron fusion in solids. *Fusion Sci Tech*. 1996; 29: 379–384.
[13] Bom VR, Demin DL, van Eijk CWE, Filchenkov VV, Grafov NN, Grebinnik VG, Gritsaj KI, Konin AD, Kuryakin AV, Nazarov VA, Perevozchikov VV, Rudenko AI, Sadetsky SM, Vinogradov YI, Yukhimchuk AA, Yukhimchuk SA, Zinov VG, Zlatoustovskii SV. Measurement of the temperature dependence of the ddμ molecule formation rate in dense deuterium at temperatures of 85-790 K. *J Experimental Theor Phys*. 2003; 96: 457–464.
[14] Takahashi A, Maruta K, Ochiai K, Miyamaru H, Iida T. Anomalous enhancement of three-body deuteron fusion in titanium-deuteride with low-energy $D^+$ beam implantation. *Fusion Sci Tech*. 1998; 34: 256–272.
[15] Flynn C.P, Stoneham AM. Quantum theory of diffusion with application to light interstitials in metals. *Phys Rev B*. 1970; 1: 3966–3978.
[16] Hirth John P. Effects of hydrogen on the properties of iron and steel. *Metall Trans A*. 1980; 11: 861–890.
[17] Sugimoto H. Diffusion mechanism of hydrogen in solids. *J Vac Soc Jpn*. 2006; 49: 17–22.
[18] Oku T, Kitao T. Hydrogen storage and possible condensation of deuterium in palladium. *Nanosci Nanotech Asia*. 2015; 5: 137–143.
[19] Ozawa N, Nakanishi H, Kunikata S, Kasai H. A behavior of a hydrogen atom $Pd_{0.75}Ag_{0.25}(111)$. *J Vac Soc Jpn*. 2007; 50: 440–443.
[20] Kitamura A, Nohmi T, Sasaki Y, Taniike A, Takahashi A, Seto R, Fujita Y. Anomalous effects in charging of Pd powders with high density hydrogen isotopes. *Phys Lett A*. 2009; 373: 3109–3112.
[21] Grochala W, Edwards PP. Thermal decomposition of the non-interstitial hydrides for the storage and production of hydrogen. *Chem Rev*. 2004; 104: 1283–1316.
[22] Horinouchi S, Yamanoi Y, Yonezawa T, Mouri T, Nishihara H. Hydrogen storage properties of isocyanide-stabilized palladium nanoparticles. *Langmuir*. 2006; 22: 1880–1884.
[23] Mitsui T, Rose MK, Fomin E, Ogletree DF, Salmeron M. Dissociative hydrogen adsorption on palladium requires aggregates of three or more vacancies. *Nature*. 2003; 422: 705–707.
[24] Kusada K, Kobayashi H, Ikeda R, Kubota Y, Takata M, Toh S, Yamamoto T, Matsumura S, Sumi N, Sato K, Nagaoka K, Kitagawa H. Solid solution alloy nanoparticles of immiscible Pd and Ru

elements neighboring on Rh:changeover of the thermodynamic behavior for hydrogen storage and enhanced CO-oxidizing ability. *J Am Chem Soc.* 2014; 136: 1864–1871.
[25] Bowman Jr. RC, Brent F. Metallic hydrides I: hydrogen storage and other gas-phase applications, *MRS Bull.* 2002: 688–693.
[26] Kim YE. Theory of Bose–Einstein condensation mechanism for deuteron-induced nuclear reactions in micro/nano-scale metal grains and particles. *Naturwissenschaften.* 2009; 96: 803–811.
[27] Kim YE. Generalized theory of Bose-Einstein condensation nuclear fusion for hydrogen-metal system. *J Cond Matt Nucl Sci.* 2011; 4: 188–208.
[28] Widom A, Larsen L. Ultra low momentum neutron catalyzed nuclear reactions on metallic hydride surfaces. *Eur Phys J C.* 2006; 46: 107–111.
[29] Burchell TD, Oku T. Material properties data for fusion reactor plasma-facing carbon-carbon composites, *Atomic and Plasma-Material Interaction Data for Fusion.* 1994; 5: 77–128.
[30] Oku T, Hiraoka T, Kuroda K, Improvement of thermal conductivity of carbon materials due to addition of metal particles. *J Nucl Sci Technol.* 1995; 32: 816–818.
[31] Oku T, Kurumada A, Sogabe T, Oku T, Hiroka T, Kuroda K, Effects of titanium impregnation on the thermal conductivity of carbon/copper composite materials. *J Nucl Mater.* 1998; 257: 59–66.
[32] Su YC, Yan J, Lu PT, Su JT. Thermodynamic analysis and experimental research on Li intercalation reactions of the intermetallic compound $Al_2Cu$. *Solid State Ion.* 2006; 177: 507–513.
[33] Reibold M, Levin AA, Meyer DC, Paufler P, Kochmann W. Microstructure of a Damascene sabre after annealing. *Int J Mater Res.* 2006; 97: 1172–1182.
[34] de Boer FR, Boom R, Mattens WCM, Miedema AR, Niessen AK. Cohesion in Metals, Transition Metal Alloys. North-Holland, Amsterdam. 1988.
[35] Oku T, Oku T. Effects of titanium addition on the microstructure of carbon/copper composite materials. Solid State Commun. 2007; 141: 132–135.
[36] Massalski TB (Ed). Binary Alloy Phase Diagrams. ASM International. 1990.
[37] Oku T, Narita I, Nishiwaki A, Koi N. Atomic structures, electronic states and hydrogen storage of boron nitride nanocage clusters, nanotubes and nanohorns. *Defect Diffus Forum.* 2004; 226–228: 113–140.
[38] Oku T, Narita I, Nishiwaki A, Koi N, Suganuma K, Hatakeyama R, Hirata T, Tokoro H, Fujii S. Formation, atomic structures and properties of carbon nanocage materials. *Topics Appl Phys.* 2006; 100: 187–216.
[39] Oku T, Kawakami E, Uekubo M, Takahiro K, Yamaguchi S, Murakami M. Diffusion barrier property of TaN between Si and Cu. *Appl Surf Sci.* 1996; 99: 265–272.
[40] Uekubo M, Oku T, Nii K, Murakami M, Takahiro K, Yamaguchi S, Nakano T, Ohta T. WNx diffusion barriers between Si and Cu. *Thin Solid Films.* 1996; 286: 170–176.
[41] Oku T, Furumai M, Uchibori CJ, Murakami M. Formation of WSi-based ohmic contacts to n-type GaAs. *Thin Solid Films.* 1997; 300: 218–222.

# 9 Other energy materials

## 9.1 Hydrogen storage materials

Global warming due to carbon dioxides ($CO_2$) has become a serious point of concern, and the development of new energy sources to surpass conventional fossil fuel has become an important issue. Although solar cells represent one of the most important methods for solving current energy problems, the energy densities of solar cells are not very high. On the other hand, hydrogen is a carrier with high energy density and forms only water and heat. Therefore, clean hydrogen energy is expected to serve as a substitute for fossil fuel in the 21st century and a gas storage ability more than 6.5 wt.% is needed for car application according to the US Department of Energy. Although $LaNi_5H_6$ is already used as a $H_2$ gas storage material, its storage ability is ~ 1 wt.% because of the large atomic numbers of La and Ni [1–3]. On the other hand, fullerene-like materials, which consist of light elements such as boron, carbon and nitrogen, may store more $H_2$ gas compared to metal hydrides. In this chapter, hydrogen gas storage in boron nitride (BN) and carbon (C) nanomaterials are reviewed.

Many works have reported on the hydrogen storage ability of carbon nanotubes, fullerenes and carbon nanomaterials [4–9]. It has been reported that multi-walled carbon nanotubes could absorb hydrogen from 1 wt.% up to 4.6 wt.%. These results indicate that carbon nanotubes would be a possible candidate for a hydrogen storage material, although further evaluation of hydrogen storage measurements is necessary, and many other studies have reported on $H_2$ gas storage in carbon nanomaterials.

Boron nitride (BN) nanomaterials are also expected to have application because they provide good stability at high temperatures with high electronic insulation in air [10]. Hydrogen storage in BN nanomaterials has also been studied recently [11–22]. It is difficult to absorb hydrogen by physisorption either inside nanotubes or at the interstitial channels of nanotubes at room temperature, which results from the fact that the bonding of physisorbed hydrogen is too weak to achieve large-scale storage at room temperature. Therefore, chemisorption is considered to be a necessary requirement for hydrogen storage at room temperature. It has been reported that the chemisorption ratio observed at room temperature increases with an increasing alkali/carbon rate. Therefore, if the energy of chemisorption can be lowered, the hydrogen storage ability of carbon and BN nanotubes would be improved.

Various metal alloys have also been developed and investigated [23–25], as listed in Table 9.1 [2]. The crystal structures of $LaNiH_6$ [26], (b) $FeTiH_{0.06}$ [27], (c) $ZrV_2H_{5.1}$,[28] (d) $Ti_{0.35}V_{0.65}H_{1.95}$[29] and (e) $Mg_2NiH_4$[30] are shown in Fig. 9.1, indicating the existence of various crystal systems. Since hydrogen atoms are ordered at special sites in the crystal structures of the hydrogen storage alloys, high hydrogen storage densities are possible. Hydrides of magnesium alloys store ~ 7 wt.% hydrogen, and titanium, zirconium, palladium and vanadium alloys have also been studied as hydrogen storage

DOI 10.1515/9783110298505-012

Table 9.1: Materials and their hydrogen storage.

| Type | Structure | Hydride | wt.% of H | Pressure (MPa) | T (K) |
|---|---|---|---|---|---|
| A | *Fm3m* | $PdH_{0.6}$ | 0.56 | 0.0020 | 298 |
| $AB_5$ | $P_6/mmm$ | $LaNi_5H_6$ | 1.37 | 0.2 | 298 |
| $AB_2$ | *Fd3m* | $ZrV_2H_{5.5}$ | 3.01 | 10–9 | 323 |
| $AB_2$ | *Fd3m* | $TiV_2H_4$ | 2.6 | 1 | 313 |
| AB | *Pm3m* | $FeTiH_2$ | 1.89 | 0.5 | 303 |
| $A_2B$ | $P_6222$ | $Mg_2NiH_4$ | 3.59 | 0.1 | 555 |
| Nano carbon | | | ~ 1.5 | 0.1 | 77 |

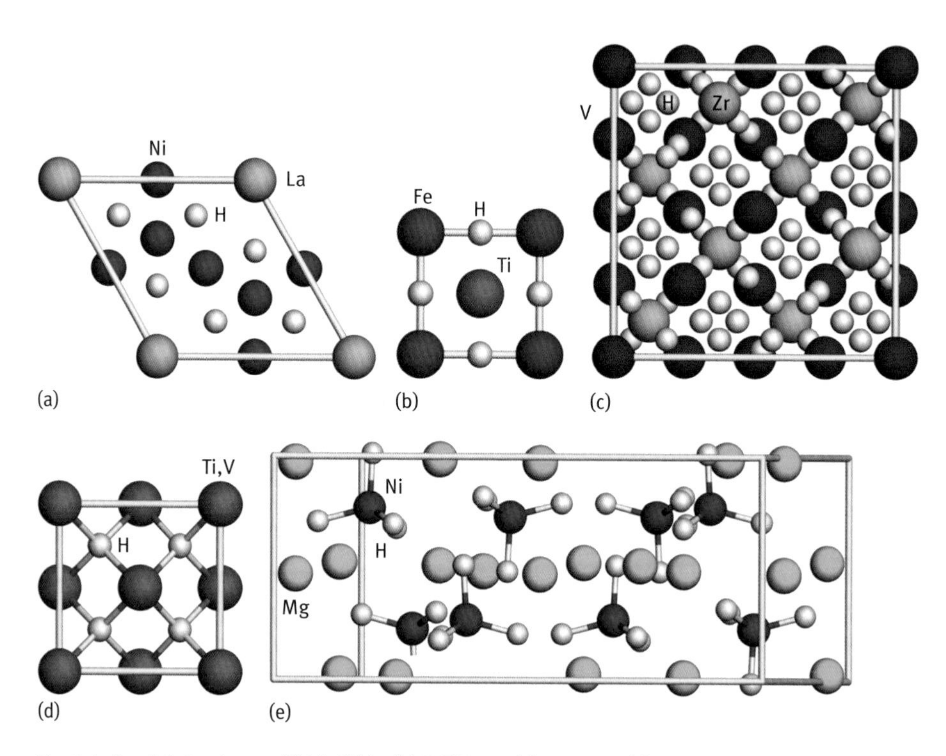


Fig. 9.1: Crystal structures of (a) $LaNiH_6$, (b) $FeTiH_{0.06}$, (c) $ZrV_2H_{5.1}$, (d) $Ti_{0.35}V_{0.65}H_{1.95}$ and (e) $Mg_2NiH_4$.

materials. Although pure palladium is not superconductive, PdH and PdD are superconductors with transition temperatures Tc of ~ 9 K and ~ 11 K, respectively, and the crystal lattice is slightly expanded [31–35]. Lithium hydride and sodium borohydride are also good candidates for hydrogen storage. Since these metal alloys bind with hydrogen strongly, high temperatures are required to release the hydrogen content. This energy cost could be reduced by using alloys which consist of elements with larger and smaller binding energies for hydrogen.

Nanostructured carbon materials reversibly absorbed 1.5 wt.% hydrogen at 77 K, which agreed with the calculation [36]. Various metals' alloys are able to absorb large amounts of hydrogen reversibly, as listed in Table 9.1. The equilibrium pressure depends on the temperature, which is related to changes in enthalpy and entropy. The entropy change corresponds to the change from molecular hydrogen to dissolved atomic hydrogen, and is ~ 130 J $K^{-1}$ $mol^{-1}$ for metal-hydrogen systems [2]. Hydrogen atoms are located at the interstitial site of the crystal lattice. The host metal lattice expands through hydrogen storage, sometimes losing the high symmetry of the crystal lattice [2]. Hydrogen positions have already been described in the chapter on nuclear fusion. The loss of crystal symmetry would also affect the deuterium condensate in Pd.

## 9.2 BN nanomaterials

BN nanotubes, nanocapsules and nanocages were prepared, and hydrogen gas storage was investigated using thermogravimetry/differential thermogravimetric analysis (TG/DTA). $LaB_6$ and Pd were selected in order to take advantage of their catalytic effects to produce the BN nanomaterials. La has shown excellent catalytic properties for producing a large number of single-walled carbon nanotubes and enlarging their diameter [37], and Pd is also expected to act as a hydrogen storage material. Although gas storage of hydrogen and argon in carbon nanotubes has been reported [4, 38], carbon nanotubes are oxidized at 600 °C in air [39]. On the other hand, BN starts to oxidize into boron oxide and nitrogen at ~ 900 °C in air, which indicates excellent heat resistance when compared to carbon materials in terms of gas storage [10]. The differences between BN and C nanomaterials are summarized, as shown in Table 9.2. To understand the formation of BN nanostructures, HREM were carried out, which are very powerful methods for atomic structural analysis [40, 41]. For the BN nanomaterials, hydrogen gas storage measurements were carried out using TG/DTA.

Mixture powder compacts made of boron particles (99 %, 40 mm, Niraco Co. Ltd), $LaB_6$ particles (99 %, 1 mm, Wako) and Pd particles (99.5 %, 0.1 mm, Niraco), with 3 mm in height and 30 mm in diameter were produced by pressing powder at 10 MPa. The atomic ratios of metal (M) to boron (B) were in the range of 1 : 1–1 : 10. The green compacts were set on a copper mold in an electric-arc furnace, which was evacuated. After introducing a mixed gas of Ar 0.025 MPa and $N_2$ 0.025 MPa, arc-melting was applied to the samples at an accelerating voltage of 200 V and an arc current of 125 A for 2 s. Arc-melting was performed with a vacuum arc-melting furnace, and the white/gray BN nanomaterial powders were collected from the surface of the bulk. To confirm the formation of BN fullerene materials, energy dispersive x-ray (EDX) spectroscopy analysis was performed using a probe size of ~ 10 nm. In order to measure hydrogen gas storage in BN nanomaterials, BN nanomaterials were extracted from the obtained powder by a supersonic dispersion method based on the Stokes equation using ethanol. Since there is a big difference in the size and density of theBN nanocap-

Table 9.2: Differences between boron nitride (BN) and C nanomaterials.

| | BN | C |
|---|---|---|
| Structure | 4-, 6-, and 8-membered rings | 5-, 6-, and 7-membered rings |
| Oxidation resistance | ~ 900 °C | ~ 600 °C |
| Electronic property | Insulator | Metal-semiconductor |
| Energy gap (eV) | ~ 6 | 0~ 1.7 |
| Band structure | Direct transition | Indirect transition |

sules/nanotubes produced and other powders, it is believed that this method would be effective for separating BN nanomaterials [42, 43]. After separation of BN nanomaterials, hydrogen storage was measured by TG/DTA at temperatures in the range of 20–300 °C in $H_2$ atmosphere [44].

For some metals, formation enthalpies with boron ($\Delta H_{MB}$) and nitrogen ($\Delta H_{MN}$) are indicated in Fig. 9.2(a) and (b), respectively. The data was derived from theoretical calculations. Differences in formation enthalpy ($\Delta H_{MN} - \Delta H_{MB}$) are also shown in Fig. 9.2(c). The difference in formation enthalpy ($\Delta H_{MN} - \Delta H_{MB}$) is important for the formation of BN fullerene nanomaterials, because reactivity with nitrogen and boron is strongly dependent on this enthalpy. Basically, BN nanotubes are formed when rare earth metals are used as catalytic metals, such as Y, Zr, Nb, Hf, Ta, W and La. These elements have minus enthalpy, as shown in Fig. 9.2(c). This means that catalytic elements for synthesis of BN nanotubes should be selected from those with a minus formation enthalpy ($\Delta H_{MN} - \Delta H_{MB}$). According to these guidelines, Sc could be a good catalytic

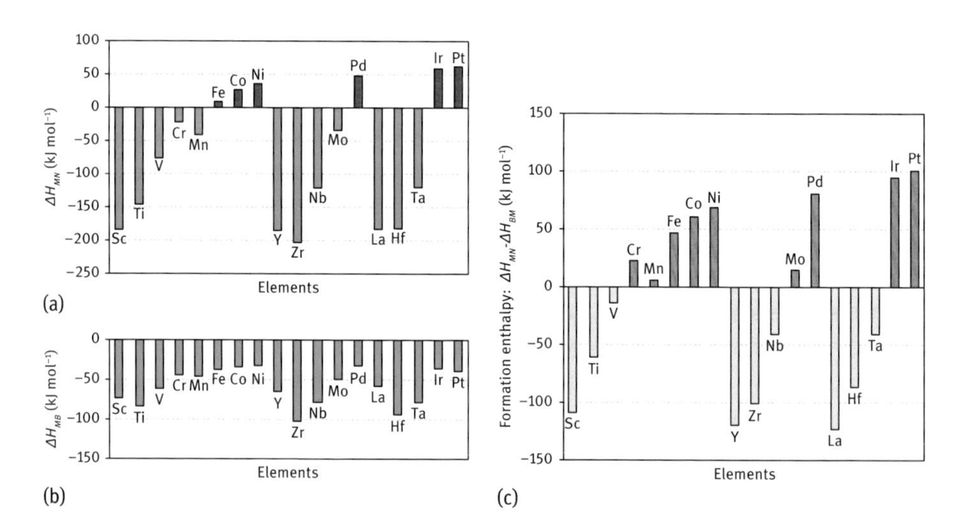


Fig. 9.2: Formation enthalpy with (a) nitrogen $\Delta H_{MN}$ and (b) boron $\Delta H_{MB}$. (c) Difference of formation enthalpy between boron $\Delta H_{MN}$ and nitrogen $\Delta H_{MB}$.

element for forming BN nanotubes. The detailed formation mechanism of the nanostructures with different metals should be investigated further.

A HREM image of a BN nanotube produced using $LaB_6$ powder is shown in Fig. 9.3(a) [19]. In Fig. 9.3(a), the diameter of the five-layered BN nanotube changes from right to left. Low magnification and HREM images of BN nanostructures were reported in detail [40]. A HREM image of a BN nanocage produced from $LaB_6$/B powder is shown in Fig. 9.3(b), which indicates a square-like shape, and four-membered rings of BN exist at the corner of the cage, as indicated by an arrow. The BN nanocage has a network-like structure, whose atomic arrangement is basically consistent with the $B_{36}N_{36}$ cluster structure [10]. BN nanocapsules with Pd nanoparticles were also produced as shown in Fig. 9.3(c), and Pd nanoparticles were covered by a few BN layers, as indicated by arrows. In order to confirm the formation of BN nanocapsules, EDX analysis was carried out, which showed the atomic ratio of B:N ~ 1.

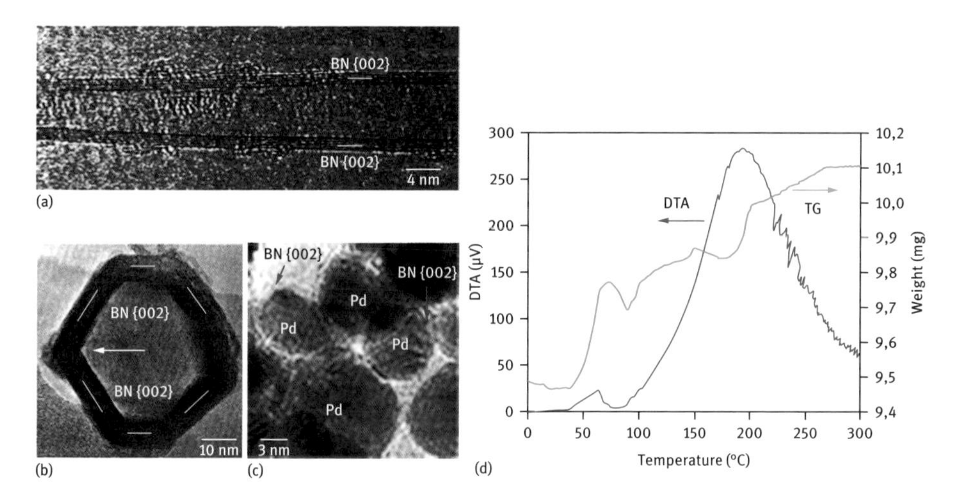


Fig. 9.3: HREM images of BN nanotube and BN nanocapsules produced using powder with ratios of: (a) La:B = 1 : 6; (b) La:B = 1 : 4; and (c) Pd:B = 1 : 4. (d) Differential thermogravimetric analysis and thermogravimetry curve of BN nanocapsules and nanotubes produced using $LaB_6$ powder.

The DTA and TG curve of BN nanomaterials produced from $LaB_6$ powder is shown in Fig. 9.3(d) [44]. At a temperature of around 70 °C, an increase in sample weight of 0.3 mg is observed. Weight change for this sample was almost reversible, which indicates the reversibility of hydrogen adsorption. It also suggests that the hydrogen atoms can be physically absorbed. For the samples of La:B = 1 : 6 and Pd:B = 1 : 4, weight increases of 3.2 and 1.6 % were observed, respectively, as listed in Table 9.3.

One formation mechanism of the BN nanotubes and nanocapsules synthesized in the present work is described below. Metal and boron particles are melted by arc-melting, and during the solidification of the liquid into metal and/or boride

Table 9.3: Atomic ratio of starting materials, formed structures and weight change by hydrogen storage.

| M:B | Nanostructures | Weight change (wt.%) |
|---|---|---|
| La:B = 1 : 6 | Tubes and cages | +3.2 |
| La:B = 1 : 10 | Tubes | +0.58 |
| Pd:B = 1 : 4 | Capsules | +1.6 |

nanoparticles, excess boron can react with nitrogen to form BN layers at the surface of the nanoparticles. Because of their electrical insulation, BN fullerene materials are usually fabricated using an arc-discharge method with specific conducting electrodes such as $HfB_2$ and $ZrB_2$. The arc-melting method presented here, which uses mixed powder, has two advantages for BN nanomaterial production. Since the powder becomes conductive upon pressing, special electrodes are not needed. In addition, ordinary, commercial arc-melting furnaces can be used. These advantages lead to a simpler fabrication method compared to ordinary arc-discharge methods.

Although gas storage of hydrogen and argon in carbon nanotubes has been reported, there are few reports of gas storage in BN fullerene materials or theoretical calculations on gas storage in BN fullerenes. A weight increase of the sample in TG measurements was observed, as shown in Fig. 9.3(d). This might be due to the hydrogen gas storage in the BN nanomaterials. Since there might be metal and boron nanoparticles in the separated BN nanomaterials even after the separation, further qualification and evaluation of the samples are needed for hydrogen storage.

Carbon fullerenes and boron nitride fullerenes are sublimed at 600 and 1000 °C, respectively. Boron nitride fullerenes could store $H_2$ molecules using a smaller amount of energy than carbon fullerenes, and would maintain good stability at high temperature. Boron nitride fullerene materials would be better candidates for $H_2$ storage materials.

## 9.3 Calculations of hydrogen storage

The $H_2$ gas storage abilities of BN and C fullerene-like materials were investigated by theoretical calculation. Although a huge amount of calculation is required to do calculations on nanotubes, it is believed that $H_2$ molecules enter from the cap of nanotubes. The chemisorption energies and stable hydrogen positions inside the clusters were investigated as cap structures of nanotubes. The influence of the endohedral element on the $H_2$ gas storage ability of BN fullerene-like materials was also investigated using theoretical calculations. The chemisorption energies were investigated for cage clusters as a cap structure of nanotubes. $B_{24}N_{24}$ clusters, which were reported by mass spectrometry [10, 45], were selected as a gas storage material. Li, K and Na elements were also selected for use as doping atoms because these elements have an electropositive character. The present study will provide us with a guideline for the hydrogenation of BN nanomaterials as hydrogen storage materials.

To research the optimized structure of $B_{99}N_{99}$ and $B_{36}N_{36}$ with $H_2$ molecules, semi-empirical molecular orbital calculations (parameterized model revision 3, PM3) were performed. The energies of $B_{36}N_{36}$ with $H_2$ molecules were calculated using a first principle single point energy calculation using Gaussian. In the calculation, a basis set of 3-21G was used in a Hartree–Fock method. $B_{24}N_{24}$, $B_{36}N_{36}$, $B_{60}N_{60}$ and $C_{60}$ were selected for cluster calculations. To investigate the optimized structures, semi-empirical molecular orbital calculations (parameterized model revision 5, PM5) were performed using MOPAC. The Eigenvector following (EF) method was used for geometry optimization, and the charge of the molecule is 0. The default self-consistent method was restricted Hartree–Fock (HF). Multi-electron configuration interaction (MECI) was used in order to prevent the repulsion between electrons from becoming excessive. Chemisorption calculations for hydrogen atoms were performed for $B_{24}N_{24}$, $B_{36}N_{36}$ and $C_{60}$ by PM5 calculations [46, 47].

BN nanostructures would have energy barriers for $H_2$ molecules to pass through tetragonal and hexagonal rings. Fig. 9.4(a–c) show structural models in which a $H_2$ molecule passes through the hexagonal rings of $B_{99}N_{99}$ and $B_{36}N_{36}$ [19]. Single point energies were calculated with changing set point of $H_2$ molecule from the center of the cage at intervals of 0.1 nm. There is an energy barrier that is given for $H_2$ molecules to pass through the hexagonal rings. The energy barrier of the hexagonal rings in $B_{36}N_{36}$

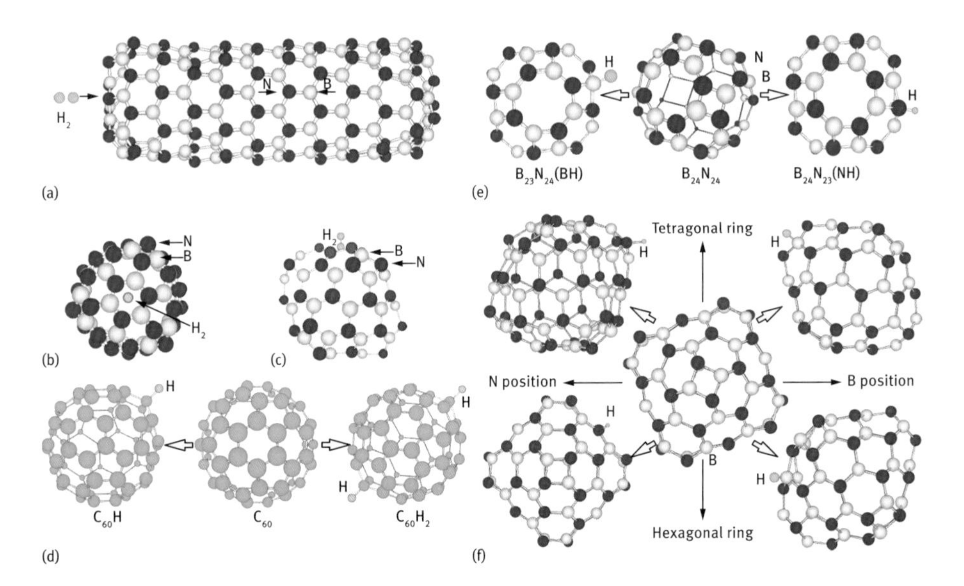


Fig. 9.4: Structure models that $H_2$ molecule passes through hexagonal BN rings of (a,b) tip of $B_{99}N_{99}$ nanotube and (c) $B_{36}N_{36}$ cluster. Structure models of H atoms chemisorbed on (d) $C_{60}$ and (e) $B_{24}N_{24}$. (f) Structure models for hydrogen chemisorption on $B_{36}N_{36}$ on N and B positions of tetragonal and hexagonal BN rings.

showed the smallest value of 14 eV in the present calculation, which is smaller than that of 27 eV for tetragonal rings. The energy barrier of hexagonal rings in $C_{60}$ was also calculated to be 16 eV, for comparison. This value is higher than that of the $B_{36}N_{36}$ hexagonal rings, and the $H_2$ molecule could pass through the hexagonal rings of $B_{36}N_{36}$ more easily than through the hexagonal rings of $C_{60}$. It was reported that $H_2$ molecules are adsorbed on the walls of single-walled carbon nanotubes over 7 MPa as an experimental result [48]. What can be concluded from this comparison is that $H_2$ molecules enter into $B_{36}N_{36}$ through hexagonal rings more easily than through tetragonal rings of $B_{36}N_{36}$ or through the hexagonal rings of $C_{60}$.

Figure 9.4(d) and (e) show structural models of hydrogen atoms chemisorbed on carbon clusters and boron and nitrogen for BN clusters, respectively. Atoms bonded with hydrogen are moved outside of the clusters. The energies for hydrogen chemisorption on each position are summarized in Table 9.4. Hydrogen bonding with nitrogen is more stable than with boron, and hydrogen bonding with carbon is more stable than with $C_{60}$. When two hydrogen atoms were chemisorbed on carbon clusters, the energies of the carbon clusters increased.

Table 9.4: Chemisorption energy of hydrogen (H) atoms on $C_{60}$ and $B_{24}N_{24}$. $\Delta E$ = (Heat of formation after hydrogen addition) – (Heat of formation before hydrogen addition).

| Cluster | Number of H atoms | Additional position of H | Heat of formation (eV) | | $\Delta E$ (eV) |
|---|---|---|---|---|---|
| | | | Before addition | After addition | |
| $C_{60}$ | 1 | C | 35.21 | 35.03 | −0.18 |
| | 2 | C | 35.21 | 35.81 | 0.6 |
| $B_{24}N_{24}$ | 1 | B | −36.12 | −34.66 | 1.46 |
| | 1 | N | −36.12 | −35.67 | 0.45 |
| $B_{36}N_{36}$ | 1 | B of tetragonal ring | −69.33 | −67.83 | 1.50 |
| | 1 | N of tetragonal ring | −69.33 | −69.16 | 0.17 |
| | 1 | B of tetragonal ring | −69.33 | −67.51 | 1.83 |
| | 1 | N of tetragonal ring | −69.33 | −68.83 | 0.51 |

According to calculations performed during this study, chemisorption of hydrogen was performed on the outside of the cage, and on boron, nitrogen and carbon. The $B_{36}N_{36}$ cluster has tetragonal and hexagonal BN rings, and there are four kinds of boron and nitrogen positions for hydrogen chemisorption, as shown in Fig. 9.4(f) [19]. Figure 9.4(f) shows a structural model of hydrogen atoms chemisorbed on boron and nitrogen for $B_{36}N_{36}$. The energies for hydrogen chemisorption on each position are summarized in Table 9.4. Hydrogen bonding with nitrogen is more stable than that with boron because nitrogen atoms are more electrophilic compared to boron atoms.

In addition, hydrogen bonding on tetragonal rings is more stable than on hexagonal rings. Chemisoption of hydrogen with $C_{60}$ reduced the energy. When two hydrogen atoms were chemisorbed on carbon clusters, the energies of the carbon clusters increased.

To investigate the stability of $H_2$ molecules in clusters, energies were calculated for $H_2$ molecules introduced in the center of clusters. The corresponding structural models are shown in Fig. 9.5. The energies of $C_{60}$, $B_{24}N_{24}$, $B_{36}N_{36}$ and $B_{60}N_{60}$ were calculated to be 0.58, 20.71, 20.93 and 20.82 eV/mol atom. This result indicates that $B_{24}N_{24}$, $B_{36}N_{36}$ and $B_{60}N_{60}$ with $H_2$ molecules are more stable than $C_{60}$ with $H_2$ molecule, and that $B_{36}N_{36}$ is the most stable of these BN clusters, as is shown in Table 9.5.

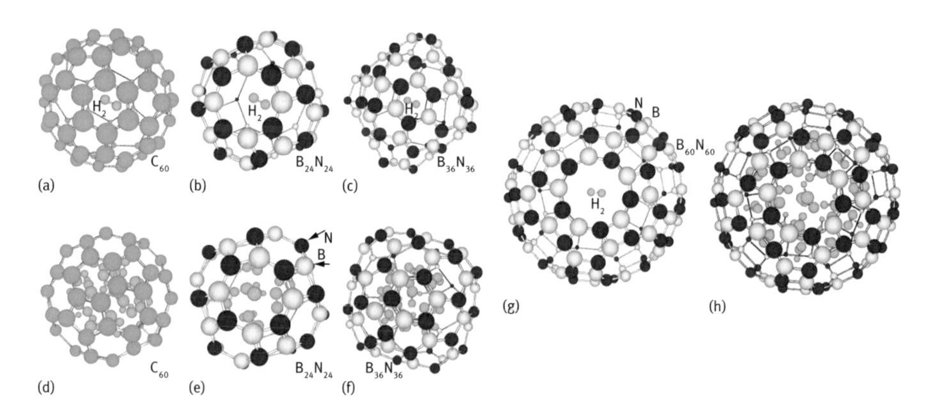


Fig. 9.5: Optimized structural models of $H_2$ molecules in the clusters. An $H_2$ molecule in the center of (a) $C_{60}$; (b) $B_{24}N_{24}$; (c) $B_{36}N_{36}$; (d) 22 $H_2$ molecules inside $C_{60}$ and 8 atoms chemisorbed; (e) 9 $H_2$ molecules in $B_{24}N_{24}$; and (f) 20 molecules in $B_{36}N_{36}$; (g) 1 and (h) 38 $H_2$ molecules in $B_{60}N_{60}$.

$M@B_{24}N_{24}$ (M@: element encaged in $B_{24}N_{24}$) was selected for cluster calculations to investigate the effect of endohedral element in $B_{24}N_{24}$ clusters on hydrogenation. To investigate the optimized structures and the chemisorption calculations of hydrogen atoms, molecular orbital calculations were performed by using PM5 in MOPAC. The EF method was used for geometry optimization. The default self-consistent method was restricted as HF, and the MECI was used. In the calculation, chemisorption of hydrogen was performed on boron and nitrogen positions outside the cage, and their heats of formation were calculated. The energy levels and densities of states (DOS) of $B_{24}N_{24}$ and $Li@B_{24}N_{24}$ were also calculated by the first principles calculation with discrete variational (DV)-X$\alpha$ method.

Figure 9.6(a) shows a structural model of hydrogen atoms chemisorbed to nitrogen for $M@B_{24}N_{24}$.[49] The energy levels and density of states for $B_{24}N_{24}$ and $Li@B_{24}N_{24}$ are shown in Fig. 9.6(b). The Fermi levels in the energy level diagrams and DOS diagrams correspond to 0 eV. $B_{24}N_{24}$ and $Li@B_{24}N_{24}$ show energy gaps of 4.8873 and 0.0247 eV

Table 9.5: Energy of clusters with hydrogen. *A C-C bond was broken.

| Cluster | Introduced $H_2$ | Heat of formation (eV) | H atoms chemisorbed inside cluster | Hydrogen storage (wt.%) | Heat of formation per added H atom (eV/H atom) |
|---|---|---|---|---|---|
| $C_{60}$ | 0 | 35.21 | 0 | 5.8–6.5 | 4.9–5.2 |
| | 22 | 143.01 | 0 | | |
| | 25 | 164.87 | 4 | | |
| | 26 | 169.63 | 8* | | |
| $B_{24}N_{24}$ | 0 | −36.12 | 0 | 2.9 | 3.0 |
| | 9 | −9.44 | 0 | | |
| $B_{36}N_{36}$ | 0 | −69.33 | 0 | 4.3 | 3.1 |
| | 20 | −6.66 | 0 | | |
| $B_{60}N_{60}$ | 0 | −100.28 | 0 | 4.9 | 1.0 |
| | 38 | −61.74 | 0 | | |

between the highest occupied molecular orbital (HOMO) and the lowest unoccupied molecular orbital (LUMO), respectively. This means that the electron transfered from Li to the $B_{24}N_{24}$ cage, and that the electronic state of the BN cluster could change from semiconductive to metallic by Li doping in BN clusters.

The energies for hydrogen chemisorption to boron and nitrogen were calculated, and heats of formation of M@$B_{24}N_{24}$ by hydrogenation are indicated in Fig. 9.6(c). Hydrogen bonding with nitrogen is more stable than that with boron because nitrogen atoms are more electrophilic compared to boron atoms. Fig. 9.6(c) also indicates that the energies of chemisorption on M@$B_{24}N_{24}$ are much lower than that of $B_{24}N_{24}$. From these results it is possible to conclude that Li atom works as a good endohedral element for hydrogen chemisorption. In the earlier findings of this work, metal catalysts were reported to generate hydrides such as LiH because of the strong interaction between hydrogen and metal atoms. I then clarified that Li doping and nitrogen positions are suitable for the hydrogenation of the $B_{24}N_{24}$ clusters. The bond lengths of B–H and N–H for M@$B_{24}N_{24}$ were also calculated. B–H and N–H distance decreased by doping elements in the BN cluster. The bond lengths of N–H are shorter than those of B–H for these BN clusters, and the bond length of N–H for Li@$B_{24}N_{24}$ is the shortest. The endohedral atoms appear to decrease the repulsion energy between the electrons of the hydrogen atom and the p-electrons of $B_{24}N_{24}$. Fig. 9.6(a) shows a structural model of M@$B_{24}N_{24}H_{24}$, which indicates hydrogenation on all nitrogen positions for M@$B_{24}N_{24}$, and that the hydrogen storage capacity for Li@$B_{24}N_{24}$, Na@$B_{24}N_{24}$ and K@$B_{24}N_{24}$ are 3.86. 3.76 and 3.67, respectively. Li is also a good element for hydrogen storage capacity because Li is the lightest element among these elements.

Molecular dynamics calculations were carried out to confirm the stability of $H_2$ molecules into $C_{60}$ at 298 K and 0.1 MPa [50]. NTP ensembles were used in the calculation as follows: number of atoms ($N$), temperature ($T$) and pressure ($P$) are constant.

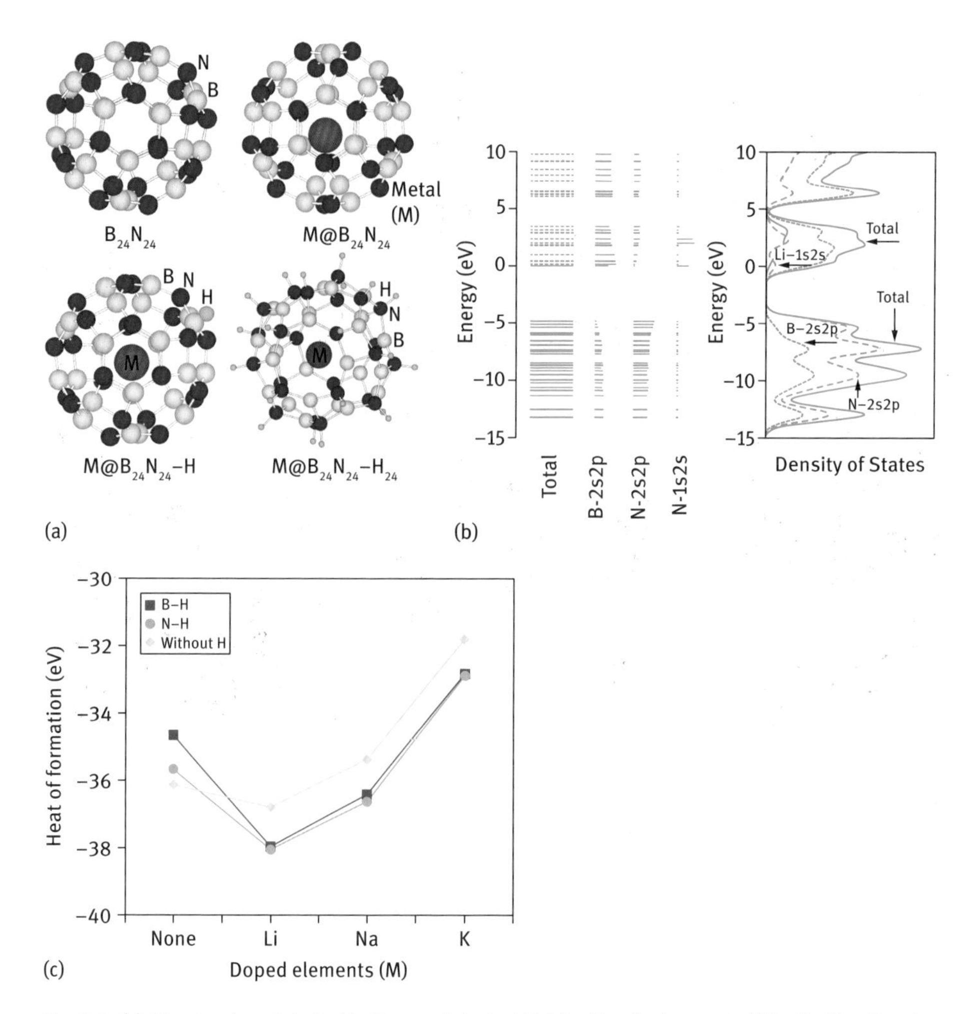


Fig. 9.6: (a) Structural models for $B_{24}N_{24}$, endohedral $M@B_{24}N_{24}$, hydrogenated $M@B_{24}N_{24}$-H and $M@B_{24}N_{24}$ which chemisorbed 24 hydrogen atoms. (b) Energy levels diagrams and density of states of $Li@B_{24}N_{24}$. (c) Heats of formation of hydrogenation for $M@B_{24}N_{24}$ clusters by endohedral elements.

The conditions for $H_2$ gas storage were also calculated. NPH ensembles were used in the calculation as follows: number of atoms ($N$), pressure ($P$) and enthalpy ($H$) are constant. These molecular dynamics were calculated by organic potential.

$C_{60}$ included a $H_2$ molecule kept in stable state at $T = 298$ K and $P = 0.1$ MPa. The model was calculated with $N = 62$ atoms, $T = 298$ K and $P = 0.1$ MPa. Although the $H_2$ molecule vibrated in the $C_{60}$ cage, it was not discharged from the cage.

Some $H_2$ molecules were stored in the $C_{60}$ cage when the pressure was 5 MPa. The unit cell of $C_{60}$ has a face-centered cubic (fcc) structure, as shown in Fig. 9.7 [19], which is composed of $32C_{60}$ with 288 $H_2$ molecules. This model was calculated with

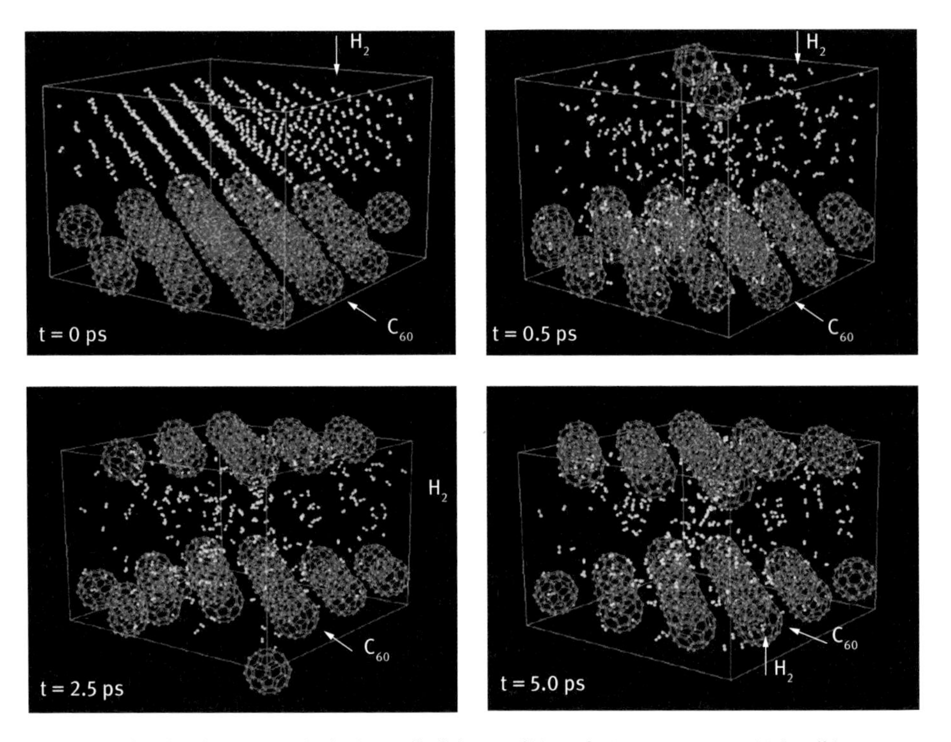


Fig. 9.7: Molecular dynamics calculation to find the conditions for $H_2$ gas storage. Unit cell is composed from 32 $C_{60}$ (fcc) with 288 $H_2$ molecules (fcc). NPH ensembles and organic potential were used in the calculation.

$N = 2496$ atoms under a pressure of $P = 5$ MPa. $H_2$ molecules pass through the hexagonal rings of the $C_{60}$ cage after 0.5 ps, as observed in Fig. 9.7.

After introducing $H_2$ molecules into the $C_{60}$ cage at 2.5 ps, they are stored and become stable in $C_{60}$. Fig. 9.8 shows a calculated model of a single $H_2$ molecule stored in a hexagonal ring of $C_{60}$. These results indicate that fullerene-like materials can store $H_2$ gas in a cage at $T = 298$ K and $P = 0.1$ MPa, and some $H_2$ molecules were stored in the $C_{60}$ cage when the pressure was greater than 5 MPa. $H_2$ molecules were reported to be adsorbed on the walls of single-walled carbon nanotubes over 7 MPa [50], which is almost comparable with the present calculation. Since a $C_{60}$ cluster has a large curvature and $H_2$ molecules are very small compared to $C_{60}$, it is considered that the adsorption and storage of $H_2$ gas occur at the same time. Detailed theoretical calculations on hydrogen storage in BN nanomaterials have also been performed [51–54].

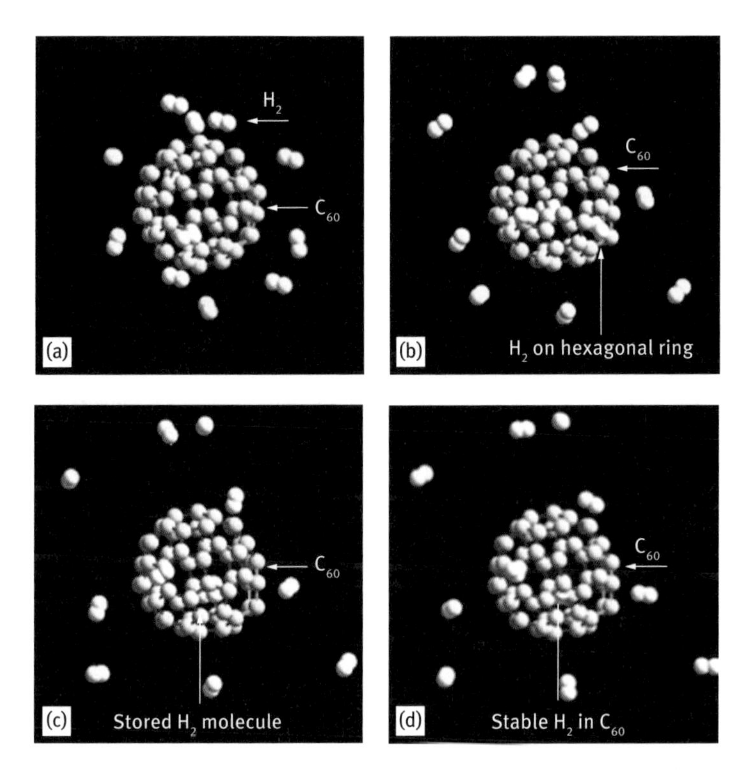


Fig. 9.8: Enlarged calculated model of $H_2$ molecule stored in a hexagonal ring in the order of (a) → (b) → (c) → (d).

## 9.4 Thermoelectric power generation

A closed circuit is fabricated with two different materials such as a *p*-type semiconductor and an *n*-type semiconductor, and the junction part is exposed to different temperatures. Then, electromotive force occurs at both electrodes by the Seebeck effect, thermoelectric power is generated, and current flows by connecting a load in the circuit, as shown in Fig. 9.9. When high and low temperatures at the junction are $T_h$ and $T_c$, the electromotive force is expressed by a following equation. The $\alpha$ is a Seebeck coefficient, and the unit is V $K^{-1}$.

$$V = \alpha\,(T_h - T_c) \tag{9.1}$$

Thermoelectric power generation refers to a method for transforming thermo-energy to electric energy using the Seebeck effect. Since thermoelectric devices have no moving parts, they have long lives and don't require maintenance for a long time. Thermoelectric devices are suitable as an electric power source for artificial satellites, and have been studied as a power source for space exploration. Thermoelectric power

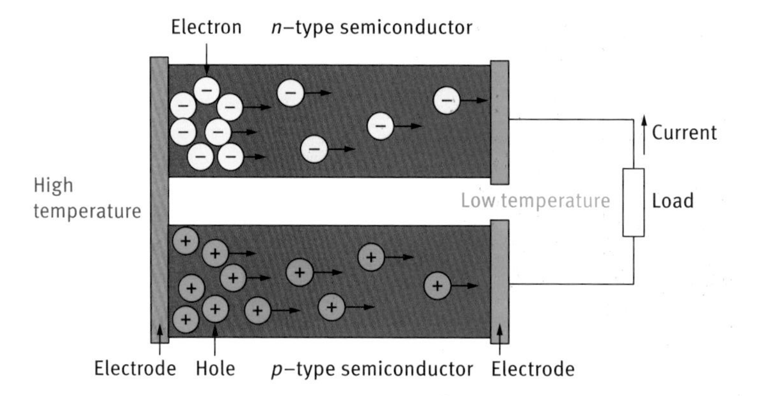


Fig. 9.9: Schematic illustration of a thermoelectric device.

generation is an expected method for generating electric power using wasted heat such as that from automobiles, factories, the heat of the earth and hot springs.

Various systems such as Bi-Te system (300 ~ 500 K), Pb-Te system (300 ~ 800 K) and Si-Ge system (300 ~ 1000 K) are used for materials combination at different temperatures. The crystal structures of $Bi_2Te_3$ [55], PbTe [56], $CoSb_3$ [57], $Mg_2Si$ [58], $Ge_{0.5}Si_{0.5}$ [59], and $Mn_4Si_7$ [60] are shown in Fig. 9.10, indicating the existence of various crystal systems. To solve some problems such as oxidation at high temperatures and limited resources, other oxide materials and quantum structures are also being studied. Since the voltage obtained from one thermoelectric device is low (0.01 ~ 0.1 V),

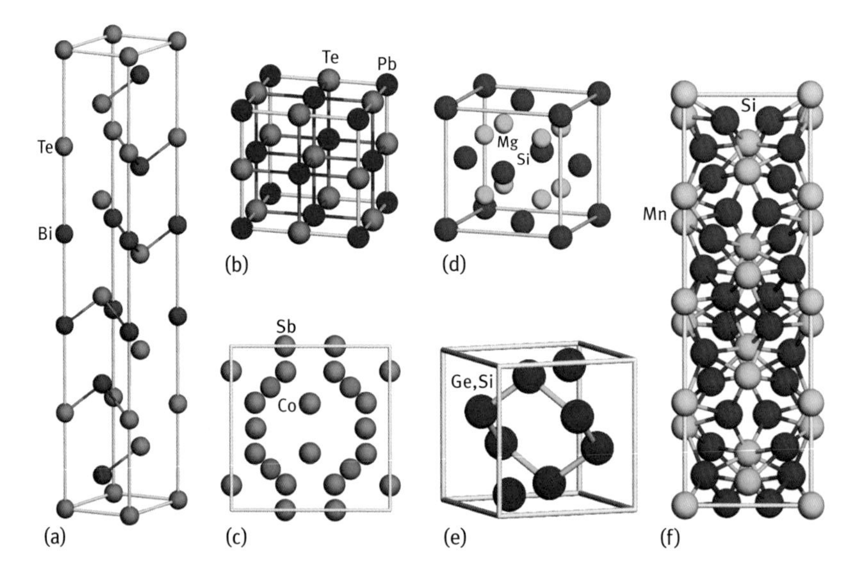


Fig. 9.10: Crystal structures of (a) $Bi_2Te_3$, (b) PbTe, (c) $CoSb_3$, (d) $Mg_2Si$, (e) $Ge_{0.5}Si_{0.5}$ and (f) $Mn_4Si_7$.

plural devices are connected in series to obtain high voltage in current thermoelectric power generation modules [61], as shown in Fig. 9.11.

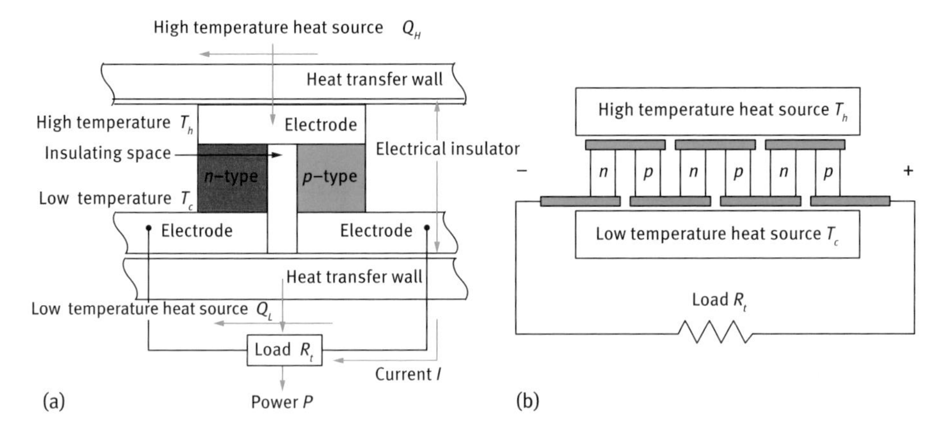


Fig. 9.11: (a) Thermoelectric mechanism and (b) module structure.

Thermal energy generated by self-nuclear decay was utilized by thermoelectric devices called nuclear batteries, which have been used as the power sources of many artificial satellites. Radioisotopes with long radioactive half-lives are used in nuclear batteries. $\alpha$-decay occurs for radioisotopes such as plutonium 238 and polonium 90, and the $\alpha$ particles emitted from these radioisotopes are absorbed in the materials, causing heat, which is then utilized in thermoelectric devices. Isotopes with long radioactive half-lives such as strontium 90 are used for long-life nuclear batteries. When sunlight cannot be used for the power source of artificial satellite, radioisotopes have been used as a heat source, and thermoelectric devices with a SiGe system were used.

Although nuclear batteries have largely been replaced by solar cells, they are still used in exploration satellites outside of Jupiter and for the night operation of the Mars rover. Since constant electric power can be obtained regardless of day or night, a nuclear battery utilizing the nuclear decay heat of plutonium 238 is used in the Mars rover. The atmospheric temperature of Mars is in the range of −127 to +30 °C, and its residual heat can be utilized for retaining the warmth of the space probe system. A nuclear battery used in the Mars rover Curiosity is the newest model, and has a weight of 50 kg, including a 4.8 kg load of plutonium 238. The nuclear battery is covered by a protective layer to protect the plutonium in case of accident such as collision and explosion. An electric power of 125 W is obtained from the generated heat of 2000 W when the battery is new, and even 14 years later an electric power of 100 W. The amount of power generated for the Mars rover per day is 2.5 kWh.

## 9.5 Icosahedral boron

Boron (B) is one of the elements that are essential for plants and is found in the cell walls of most plant life. Boron could play a role in the synthesis of cell walls, membrane transport of glucose, nucleic acid synthesis, etc. For humans, boron is a trace element that activates the vitamin D necessary for bone formation. It is sold as a supplement for the prevention of osteoporosis and is also found in fruits such as apples, pears, peaches and grapes.

Boron has three structure types: $\alpha$-rhombohedral [62], $\beta$-rhombohedral [63] and tetragonal, as shown in Figs. 9.12 and 9.13. They all consist of various arrangements of icosahedral $B_{12}$ clusters with 12 boron atoms. The icosahedral structure is one of Plato's regular polyhedra. A single boron element indicates a semiconductive property and its hardness nearly rivals that of diamond. $\beta$-rhombohedral boron is the form of boron usually found in nature.

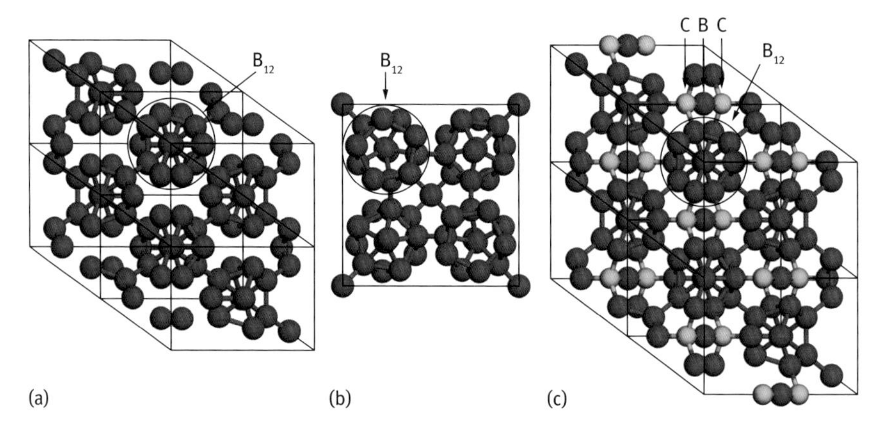


Fig. 9.12: Structural models of (a) $\alpha$-rhombohedral, (b) tetragonal boron and (c) boron carbide ($B_{13}C_2$).

$\beta$-rhombohedral boron has a particularly unique structure based on the $B_{84}$ clusters, as shown in Fig. 9.13. $B_{84}$ consists of a $B_{12}$ center encapsulated in large $B_{12}$ bonding to outside $B_{60}$, which has the same structure as fullerene $C_{60}$. The $B_{60}$ is not an isolated structure like a $C_{60}$, but a cluster solid that exist in solid as a cage structure. The icosahedral structure has 2-, 3- and 5-fold symmetry as shown in Fig. 9.13. When a small amount (~ 4 %) of element such as vanadium or copper was doped to the $\beta$-rhombohedral boron, which was expected to function as a thermoelectric material, the semiconducting properties changed drastically as a result of the doping.

The $\alpha$-rhombohedral boron has vacant space in its crystal structure. When some carbon and boron atoms occupy this vacant space, boron carbide ($B_{13}C_2$) is formed [64]. $B_{13}C_2$ is the hardest material after diamond and cubic boron nitride, and is uti-

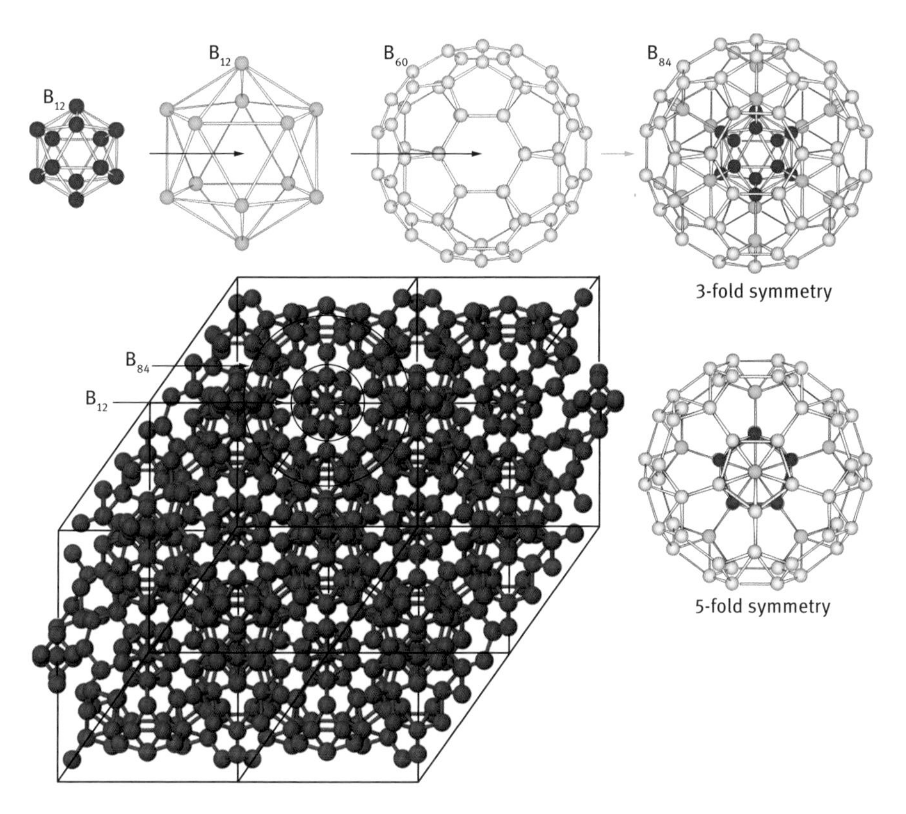


Fig. 9.13: Structural models of $B_{84}$ and $\beta$-rhombohedral boron.

lized as a super hard material. During the nuclear fission of uranium and plutonium in nuclear reactors, lots of neutrons are generated, which is dangerous in terms of the fission chain. Since boron absorbs neutrons effectively, control rods made of $B_{13}C_2$ are used in nuclear reactors.

## 9.6 Titanium dioxide

Titanium dioxide ($TiO_2$) has recently attracted attention for its potential in perovskite solar cells, dye-sensitized solar cells and as a photocatalyst. $TiO_2$ has three main structures, rutile [65], anatase [65] and brookite [66], in addition to some other forms, ramsdellite [67], II [68] and baddeleyite [69], as shown in Fig. 9.14. TiO's anatase structure has photocatalytic properties and photovoltaic behavior. Rutile is mainly used for cosmetics, and $TiO_2$ nanotube have also been produced.

When ultraviolet rays from the sun and fluorescent lights are irradiated onto $TiO_2$, various materials around $TiO_2$ can become oxidized. Since $TiO_2$ does not change upon radiation, it can be used for long time.

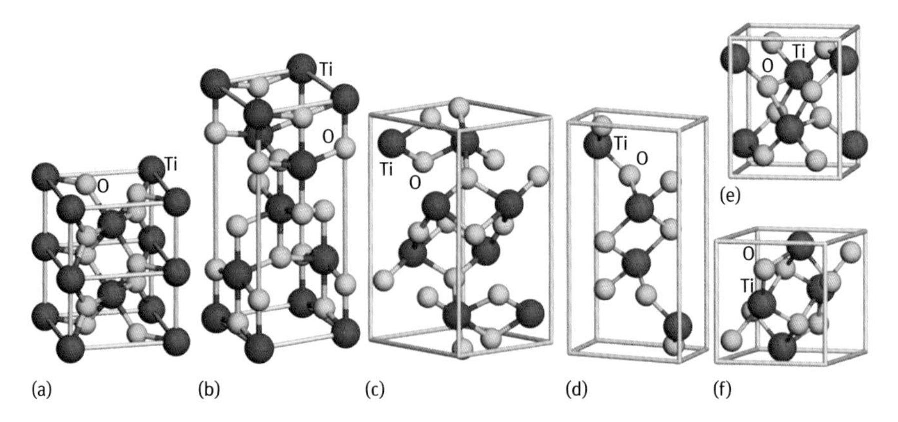


Fig. 9.14: Crystal structures of (a) rutile, (b) anatase, (c) brookite, (d) ramsdellite, (e) II and (f) baddeleyite-types of $TiO_2$.

$TiO_2$ is expected to serve as a photocatalyst that enables the decomposition of water without the use of an electric current [70]. When $TiO_2$ is dipped in water and irradiated by ultraviolet rays (< 380 nm) with larger energies compared with the energy gap of $TiO_2$ (3.2 eV), $H_2O$ molecules in the water decompose into $H_2$ and $O_2$. Since hydrogen gas $H_2$ is obtained without decomposition of $TiO_2$, the $TiO_2$ is expected to have potential as an energy material.

$TiO_2$ has a bactericidal effect and can decompose some of the harmful substances polluting the environment. $TiO_2$ is changeless, and can be used semi-permanently. More specifically, harmful nitrogen oxides in the exhaust gas of cars are decomposed by $TiO_2$ into harmless materials, and harmful materials in industrial sewage and oil that has leaked into the sea can also be decomposed. Because of this, $TiO_2$ has attracted attention for its potential as an environmental material for cleaning the air and water.

The decomposition mechanism is shown in Fig. 9.15. Hydroxyl radical (OH) is formed, which decomposes various organic materials. Various forms of pollution such

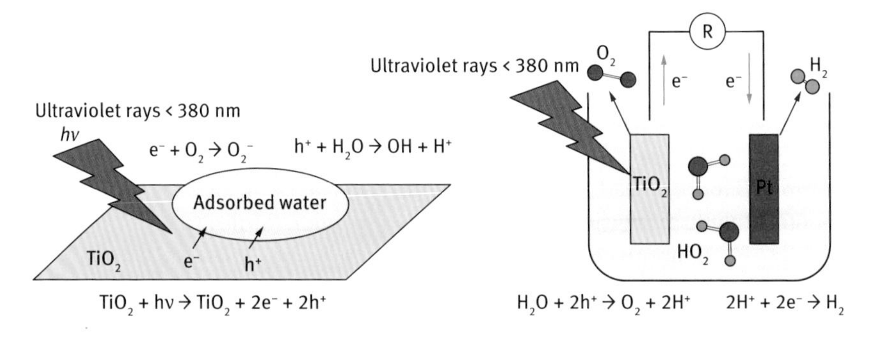


Fig. 9.15: Photocatalysis and decomposition of water by $TiO_2$.

as the exhaust gas of cars build up on roofs and walls of buildings as time passes. When the pollution is irradiated by the sunlight, it is decomposed automatically. If it rains, the decomposed materials are washed away, and the roofs and walls are cleaned automatically.

If $TiO_2$ is used in walls and floors in hospitals, they can be sterilized, and smells from the hospitals can be reduced by its deodorant effects. Since $TiO_2$ nanoparticles are white, they are utilized in cosmetics to reflect ultraviolet rays, as well as in some toothpastes.

## 9.7 Zeolite

Zeolites consists of an ordered arrangement of vacancies with a size of ~1 nm, as shown in Fig. 9.16. Zeolite is a general term for crystals consisting of silicon, aluminum and oxygen, and more than 150 kinds of zeolites are known. Since natural mineral zeolites contain water, they appear to boil when heated. When the water is removed by heating, the cavities are preserved, and gases and water are absorbed into them again. Various artificial zeolites have recently been synthesized, and they are used as catalysts, molecular sieves, adsorption materials and ion-exchange materials by utilizing the zeolite's ordered cavities. Molecules with smaller sizes than the size of the zeolite's cavities can be absorbed by the molecular sieve effect. Since water molecules can be absorbed and desorbed in the vacancies of zeolites, they are used for wallpapers to control the humidity.

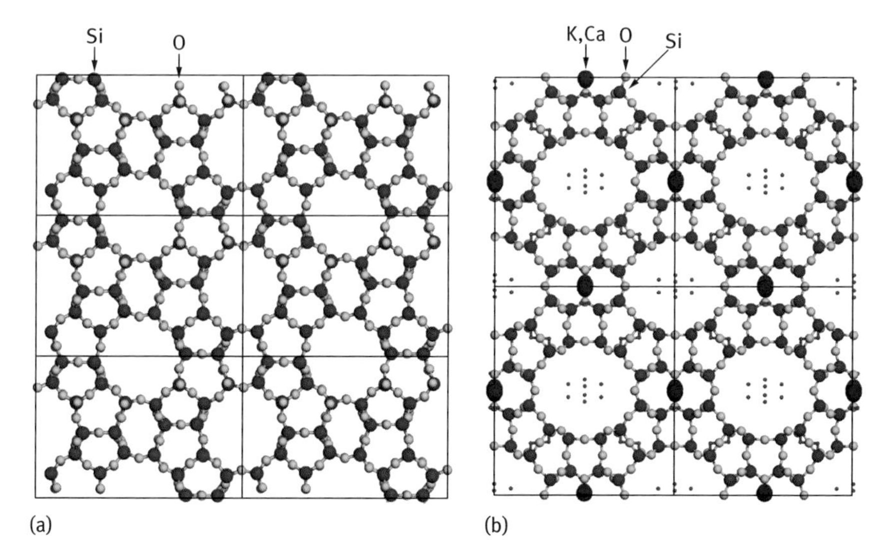


Fig. 9.16: Structural models of (a) ZSM-5 and (b) mordenite.

Gasoline can be synthesized from ethanol by using the ZSM-5 zeolite developed by the Mobil Oil Company. A structural model of ZSM-5 is shown in Fig. 9.16(a) [71]. Molecules in the vacancies of the zeolite play a role in catalyzing the synthesis. In New Zealand, gasoline is synthesized from methane gas by using the ZSM-5. In addition, the zeolite is studied as a catalyst that remove harmful nitrogen oxides from the exhaust gas of cars. Mordenite is one natural zeolite and is shown in Fig. 9.16(b) [72].

## 9.8 Lithium-ion batteries

Lithium-ion batteries (LIB) are one form of rechargeable batteries, and lithium ions in the electrolytes carry electric charge from the negative electrode to the positive electrode for storage. Lithium cobalt oxide ($LiCoO_2$) and graphite are mainly used as cathodes and anodes [73]. Since non-liquid electrolytes are used in lithium-ion batteries, a high voltage of ~ 4 V beyond the decomposition voltage of water can be obtained, achieving a high energy density. The cathode includes lithium, and the graphite of the anode stores lithium. Therefore, no metal lithium exists intrinsically in the battery, and it is safe. The electrolyte includes no water, and the LIB can be used in environments below the freezing point.

Lithium-ion batteries are one of the most popular types of rechargeable batteries for portable electronics such as notebook computers and mobile phones. They are also used as electric batteries for vehicles. Lithium-ion batteries are becoming a replace-

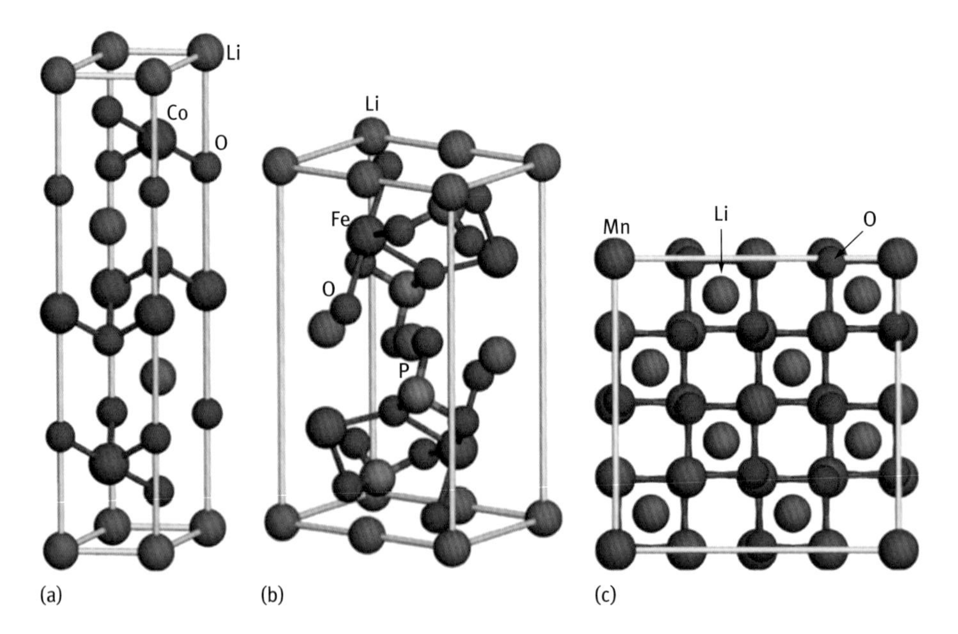


Fig. 9.17: Structural models of (a) $LiCoO_2$, (b) $LiFePO_4$, and (c) $LiMn_2O_4$.

ment for lead acid batteries with heavy lead plates and acid electrolytes. Lightweight lithium-ion batteries can provide the same voltage as lead-acid batteries. In addition to $LiCoO_2$, lithium iron phosphate ($LiFePO_4$) [74], Lithium ion manganese oxide battery ($LiMn_2O_4$)[75] and lithium nickel manganese cobalt oxide ($LiNiMnCoO_2$) with lower energy density but longer lives are also utilized, as shown in Fig. 9.17.

## 9.9 Si clathrate

lathrate is a structure that encapsulates other elements in cage structures. For silicon (Si) clathrate, various elements such as barium (Ba), potassium (K) and sodium (Na) are incorporated in the cage structures such as $Si_{20}$ and $Si_{24}$.[76] Germanium (Ge) clathrate has also been reported [77]. Ordinary Si has a diamond structure, and is widely used in semiconductor devices and solar cells. The present Si clathrate is completely different from the diamond structure, and the structural models of $Si_{46}$ and $Ba_8Si_{46}$ are shown in Fig. 9.18(a) and (b), respectively. Ba atoms are encapsulated in $Si_{20}$ and $Si_{24}$ clusters. The crystal structure is a cubic system with a lattice constant of 10.19 Å (space group No. 233, $Pm\bar{3}n$).

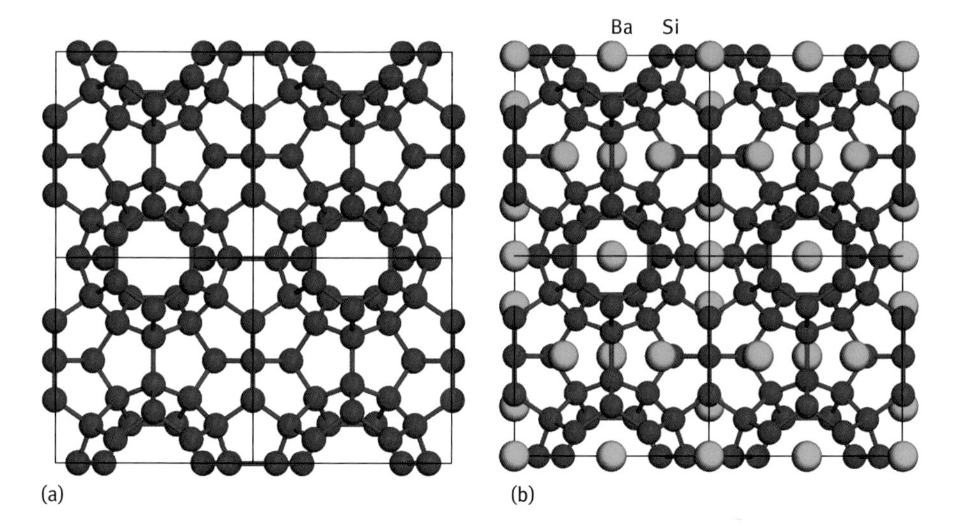


Fig. 9.18: Structure models of (a) $Si_{46}$ and (b) $Ba_8Si_{46}$.

The $Ba_8Si_{46}$ was found to provide superconductivity at 9 K [76], and many attentions are attracted to the property. Thermoelectric properties are also predicted from the theoretical calculations, and the bandgap energy seems to be larger than that of Si. Various kinds of studies on these group 14 elements have been performed because of their unique structures.

## 9.10 Ice and hydrate

Water is the source of life, and 60 ~ 70 % of the human body consists of water. Molecules of water ($H_2O$) are dispersed randomly in liquid. The bonding angle of H-O-H is 104.45 °, which is close to 109.28 ° of tetrahedron angle. Oxygen and hydrogen atoms polarized as negative and positive charge, respectively, which form electric dipole. The oxygen atom has a covalent bond with two hydrogens, and other hydrogens are attracted by two remained electron lone pair.

When the temperature of water is 0 °C, water becomes ice, and $H_2O$ molecules become arranged regularly. Ice has a hexagonal structure with the space group is $P6_3/mmc$ [78], and the lattice constants are $a = 4.5$ Å and $c = 7.3$ Å, as shown in Fig. 9.19. Crystals of snow have hexagonal shapes due to the hexagonal structure of $H_2O$. As observed in the model of ice, hydrogen atoms are bonded to the tetrahedral directions of an oxygen atom, and the hydrogens are also attracted to lone pair of the oxygen. Note that two hydrogens are drawn between the oxygen in the atomic model, and that actually only one exists there.

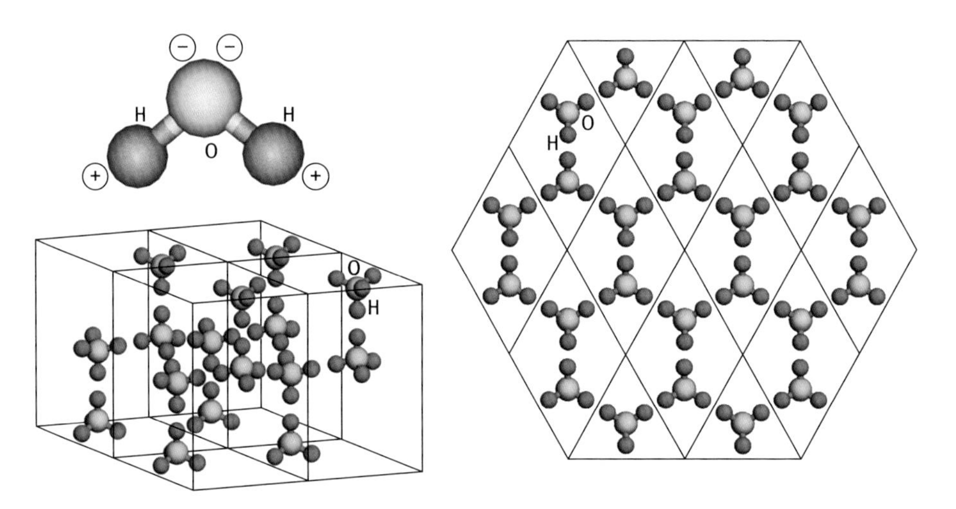


Fig. 9.19: Atomic structure of ice.

Ice crystals contain a lot of space between oxygen atoms. Ice floats on water because of the volume of this space. When pressures are applied to the ice, various other structures are formed, and 12 kinds of ice structures have been found, 10 kinds of which sink in water. Such sinking ices actually exist within the inner part of the earth and on planets outside of Jupiter.

Hydrates are known as new types of energy materials. Water molecules encapsulating methane ($CH_4$), called methane hydrate, are shown in Fig. 9.20(a). The crystal structure is a cubic system with a lattice constant of 11.82 Å (space group No. 233,

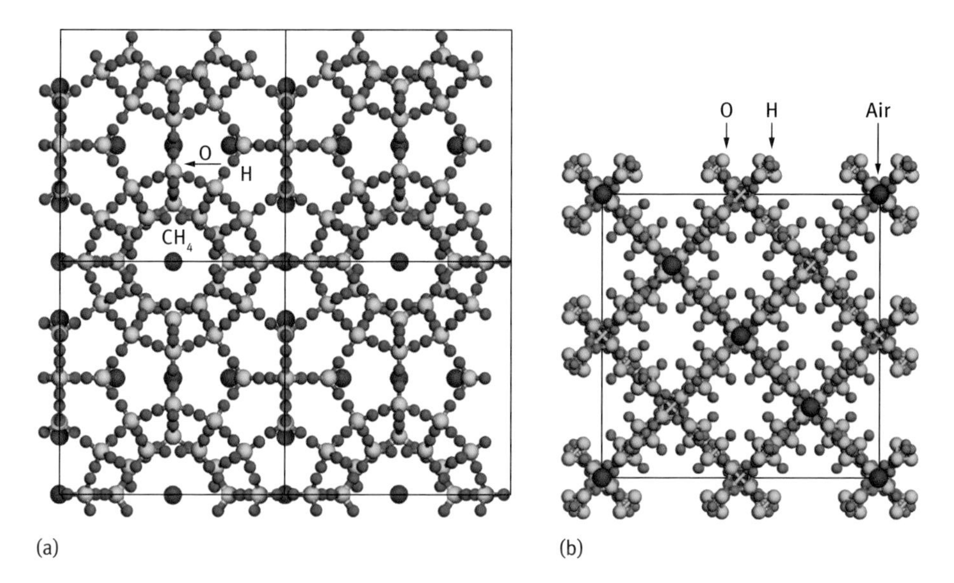


Fig. 9.20: Crystal structure of (a) methane hydrate and (b) air hydrate.

$Pm\bar{3}n$) [79]. In addition to the methane, various gas molecules such as nitrogen and oxygen are also encapsulated in the materials, as shown in Fig. 9.20(b).

The basic structure of hydrates is the cage structure of $H_2O$, and several structures have been found that are similar to those of clathrate materials. Hydrates are formed under high pressure and low temperatures, and a large amount of gas molecules are encapsulated in the cage structure. They were found at the bottom of the sea and on planets outside of Saturn.

Huge amount of methane hydrates were found at the bottom of the deeper sea than 500 m, and they are expected to have potential as an alternative energy source of oil. Although the hydrates look like ordinary ice, encapsulated methane molecules burn, and only water remains after burning. Therefore, they are called "burning ice". Fuels for taxis are natural gas, and the main constituent of natural gas is methane. If the methane hydrate is utilized wisely, it would be possible to transport the methane confined in the hydrate, and to remove carbon dioxide causing global warming.

## 9.11 Fuel cells

Fuel cells are chemical batteries that generate electricity from a supply of fuel, such as hydrogen and an oxidizing agent. Since chemical energy is directly transformed into an electric energy, the conversion efficiencies are very high, and fuel cells are promising as energy resources. For the fuel cells, electricity and water are obtained by the electrochemical reaction of hydrogen and oxygen ($2H_2 + O_2 \rightarrow 2H_2O + e^-$) is shown in

Fig. 9.21, and the fuel cells produce clean and pollution-free energy. In the electrolysis of water, hydrogen and oxygen are obtained by the use of an electric current ($2H_2O + e^- \rightarrow 2H_2 + O_2$), as shown in Fig. 9.21. Fuel cells have an opposite mechanism to that of electrolysis, and it is more accurate to refer to them as power generators than as batteries.

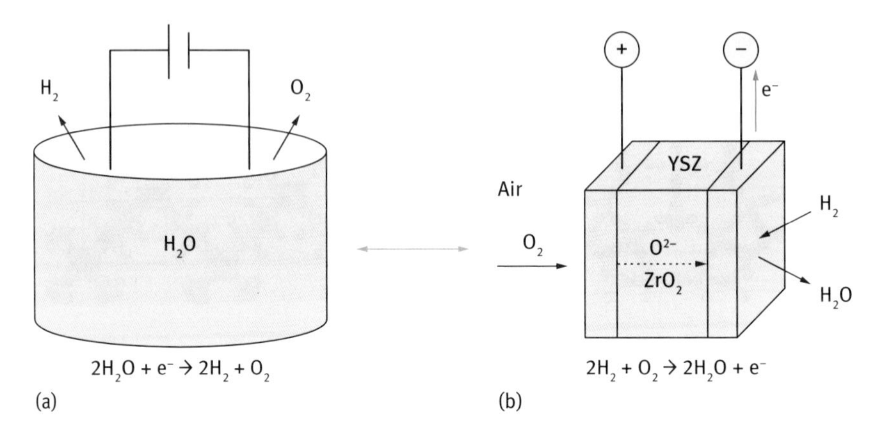


Fig. 9.21: Comparison of the (a) electrolysis of water and a (b) fuel cell.

Fuel cells are classified according to the kinds of electrolytes they contain, and four types of fuel cells,polymer electrolyte fuel cells (PEFC), molten carbonate fuel cells (MCFC), phosphoric acid fuel cell (PAFC) and solid oxide fuel cells (SOFC), have been developed, as listed in Table 9.6. Although alkali-type fuel cells (AFC) that can be used at low temperatures were also developed as batteries for the Apollo spacecraft and Space shuttle, they need hydrogen with high purity.

Since the electrolytes of the polymer electrolyte fuel cells are solid and can be downsized and thereby lightened, they are used in motor vehicles. Phosphoric acid fuel cells are used for usual power generation. Solid oxide fuel cells and molten carbonate fuel cells, which are operated at ~ 1000 °C and ~ 650 °C, provide high power generation efficiencies and effective utilization of thermal energy. However, they need some time (~ 10 min) for starting.

A typical material for solid oxide fuel cells is zirconium dioxide ($ZrO_2$), as shown in Fig. 9.22 [80–83]. Solid electrolytes are materials that conduct electric currents due to ion transfer in the solid, and the electric current flows by oxygen ion ($O^{2-}$) transfer for the $ZrO_2$. Generally, the energy conversion efficiencies of thermal power generation and nuclear power generation are ~ 35 % since the electricity is obtained by turning turbine using water vapor generated by heat from the fuels. On the other hand, the energy conversion efficiency of solid oxide fuel cells using $ZrO_2$ is ~ 50 %. These cells generate electric power at temperatures of 900~ 1000 °C, and are utilized as small electric power generators for home use.

Table 9.6: Various fuel cells and their properties.

| Type of cells | Alkali-type (AFC) | Polymer electrolyte (PEFC) | Phosphoric acid (PAFC) | Molten carbonate (MCFC) | Solid oxide (SOFC) |
|---|---|---|---|---|---|
| Electrolyte | KOH | Polymer membrane | Phosphoric acid | Molten carbonate | YSZ |
| Operation temperature (°C) | < 100 | < 100 | 200 | 650 | 1000 |
| Fuel | High purity $H_2$ | $H_2$ | $H_2$ | $H_2$ | $H_2$ |
| Efficiency (%) | 60 | 40 | 35 ~ 45 | 45 ~ 55 | 50 |
| Application | Special environment (Space, abyss) | Decentralized power supply (motor vehicle) | Cogeneration power plant (Medium-scale) | Cogeneration power plant (Large-scale) | Cogeneration power plant (Medium-scale) |

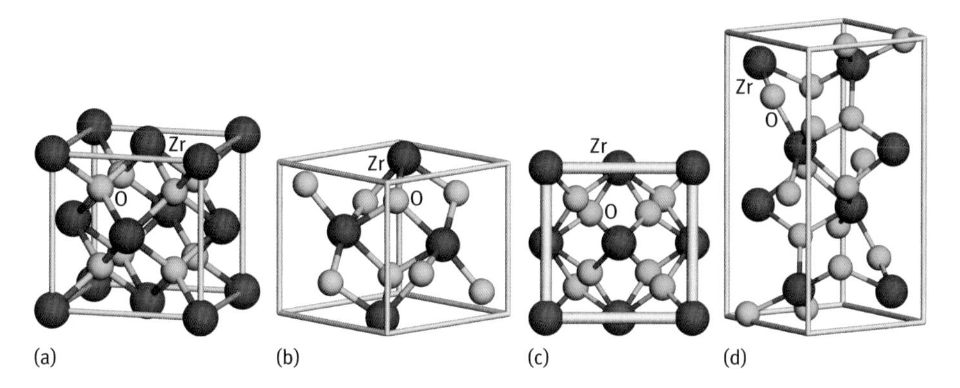


Fig. 9.22: Structures of (a) cubic, (b) monoclinic, (c) tetragonal and (d) orthorhombic $ZrO_2$.

The crystal structures of $ZrO_2$ change from monoclinic, tetragonal to cubic as temperature increases. Since the structures tend to change along with temperature changes, a stable cubic structure at high temperature can be preserved by adding yttrium oxides, which is called yttria-stabilized zirconia (YSZ). Since the $ZrO_2$ is hard, tough and stable at high temperatures, it is also used for ceramic knives, scissors and refractory materials.

The appearance of $ZrO_2$ is very similar to that of diamond. The brilliance of diamonds results from their high refractive index of light, and the $ZrO_2$ has a refractive index close to that of the diamond. In addition, $ZrO_2$'s color can be changed to blue, green, orange, pink and red by adding various elements to the $ZrO_2$. Since bigger $ZrO_2$ crystals can be synthesized at low cost, they are also used for jewelery.

Bio fuel cells using enzymes and microorganisms have attracted much attention. Electric power is obtained by decomposing sugar through the action of the enzymes and microorganisms. A photovoltaic voltage of 1.24 V can be obtained for glucose-air fuel cells by the reaction of ($1/6C_6H_{12}O_6 + H_2O \rightarrow CO_2 + 4H^+ + 4e^-$). Bio fuel cells in a body that generated electric power have also been developed. One of the electrodes of this cell is a pressed pellet of carbon nanotubes and glucose oxidase, and the other electrode consists of carbon nanotubes, glucose and polyphenol oxidase, and a current flows in a platinum wire. Electrons are removed from glucose of the electrode, and the electrons are transferred to oxygen and hydrogen to form water at the other electrode. The current generated from the electrode in the circuit is expected to have potential as an energy resource for body equipment such as pacemakers. Since glucose and oxygen always exist in the body, the cells could continue to function without supplying fuel from outside the body. The electrodes are protected by mesh materials to protect the immune system.

## 9.12 SiC FET

Most of the energy consumed by human society in its current form is derived from finite fossil fuels, and it is therefore mandatory to develop clean alternative energy sources. Renewable energy sources such as a photovoltaic power systems, consisting of solar cells connected with inverters, are constitute some of the most promising sources of alternative energy power. Solar cells provide electricity, and produce no carbon dioxide or other hazardous waste gases involved in global warming. Most power electronic converters are based on a silicon (Si) semiconductor, and the performance of the Si devices is approaching the theoretical limits. Therefore, power devices equipped with wide bandgap semiconductors such as silicon carbide (SiC) and gallium nitride (GaN) are needed. The crystal structures of the SiC [84–87] and GaN [88] are shown in Fig. 9.23.

The properties of semiconductors are listed in Table 9.7. Wide gap semiconductors have superior features compared to Si for use in power electronic systems. The higher electrical breakdown voltage, lower 'on'-resistance ($R_{on}$) and higher thermal conductivity provide small size, improved efficiency of components for high power switching and less heating, respectively. Therefore, the SiC and GaN devices have smaller size and weight [89], in comparison to ordinary Si devices, as shown in Fig. 9.24. In addition, the SiC device has higher efficiencies for high power switching devices with high frequencies. SiC devices can be operated at temperatures up to ~ 600 °C without much change in their electrical properties, which is a higher temperature than Si devices can withstand (~ 200 °C). SiC devices are one of the leading candidates for next generation power devices [90–93].

Si consumption can be reduced by using spherical Si solar cells, in comparison to conventional crystal Si solar cells [94]. Flexible and light solar cells were also manufac-

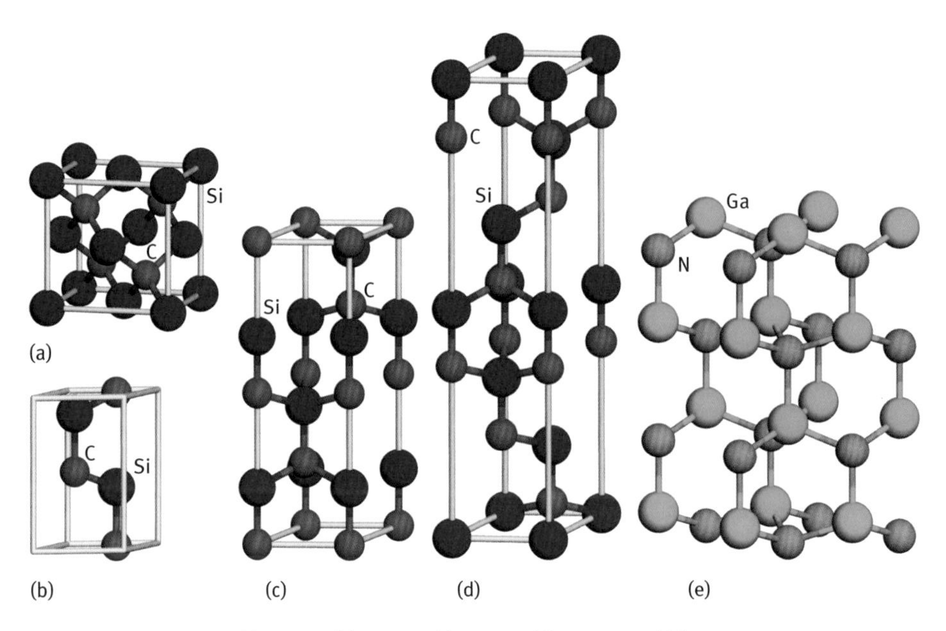


Fig. 9.23: Structures of (a) 3C-SiC, (b) 2H-SiC, (c) 4H-SiC, (d) 6H-SiC and (e) GaN.

Table 9.7: Properties of semiconductor materials.

| **Semiconductors** | **Si** | **GaAs** | **4H-SiC** | **GaN** |
|---|---|---|---|---|
| Energy gap (eV) | 1.12 | 1.42 | ~ 3.3 | 3.4 |
| Electron mobility ($cm^2\ V^{-1}\ s^{-1}$) | 1400 | 9000 | ~ 500 | 1200 |
| Hole mobility ($cm^2\ V^{-1}\ s^{-1}$) | 500 | 300 | 100 | 800 |
| Dielectric constant | 11.9 | 12.4 | 10 | 9 |
| Electric breakdown field ($MV\ cm^{-1}$) | 0.30 | 0.40 | 3.0 | 2.0 |
| Thermal conductivity ($W\ cm^{-1}\ K^{-1}$) | 1.5 | 0.46 | 4.9 | 1.3 |

tured using silicon spheres with a diameter of ~ 1 mm with a *pn* junction at the sphere surface. Spherical Si solar cells are one of the candidates for creating a portable source of electricity.

The construction and characterization of photovoltaic power storage systems using spherical Si solar cells, maximum power point tracking (MPPT) charge controller, lithium-ion battery and a direct current-alternating current (DC-AC) inverter using SiC field-effect transistor (FET), considering self-standing, portable emergency power sources with low electric power ranges of ~ 100 W are presented here. SiC-FET devices were introduced into a DC-AC converter [92, 95], which was connected with Li ion batteries. Since spherical Si panels are lighter and more flexible in comparison to ordinary flat Si solar panels [94], spherical Si solar cells were selected as the power source in the present work. To accumulate the generated electricity effectively, MPPT and Li ion

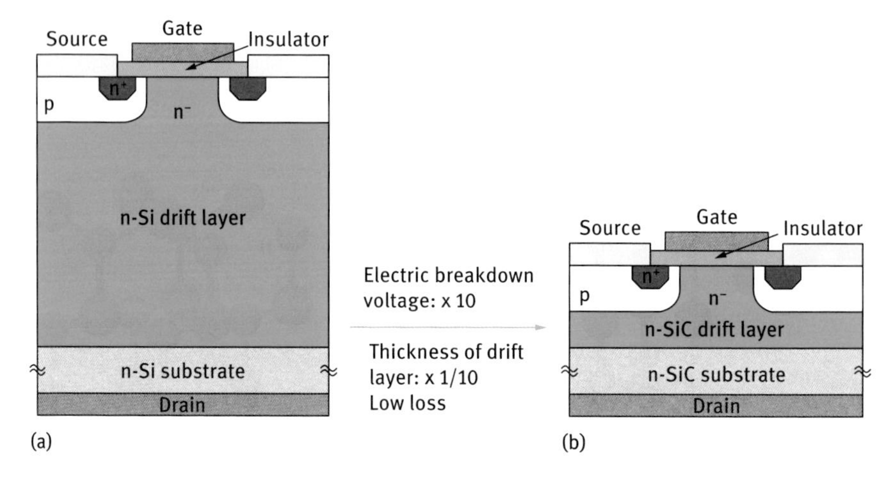


Fig. 9.24: Schematic illustration of (a) Si-FET and (b) SiC-FET.

batteries were also connected with the system. The conversion efficiencies and device losses were analyzed and discussed, and then compared to those of an ordinary Si-based converter.

The experimental setup for the solar cell and power storage systems presented here is shown in Fig. 9.25 [96]. Spherical silicon solar cells (Clean venture 21, CVFM-0540T2-WH) were used as a power source for the inverter operation, and 4 solar cell panels were connected to the MPPT in parallel. Li ion batteries (O'Cell, IFM12-200E2) and MPPT controller (EPsolar, Tracer-2215BN) were used to accummulate electricity storage and control charge, respectively. The generated voltage was almost constant (~ 14 V), and the power and current strongly depended on the intensity of the natural sunlight. The SiC-FET power inverter was prepared by replacing Si-FET (Fairchild, FQPF16N25C) in a DC-AC inverter (Daiji Industry, SXCD-300) with SiC-FET (Rohm, SCT2120AF), as shown in Fig. 9.25. Measurements using conventional SXCD-300 were also performed at the same time for reference. The measurement interval was set as 200 ms, and the measured data were averaged for each minute. The load was controlled in response to changes in solar radiation. The power efficiencies, temperature, humidity and solar radiation were measured at the same time by synchronizing them. The conversion efficiencies of each inverter were calculated from the input power and output power. Since the $R_{on}$ of SiC-FET power devices is smaller (156 mΩ) compared with that of the Si power device (270 mΩ) in the DC-AC part, an improvement in the conversion efficiency of the inverter was expected to result from this addition. On the other hand, the $R_{on}$ of the Si-FET in the DC-DC part provided a very low value of 4 mΩ, and the SiC-FET was introduced only in the DC-AC part.

The data measured is shown as a function of time in Fig. 9.26. The load was 90 W. The amount of solar radiation during the measurements is shown in Fig. 9.26(a). Temperatures and humidity during the measurements are shown in Fig. 9.26(b). The

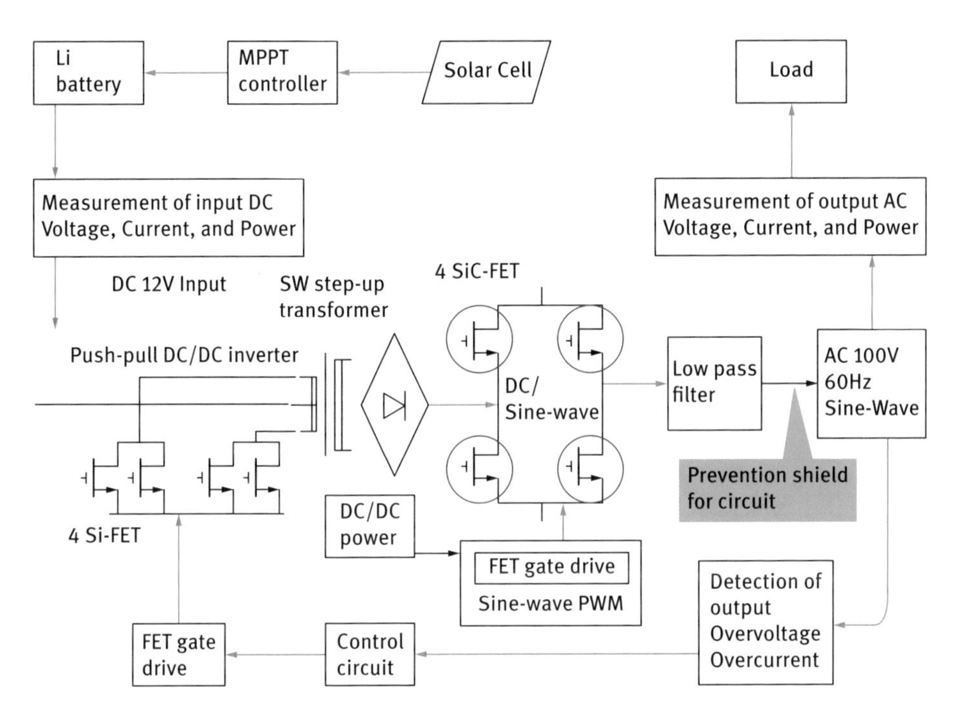


Fig. 9.25: Schematic illustration of the present photo-generation system with SiC-FET inverter.

amount of solar radiation between 11:00 and 13:00 decreased because of cloudy weather. Fig. 9.26(c) shows the input and output voltage measured for the inverter. Fluctuation in the voltage is due to capacitance of the batteries. When the electric power generated was larger or smaller than the load, the voltage increased or decreased, respectively. The input/output current and input/output power of the SiC-FET inverter supplied by spherical Si solar cells are shown in Fig. 9.26(d) and (e), respectively. The input and output power of the inverters were almost constant for the load. The current change was due to fluctuations in voltage caused by the capacity of the batteries. The difference between the input and output power of the inverters is due to power loss in the inverters. This difference is DC-AC conversion efficiency, and a large efficiency means a high-performance DC-AC inverter. Changes in the input electric voltage and current are shown in Fig. 9.26(c) and (d), respectively, which depended on the amount of solar radiation. Although drastic change of the temperatures and humidity are observed at ~ 14 in Fig. 9.26(b), DC-AC conversion efficiencies were not affected by the temperatures and humidity, as observed in Fig. 9.26(e).

The DC-AC conversion efficiencies and the difference in the DC-AC conversion efficiencies of SiC-FET and Si-FET inverters supplied by spherical Si solar cells are shown in Fig. 9.27(a) and (b), respectively. The efficiencies of both SiC and Si inverters increased for several minutes after the measurement started, and were almost stable

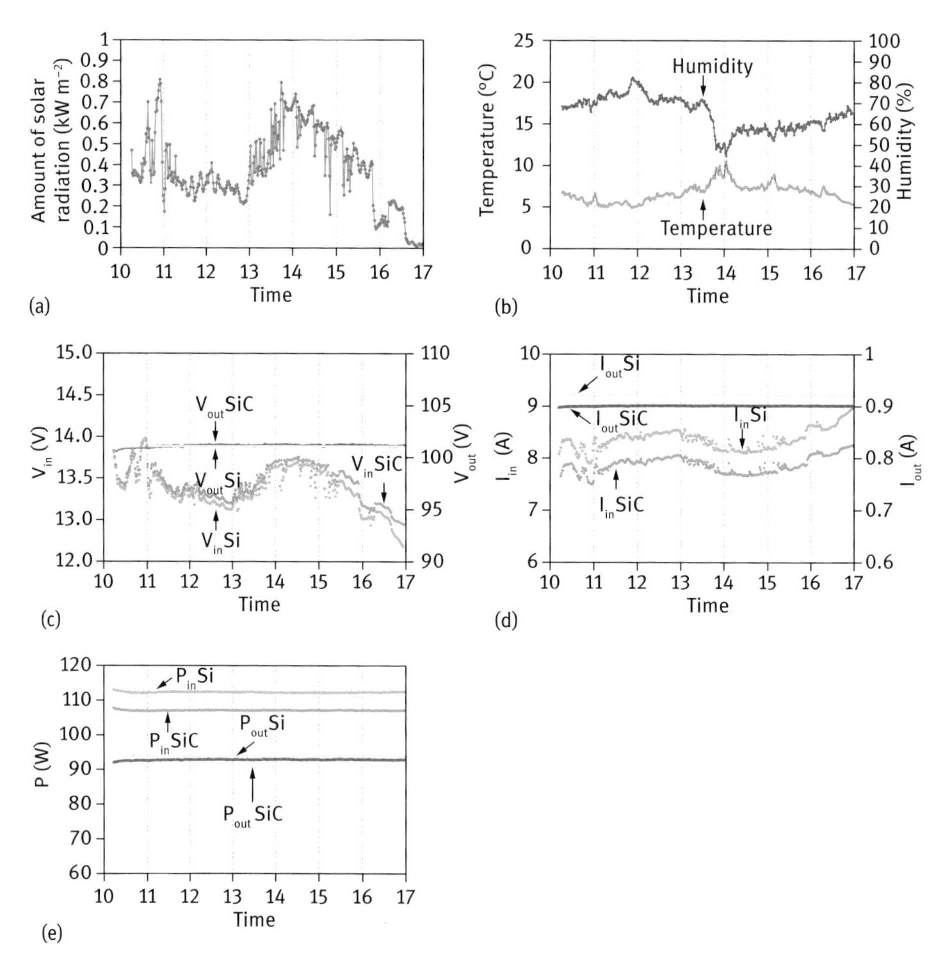


Fig. 9.26: (a) Amount of solar radiation and (b) temperatures and humidity during the measurements. Measured examples of (c) input and output voltage, (d) input and output current, and (e) input and output power of SiC-FET inverter. The load was ~ 90 W.

after that. The efficiency of the SiC inverter was higher than that of the Si inverter for the load of ~ 90 W, and the difference was ~ 3 %.

The DC-AC conversion efficiencies and the difference in DC-AC conversion efficiencies for SiC-FET and Si-FET inverters are shown in Fig. 9.28(a) and (b), respectively. Fig. 9.28(a) shows the DC-AC conversion efficiency as a function of the output power of the inverters, and indicates that both efficiencies increased as the output power of the inverters increased. Figure 9.28(b) shows the difference in the DC-AC conversion efficiencies of SiC-FET and Si-FET inverters, indicating that the conversion efficiency of the inverter with SiC-FET was much larger compared to that of conventional Si-FET inverters when the output power is over ~ 20 W.

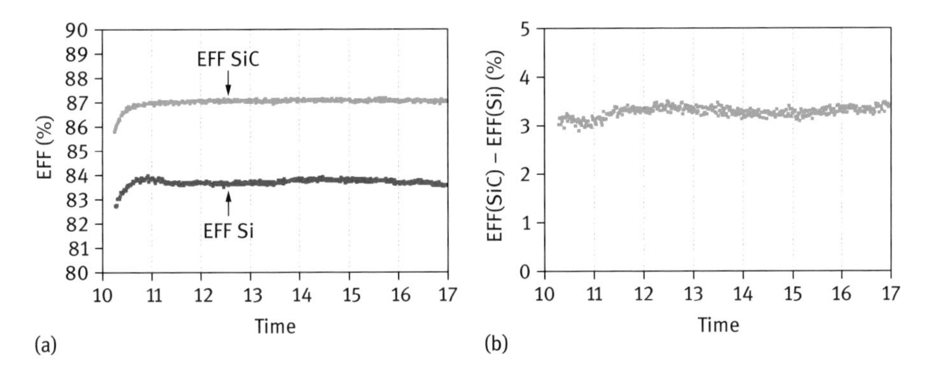


Fig. 9.27: (a) Measured DC-AC conversion efficiencies and (b) the difference of DC-AC conversion efficiencies of SiC-FET and Si-FET inverters. The load was ~ 90 W.

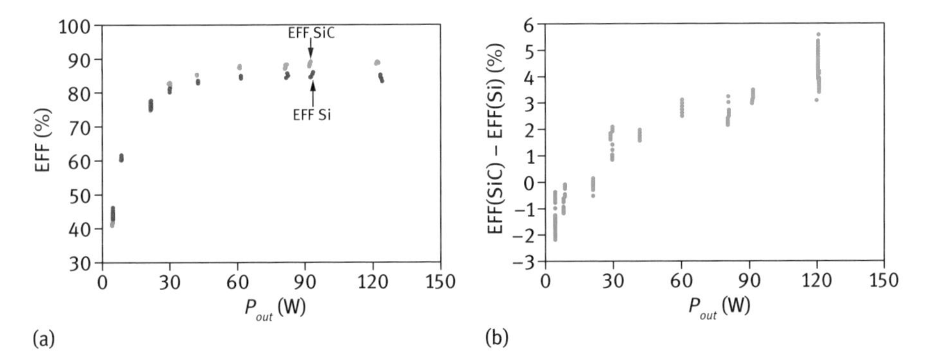


Fig. 9.28: (a) DC-AC conversion efficiencies and (b) difference of DC-AC conversion efficiencies of SiC-FET and Si-FET inverters. The power source was spherical Si solar cells.

Electric power loss in the inverters strongly depends on both conduction loss by the on-resistance and switching loss affected by accumulated carrier. The findings presented here suggest that the $R_{on}$ values were decreased by the SiC-FET instead of by the Si-FET in the conventional inverter, which resulted in an increase of the current value. As the output power increases, the difference in the conversion efficiencies of the inverters with the SiC-FET power devices and a conventional inverter with Si-FET increased up to ~ 5 % for 120 W, as observed in Fig. 9.28(b). When the output power of the inverters are below ~ 20 W, the DC-AC conversion efficiencies decrease to 70 ~ 40 %, and the efficiencies of the inverter with Si-FET seems to be higher than that with SiC-FET. This could be due to the decreased effectiveness of $R_{on}$ of SiC-FET for low output power and/or due to the originally existing resistances in the circuit. The electrical circuit presented here should be designed and optimized for SiC-FET in future work.

With the aim of developing self-standing, portable emergency power sources at low electric power of ~ 100 W, a power storage system using spherical Si solar cells,

MPPT charge controller, lithium-ion battery and a DC-AC inverter with SiC-FET was constructed and evaluated, and were compared with an ordinary Si-FET converter. Spherical Si solar cells were used as the power source in the present work because of the lightness and flexibility of the panels compared with those of ordinary flat Si solar panels. The DC-AC conversion efficiencies were improved by using the SiC-FET, which indicates the effectiveness of the SiC device. When the output power of the SiC inverter increased, the difference in the DC-AC conversion efficiencies of SiC-FET and Si-FET inverters increased to ~5 %. In addition, the efficiencies of the inverters were not affected by fluctuations in the charge condition of Li batteries, and almost constant efficiencies were maintained. This SiC-based power system, equipped with MPPT, Li ion battery and spherical Si solar cells, is expected to serve as an emergency, portable power source with good efficiency.

## 9.13 Two Concluding Asides

### 9.13.1 The energy of a vacuum

When a supernova was observed in detail in order to measure expansion rate of the universe, the accelerating expansion of the universe was discovered. This indicated that something other than gravity was expanding the universe, and this unknown energy was named dark energy. This great scientific discovery in 1998 was awarded the Nobel Prize in physics.

One of the candidates for the dark energy accelerating the expansion of the universe against gravity is the energy of the vacuum. The vacuum contains a world of pair creation and annihilation of particles at the Planck scale. This pair creation and annihilation is continuously repeated, and the vacuum's energy fluctuates unceasingly. If the universe expands, the volume of vacuum increases. Then, the ratio of vacuum energy to gravity increases, and dark energy expand the universe at an accelerated pace. Dark energy has the properties of negative energy and antigravity. The cosmological constant of Einstein's equation shows antigravity and negative pressure, and provides a peculiar quantity of finite energy density even in the vacuum. Thus, the energy of vacuum is considered to be one of the cosmological constants.

Inventors have tried to obtain vacuum energy from the vacant space. The Casimir effect, indicating the zero-point energy of a vacuum, was demonstrated and reported on in 1997 [97]. According to the static Casimir effect, two conducting plates at a very short distance in a vacuum have attractive force. The dynamic Casimir effect is that photons are generated by vibrating the two conducting plates.

The electromagnetic field between the two metal plates is expressed by the overlapping of modes of waves with integer numbers, and the zero-point vibration of each mode has zero-point energy from the quantization. When the distance between the conductors is made extremely small, the frequencies of the modes are changed, and the energy of the vacuum between the two metal plates is reduced to produce an attractive force. If the conducting plates are set at separations of 10 nm, the Casimir force would be ~ 1 atm.

The attracting force of the Casimir effect is due to the energy difference between the inside and outside of the metal plates, and the vacuum energy between the metal plates provides a negative value. Although this experiment indicates the existence of the negative energy of the vacuum, no negative energy could be extracted as an isolated energy. To produce antigravity, negative energy is needed, and the Casimir effect is often referred as an example of the detection of the negative energy.

### 9.13.2 The energy of the mind

The mind has energy. Most readers may agree with this. When our mind is full of energy, we feel cheerful and our heart seems light. When our mind lacks energy, we feel depressed. Although we cannot see the "energy of the mind" directly, readers may have a feeling that this idea corresponds with reality. What is the energy of the mind? Is it the same type of energy referred to in physics?

The density of an atomic nucleus is ~ $10^{17}$ kg $m^{-3}$. Let us transform this density into energy: When the density value is introduced to mass ($m$) in Einstein's theory of relativity $E = mc^2$ ($c = 3\times 10^8$ m $s^{-1}$), an energy density of ~ $10^{34}$ J $m^{-3}$ ( = $10^{25}$ J $mm^{-3}$) is obtained. If this high energy density (~ $10^{25}$ J $mm^{-3}$) exists, this energy could materialize automatically in our universe.

This seems to be extremely hard work. If we could produce such a huge energy, we would have to concentrate it at only one point. There are few forms of matter with such huge energy density. Energy density is generally small because there are electrons around the nucleus. However, black holes and neutron stars have energy density close to the above value. If there is a person who can concentrate his/her energy with this huge energy density, materialization of the mind's energy would be possible. It is an interesting fact that our mind and life may be interpreted by present physics [98].

## 9.14 Bibliography


[1] Chen J, Kuriyama N, Takeshita HT, Tanaka H, Sakai T, Haruta M. Hydrogen storage alloys with $PuNi_3$-type structure as metal hydride electrodes. *Electrochem Sollid-State Lett.* 2000; 3: 249–252.

[2] Schlapbach L, Züttel A. Hydrogen-storage materials for mobile applications. *Nature.* 2001; 414: 353–358.

[3] Bowman Jr. RC, Brent F. Metallic hydrides I: hydrogen storage and other gas-phase applications, *MRS Bull.* 2002: 688–693.

[4] Dillon AC, Jones KM, Bekkedahl TA, Kiang H, Bethune DS, Heben MJ. Storage of hydrogen in single-walled carbon nanotubes. *Nature.* 1997; 386: 377–379.

[5] Chen P, Wu X, Lin J, Tan KL. High $H_2$ uptake by alkali-doped carbon nanotubes under ambient pressure and moderate temperatures. *Science*. 1999; 285: 91–93.
[6] Dujardin E, Ebbesen TW, Hiura H, Tanigaki K. Capillarity and wetting of carbon nanotubes. *Science*. 1994; 265: 1850–1852.
[7] Liu C, Fan YY, Liu M, Cong HT, Cheng HM, Dresselhaus MS. Hydrogen storage in single-walled carbon nanotubes at room temperature. *Science*. 1999; 286: 1127–1129.
[8] Nützenadel C, Züttel A, Chartouni D, Schlapbach L. Electrochemical storage of hydrogen in nanotube materials. Electrochem. *Solid-State Lett*. 1999; 2: 30–32.
[9] Wu, X.B, Chen P, Lin J, Tan KL. Hydrogen uptake by carbon nanotubes. *Int J Hydrogen Energy*. 2000; 25: 261–265.
[10] Oku T, Narita I, Koi N, Nishiwaki A, Suganuma K, Inoue M, Hiraga K, Matsuda T, Hirabayashi M, Tokoro H, Fujii S, Gonda M, Nishijima M, Hirai T, Belosludov RV, Kawazoe Y. Boron nitride nanocage clusters, nanotubes, nanohorns, nanoparticles, and nanocapsules. In: Yap YK ed. B-C-N nanotubes and related nanostructures. *Springer*. 2009: 149–194.
[11] Moussa G, Demirci UB, Malo S, Bernard S, Mielea P. Hollow core@mesoporous shell boron nitride nanopolyhedron-confined ammonia borane: a pure B–N–H composite for chemical hydrogen storage. *J Mater Chem A* 2014; 2: 7717–7722.
[12] Weng Q, Wang X, Bando Y, Golberg D. One-step template-free synthesis of highly porous boron nitride microsponges for hydrogen storage. *Adv Energy Mater*. 2014; 4: 1301525.
[13] Moussa G, Salameh C, Bruma A, Malo S, Demirci UB, Bernard S, Miele P. Nanostructured boron nitride: from molecular design to hydrogen storage application. *Inorganics*. 2014; 2: 396–409.
[14] Liu Y, Liu W, Wang R, Hao L, Jiao W. Hydrogen storage using Na-decorated graphyne and its boron nitride analog. *Int J Hydrogen Energy*. 2014; 39: 12757–12764.
[15] Lei W, Zhang H, Wu Y, Zhang B, Liua D, Qin S, Liu Z, Liu L, Ma Y, Chen Y. Oxygen-doped boron nitride nanosheets with excellent performance in hydrogen storage. *Nano Energy*. 2014; 6: 219–224.
[16] Ebrahimi-Nejad S, Shokuhfar A. Compressive buckling of open-ended boron nitride nanotubes in hydrogen storage applications. *Physica E*. 2013; 50: 29–36.
[17] Wang Y, Wang Fei, Xu B, Zhang J, Sun Q, Jia Y. Theoretical prediction of hydrogen storage on Li-decorated boron nitride atomic chains. *J Appl Phys*. 2013; 113: 064309–1–6.
[18] Lim SH, Luo J, Ji W, Lin J. Synthesis of boron nitride nanotubes and its hydrogen uptake. *Catalysis Today*. 2007; 120: 346–350.
[19] Oku T. Hydrogen storage in boron nitride and carbon nanomaterials. *Energies*. 2015; 8: 319–337.
[20] Zhang H, Tong Zhang CJY, Zhang YN, Liu LM. Porous BN for hydrogen generation and storage. *J Mater Chem A*. 2015; 3: 9632–9637
[21] Muthu RN, Rajashabala S, Kannan R. Synthesis and characterization of polymer (sulfonated poly-ether-ether-ketone) based nanocomposite (h-boron nitride) membrane for hydrogen storage. *Int J Hydrog Energy* 2015; 40: 1836–1845.
[22] Kumar EM, Sinthika S, Thapa R. First principles guide to tune h-BN nanostructures as superior light-element-based hydrogen storage materials: role of the bond exchange spillover mechanism. *J Mater Chem A*. 2015; 3: 304–313.
[23] Liu DM, Tan QJ, Gao C, Sun T, Li YT. Reversible hydrogen storage properties of $LiBH_4$ combined with hydrogenated $Mg_{11}CeNi$ alloy. *Int J Hydrog Energy*. 2015; 40: 6600–6605.
[24] Urbanczyk R, Peinecke K, Felderhoff M, Hauschild K, Kersten W, Peil S, Bathen D. Aluminium alloy based hydrogen storage tank operated with sodium aluminium hexahydride $Na_3AlH_6$. *Int J Hydrog Energy*. 2014; 39: 17118–17128.

[25] Révész Á, Kis-Tóth Á, Varga LK, Schafler E, Bakonyi I, Spassov T. Hydrogen storage of melt-spun amorphous $Mg_{65}Ni_{20}Cu_5Y_{10}$ alloy deformed by high-pressure torsion. *Int J Hydrog Energy* 2012; 37: 5769–5776.
[26] Halstead TK. Proton NMR studies of lanthanum nickel hydride: structure and diffusion. *J Solid State Chem.* 1974; 11: 114–119.
[27] Reidinger F, Lynch JF, Reilly JJ. An X-ray diffraction examination of the FeTi-$H_2$ system. *J Phys F: Met Phys.* 1982; 12: L49–L55.
[28] Bogdanova AN, André G. Phase transformations in the solid solutions $ZrV_2H_x$ at high hydrogen concentrations ($4 < x < 5$). *J Alloys Comp.* 2004; 379: 54–59.
[29] Hayashi S, Hayamizu K, Yamamoto O. X-ray diffraction and $^1H$ and $^{51}V$ NMR study of the structure of a Ti-V-H alloy in relation to preparation conditions. *J Less Common Met.* 1986; 123: 75–84
[30] Häussermann U, Blomqvist H, Noréus D. Bonding and Stability of the Hydrogen Storage Material $Mg_2NiH_4$. *Inorg Chem.* 2002; 41: 3684–3692.
[31] Skoskiewicz T. Superconductivity in the palladium-hydrogen and palladium-nickel-hydrogen systems. *Phys. Stat Sol.* 1972; 11: K123–K126.
[32] Stritzker B, Buckel W. Superconductivity in the palladium-hydrogen and the palladium-deuterium systems. *Z Phys.* 1972; 257: 1–8.
[33] Eichler A, Wühl H, Stritzker B. Tunneling experiments on superconducting palladium-deuterium alloys. *Solid State Commun.* 1975; 17: 213–216.
[34] Stritzker B. High superconducting transition temperatures in the palladium-noble metal-hydrogen system. *Z Phys.* 1974; 268: 261–264.
[35] Tripodi P, Di Gioacchino D, Vinko JD. A review of high temperature superconducting property of PdH system. *Int J Mod Phys. B* 2007; 21: 3343–3347.
[36] Nijkamp MG, Raaymakers JEMJ, Van Dillen AJ, De Jong KP. Hydrogen storage using physisorption-materials demands. *Appl Phys A.* 2001; 72: 619–623.
[37] Saito Y, Kawabata K, Okuda M. Single-layered carbon nanotubes synthesized by catalytic assistance of rare-earths in a carbon arc. *J Phys Chem.* 1995; 99: 16076–16079.
[38] Gadd GE, Blackford M, Moricca S, Webb N, Evans PJ, Smith AM, Jacobesen G, Leung S, Day A, Hua Q. The world's smallest gas cylinders? *Science.* 1997; 277: 933–936.
[39] Ebbesen TW, Ajayan PM, Hiura H, Tanigaki K. Purification of nanotubes. *Nature* 1994, 367, 519.
[40] Oku T. Direct structure analysis of advanced nanomaterials by high-resolution electron microscopy. *Nanotechnol Rev.* 2012; 1: 389–425.
[41] Oku T. High-resolution electron microscopy and electron diffraction of perovskite-type superconducting copper oxides. *Nanotechnol Rev.* 2014; 3: 413–444.
[42] Koi N, Oku T, Narita I, Suganuma K. Synthesis of huge boron nitride cages. *Diam Relat Mater.* 2005; 14: 1190–1192.
[43] Koi N, Oku T, Inoue M, Suganuma K. Structures and purification of boron nitride nanotubes synthesized from boron-based powders with iron particles. *J Mater Sci.* 2008; 43: 2955–2961.
[44] Oku T, Kuno M, Narita I. Hydrogen storage in boron nitride nanomaterials studied by TG/DTA and cluster calculation. *J Phys Chem Solids.* 2004; 65: 549–552.
[45] Oku T, Nishiwaki A, Narita I, Gonda M. Formation and structure of $B_{24}N_{24}$ clusters. *Chem Phys Lett.* 2003; 380: 620–623.
[46] Oku T, Narita I. Calculation of $H_2$ gas storage for boron nitride and carbon nanotubes studied from the cluster calculation. *Physica B.* 2002; 323: 216–218.
[47] Koi N, Oku T. Hydrogen storage in boron nitride and carbon clusters studied by molecular orbital calculations. *Solid State Commun.* 2004; 131: 121–124.

[48] Ye Y, Ahn CC, Witham C, Fultz B, Liu J, Rinzler AG, Colbert D, Smith KA, Smalley RE. Hydrogen adsorption and cohesive energy of single-walled carbon nanotubes. *Appl Phys Lett.* 1999; 74: 2307–2309.
[49] Koi N, Oku T, Suganuma K. Effects of endohedral element in $B_{24}N_{24}$ clusters on hydrogenation studied by molecular orbital calculations. *Physica E.* 2005; 29: 54–545.
[50] Oku T, Narita I, Nishiwaki A, Koi N, Suganuma K, Hatakeyama R, Hirata T, Tokoro H, Fujii S. Formation, atomic structures and properties of carbon nanocage materials. *Topics Appl Phys.* 2006; 100: 187–216.
[51] Sun Q, Wang Q, Jena P. Storage of molecular hydrogen in B-N cage: energetics and thermal stability. *Nano Lett.* 2005; 5: 1273–1277.
[52] Wu H, Fan X, Kuo JL. Metal free hydrogenation reaction on carbon doped boron nitride fullerene: A DFT study on the kinetic issue. *Int J Hydrog Energy.* 2012; 37: 14336–14342.
[53] Wen SH, Deng WQ, Han KLi. Endohedral BN metallofullerene $M@B_{36}N_{36}$ complex as promising hydrogen storage materials. *J Phys Chem C.* 2008; 112: 12195–12200.
[54] Venkataramanan NS, Belosludov RV, Note R, Sahara R, Mizuseki H, Kawazoe Y. Theoretical investigation on the alkali-metal doped BN fullerene as a material for hydrogen storage. *Chem Phys.* 2010; 377: 54–59.
[55] Shelimova LE, Karpinskii OG, Konstantinov PP, Avilov ES, Kretova MA, Zemskov VS. Crystal structures and thermoelectric properties of layered compounds in the ATe-$Bi_2Te_3$ (A = Ge, Sn, Pb) systems. *Inorg Mater.* 2004; 40: 451–460.
[56] Bouad N, Chapon L, Marin-Ayral RM, Bouree-Vigneron F, Tedenac JC, Neutron powder diffraction study of strain and crystallite size in mechanically alloyed PbTe. *J. Solid State Chem.* 2003; 173, 189–195.
[57] He T. Chen JD, Rosenfeld HD, Subramanian MA. Thermoelectric properties of indium-filled skutterudites. *Chem Mater.* 2006; 18: 759–762.
[58] Akasaka M, Iida T, Matsumoto A, Yamanaka K, Takanashi Y, Imai T, Hamada N. The thermoelectric properties of bulk crystalline n- and p-type $Mg_2Si$ prepared by the vertical Bridgman method. *J Appl Phys.* 2008; 104: 013703.
[59] Dismukes JP, Ekstrom L, Paff RJ. Lattice parameter and density in germanium-silicon alloys. *J Phys Chem.* 1964; 68: 3021–3027.
[60] Miyazaki Y, Igarashi D, Hayashi K, Kajitani T, Yubuta K. Modulated crystal structure of chimney-ladder higher manganese silicides $MnSi_j$ (j = 1.74). *Phys Rev B.* 2008; 78: 214104.
[61] Yanabu S, Nishikawa H. Energy conversion engineering (in Japanese). Tokyo Denki University Pub. 2004.
[62] Hosoi S, Kim H, Nagata T, Kirihara K, Soga Kohei, Kimura K, Kato Kenichi, Takata M. Electron density distributions in derivative crystals of $\alpha$-rhombohedral boron. *J Phys Soc Jpn.* 2007; 76: 044602.
[63] Slack GA, Hejna CI, Garbauskas MF, Kasper JS. The crystal structure and density of $\beta$-rhombohedral boron. J Solid State Chem. 1988; 76: 52–63.
[64] Kwei GH, Morosin B. Structures of the boron-rich boron carbides from neutron powder diffraction: implications for the nature of the inter-icosahedral chains. *J Phys. Chem.* 1996; 100, 8031–8039.
[65] Khan MN, Shahzad K, Bashir J. Thermal atomic displacements in nanocrystalline titanium dioxide studied by synchrotron x-ray diffraction. *J Phys D: Appl Phys.* 2008; 41: 085409.
[66] Bokhimi X, Pedraza F. Characterization of brookite and a new corundum-like titania phase synthesized under hydrothermal conditions. *J Solid State Chem.* 2004; 177: 2456–2463.
[67] Kuhn A, Baehtz C, Garcia Alvarado F. Structural evolution of ramsdellite-type $Li_xTi_2O_4$ upon electrochemical lithium insertion-deinsertion ($0 \leq x \leq 2$). *J Power Sources.* 2007; 174: 421–427.

[68] Filatov SK, Bendeliani NA, Albert B, Kopf J, Dyuzheva TI, Lityagina LM. Crystalline structure of the $TiO_2$ II high-pressure phase at 293, 223, and 133 K according to single-crystal X-ray diffraction data. *Doklady Phys.* 2007; 52: 195–199.

[69] Swamy V, Dubrovinsky LS, Dubrovinskaia NA, Langenhorst F, Simionovici AS, Drakopoulos M, Dmitriev V, Weber HP. Size effects on the structure and phase transition behavior of baddeleyite $TiO_2$. *Solid State Commun.* 2005; 134: 541–546.

[70] Fujishima A. Honda K. Electrochemical photolysis of water at a semiconductor electrode. *Nature.* 1972; 238: 37–38.

[71] Olson DH, Kokotailo GT, Lawton SL, Meier WM. Crystal structure and structure-related properties of ZSM-5. *J Phys Chem.* 1981; 85: 2238–2243.

[72] Passaglia E, Artioli G, Gualtieri A, Carnevali R. Diagenetic mordenite from Ponza, Italy. *Eur J Mineral.* 1995; 7: 429–438.

[73] Khan MN, Bashir J. Synthesis and structural refinement of $LiAl_xCo_{1-x}O_2$ system. *Mater Res Bull.* 2006; 41: 1589–1595.

[74] Liang G, Park K, Li J, Benson RE, Vaknin D, Markert JT, Croft MC. Anisotropy in magnetic properties and electronic structure of single-crystal $LiFePO_4$. *Phys Rev B.* 2008; 77: 064414-1-12.

[75] Krins N, Hatert F, Traina K, Dusoulier L, Molenberg I, Fagnard JF, Vanderbemden P, Rulmont A, Cloots R, Vertruyen B. $LiMn_{2-x}Ti_xO_4$ spinel-type compounds ($x \leq 1$): Structural, electrical and magnetic properties. *Solid State Ion.* 2006; 177: 1033–1040.

[76] Fukuoka H, Kiyoto J, Yamanaka S. Superconductivity of metal deficient silicon clathrate compounds, $Ba_{8-x}Si_{46}$ ($0 < x \leq 1.4$). *Inorg Chem.* 2003; 42: 2933–2937.

[77] Li Y, Liu Y, Chen N, Cao G, Feng Z, Ross JHJ. Vacancy and copper-doping effect on superconductivity for clathrate materials. *Phys Lett A.* 2005; 345: 398–408.

[78] Fortes AD, Wood IG, Grigoriev D, Alfredsson M, Kipfstuhl S, Knight KS, Smith RI. No evidence for large-scale proton ordering in Antarctic ice from powder neutron diffraction. *J Chem Phys.* 2004; 120: 11376–11379.

[79] Gutt C, Asmussen B, Press W, Johnson MR, Handa YP, Tse JS. The structure of deuterated methane-hydrate. *J. Chem. Phys.* 2000; 113: 4713–4721.

[80] Miyashiro H, Wakihara M, Yamanaka A, Tabuchi M, Seki S, Nakayama M, Ohno Y, Kobayashi Y, Mita Y, Usami A. Improvement of degradation at elevated temperature and at high state-of-charge storage by $ZrO_2$ coating on $LiCoO_2$. *J Electrochem Soc.* 2006; 153: A348–A353.

[81] McCullough JD, Trueblood KN. The crystal structure of baddeleyite (monoclinic $ZrO_2$). *Acta Cryst.* 1959; 12: 507–511.

[82] Block S, Da Jornada JAH, Piermarini GJ. Pressure-temperature phase diagram of zirconia. *J Amer Ceram Soc.* 1985; 68: 497–499.

[83] Ohtaka O, Yamanaka T, Kume S, Hara N, Asano H, Izumi F. Structural analysis of orthorhombic $ZrO_2$ by high resolution neutron powder diffraction. *Proc Jpn Acad Ser B.* 1990; 66: 193–196.

[84] Palosz B, Gierlotka S, Stel'makh S, Pielaszek R, Zinn P, Winzenick M, Bismayer U, Boysen H. High-pressure high-temperature in situ diffraction studies of nanocrystalline ceramic materials. *J Alloys Comp.* 1999; 286: 184–194.

[85] Ruh R, Zangvil A. Composition and properties of hot-pressed SiC-AlN solid solutions. *J Amer Ceram Soc.* 1982; 65: 260–265.

[86] Bind JM. Phase transformation during hot-pressing of cubic SiC. *Mater Res Bull.* 1978; 13: 91–96.

[87] Capitani GC, Di Pierro S, Tempesta G. The 6H-SiC structure model: Further refinement from SCXRD data from a terrestrial moissanite. *Amer Mineral.* 2007; 92: 403–407.

[88] Kaminski M, Podsiadlo S, Wozniak K, Dobrzynski L, Jakiela R, Barcz A, Psoda M, Mizera J. Growth and structural properties of thick GaN layers obtained by sublimation sandwich method. *J Cryst Growth.* 2007; 303: 395–399.

[89] Burger B, Kranzer D, Stalter O. Efficiency Improvement of PV-Inverters with SiC-DMOSFETs. *Mater Sci Forum*. 2008; 600–603: 1231–1234.

[90] Okumura, H. Present status and future prospect of widegap semiconductor high-power devices. *Jpn J Appl Phys*. 2006; 45: 7565–7586.

[91] Li L, Li C, Cao Y, Wang F. Recent progress of SiC power devices and applications, *IEEJ Trans Electric Electronic Eng*. 2013; *8*: 515–521.

[92] Oku T, Matsumoto T, Hiramatsu K, Yasuda M, Shimono A, Takeda Y, Murozono M. Construction and characterization of spherical Si solar cells combined with SiC electric power inverter: *AIP Conf Proc*. 2015; 1649: 79–83.

[93] Pushpakaran BN, Subburaj AS, Bayne SB, Mookken J. Impact of silicon carbide semiconductor technology in photovoltaic energy system. *Renewable Sustainable Energy Rev*. 2016, 55, 971–989.

[94] Matsumoto T, Oku T, Hiramatsu K, Yasuda M, Shirahata Y, Shimono A, Takeda Y, Murozono M. Evaluation of photovoltaic power generation system using spherical silicon solar cells and SiC–FET inverter. *AIP Conf Proc*. 2016; 1709: 020023–1–6.

[95] Oku T, Kanayama M, Ono Y, Akiyama T, Kanamori Y, Murozono M. Microstructures, optical and photoelectric conversion properties of spherical silicon solar cells with anti-reflection $SnO_x$:F thin films. *Jpn J Appl Phys*. 2014; 53: 05FJ03.

[96] Oku T, Matsumoto T, Hiramatsu K, Yasuda M, Ohishi Y, Shimono A, Takeda Y, Murozono M. Construction and evaluation of photovoltaic power generation and power storage system using SiC field-effect transistor inverter. *AIP Conf Proc*. 2016; 1709: 020024–1–10.

[97] Lamoreaux SK. Demonstration of the Casimir force in the 0.6 to 6 mm range. *Phys Rev Lett*. 1997; 78: 5–8.

[98] Oku T. Science towards reality and meaning. *J Intl Soc Life Info Sci*. 2008; 26(1): 65–70.

# Index